MAINSTREAMING CLIMATE CHANGE IN DEVELOPMENT COOPERATION

Theory, Practice and Implications for the European Union

Climate change, development and development cooperation are, individually and jointly, three politically sensitive and complex issues, especially in the context of relations between developed and developing countries. This book tackles these issues by combining theoretical, political and practical perspectives. At the theoretical level, it analyses the dominant paradigms and explores the meaning of the concept of mainstreaming. At the political level, it highlights the sensitivities between developed and developing countries and examines the mainstreaming debate in various fora. At the practical level, it presents the results of case studies focusing on the assistance provided by the European Union and key Member States and the climate needs articulated by developing countries. This book is valuable for politicians, policymakers, academics and non-state actors working in the fields of development studies, international law, politics, international relations, economics, climate change and environmental studies.

This volume is one of the results of the three-year European Commission ADAM (Adaptation and Mitigation Strategies) research project. Three other books arise from this project, all published by Cambridge University Press:

Making Climate Change Work for Us: European Perspectives on Adaptation and Mitigation Strategies, edited by *Mike Hulme and Henry Neufeldt*

Climate Change Policy in the European Union: Confronting the Dilemmas of Mitigation and Adaption?, edited by *Andrew Jordan, Dave Huitema, Harro van Asselt, Tim Rayner and Frans Berkhout*

Global Climate Governance Beyond 2012: Architecture, Agency and Adaptation, edited by *Frank Biermann, Philipp Pattberg and Fariborz Zelli*

JOYEETA GUPTA is professor of climate change law and policy at the VU University Amsterdam and of water law and policy at the UNESCO-IHE Institute for Water Education in Delft. She is editor-in-chief of *International Environmental Agreements: Politics, Law and Economics* and is on the editorial board of the journals *Carbon and Law Review, International Journal on Sustainable Development, Environmental Science and Policy* and *International Community Law Review*. She was a lead author of the Fourth Assessment Report of the Intergovernmental Panel on Climate Change and of the Millennium Ecosystem Assessment. She is on the scientific steering committees of many international programmes, including the Global Water Systems Project and the Earth System Governance Project. She has published several books on climate change, including *The Climate Change Convention and Developing Countries: From Conflict to Consensus?* (1997, Kluwer Academic Publishers) and *Our Simmering Planet: What to Do about Global Warming* (2001, Zed Publishers). Books she has edited include *The Evolution of the Law and Politics of Water* (2009, Springer Verlag, with J. Dellapenna), *Climate Change and the Kyoto Protocol: The Role of Institutions and Instruments to Control Global Change* (2003, Edward Elgar Publishers, with M. Faure and A. Nentjes), *Issues in International Climate Policy: Theory and Policy* (2003, Edward Elgar Publishers, with E. Van Ierland and M. Kok) and *Climate Change and European Leadership: A Sustainable Role for Europe* (2000, Kluwer Academic Publishers, with M. Grubb).

NICOLIEN VAN DER GRIJP is a senior researcher at the Institute for Environmental Studies at the VU University Amsterdam, where she coordinates the European Law and Policy cluster. Her research interests are related to environmental law and policy at the international, EU and national levels. In her present work, she focuses on issues of integration, implementation and interaction at the various levels. She has worked on several projects focusing on developing countries, including the UNEP project on global product chains, the EU climate change leadership project and the EU sustainability labelling project.

MAINSTREAMING CLIMATE CHANGE IN DEVELOPMENT COOPERATION

Theory, Practice and Implications for the European Union

Edited by

JOYEETA GUPTA

and

NICOLIEN VAN DER GRIJP

Institute for Environmental Studies
VU University Amsterdam

CAMBRIDGE
UNIVERSITY PRESS

CAMBRIDGE UNIVERSITY PRESS
Cambridge, New York, Melbourne, Madrid, Cape Town, Singapore,
São Paulo, Delhi, Dubai, Tokyo

Cambridge University Press
The Edinburgh Building, Cambridge CB2 8RU, UK

Published in the United Kingdom by Cambridge University Press, UK

www.cambridge.org
Information on this title: www.cambridge.org/9780521197618

First published 2010

Printed in the United Kingdom at the University Press, Cambridge

A catalogue record for this publication is available from the British Library

ISBN 978 0 521 19761 8 Hardback

To Zubin van der Hoeven, as incorporating a bit of the North and South, and Bruno and Nena van der Grijp; all as representatives of the future generation

Contents

Contributors

Michiel van Drunen (b. Netherlands) studied environmental chemistry at Leiden University between 1986 and 1990, and received his PhD at Delft University of Technology. He has worked at the Institute for Environmental Studies (IVM) of the VU University Amsterdam since 1996. His research projects deal with making environmental information transparent for decision makers, such as the development of a decision-support system for soil remediation, sustainability indicators for Amazonia, socio-economic scenarios for climate assessments, measuring environmental performance of industries, (financing) adaptation in developing countries and many others. In addition, he developed and taught several environment-related courses at the VU University Amsterdam.

Thijs Etty (b. Netherlands) is a researcher at the Institute for Environmental Studies (IVM), and adjunct lecturer in European law at the Faculty of Law, of the VU University Amsterdam. With degrees in European and international law and in Dutch law, in recent years he has built a specialization in EU environmental law and policy, and biotechnology governance. Since 2003, he has been editor-in-chief of the *Yearbook of European Environmental Law*, published annually by Oxford University Press, and has served on its editorial board since its inception in 1998. He is an editorial board member of several law journals, and has published articles and chapters in a variety of international (peer-reviewed) journals and books. Etty is currently completing a PhD dissertation on the regulation of agricultural biotechnology law in the EU, focusing in particular on the governance of the co-existence of the cultivation of GMO and non-GMO food crops.

Joyeeta Gupta (b. India) is professor of climate change law and policy at the VU University Amsterdam and of water law and policy at the UNESCO-IHE Institute for Water Education in Delft. She is editor-in-chief of *International Environmental Agreements: Politics, Law and Economics* and is on the editorial board of *Carbon and Climate Law Review, International Journal on Sustainable Development,*

Environmental Science and Policy and *International Community Law Review*. She was lead author in the Intergovernmental Panel on Climate Change that recently shared the 2007 Nobel Peace Prize with Al Gore. She is on the scientific steering committees of, amongst others, the Global Water Systems Project and the Earth System Governance Project of the International Human Dimensions Programme.

Nicolien van der Grijp (b. Netherlands) is a senior researcher at the Institute for Environmental Studies of the VU University Amsterdam. Her research interests are related to environmental law and policy at the international, EU and national level. In her present work, she focuses on the international and EU regulation of pesticide risks, the mainstreaming of climate change concerns into development cooperation and the evaluation of Dutch environmental law and policy in various areas. She has worked on several projects focusing on developing countries, including the UNEP project on global product chains, the EU climate change leadership project and the EU sustainability labelling project. Last year, she finalized a dissertation about the regulation of pesticide risks considered from the perspective of legal pluralism, in which she focused on interactions between state and non-state actors in regulatory processes.

Joanne Linnerooth-Bayer (b. USA) is leader of the Risk, Modeling and Society Project. She received her PhD in economics from the University of Maryland. At the IIASA, she has worked on interdisciplinary teams exploring the social and economic issues related to environmental and technological risks, including issues of risk estimation, risk–benefit analysis, risk perception, culturally determined risk construction and risk burden sharing. Her current interest is global change and the risk of catastrophic natural disasters, and she is investigating options for improving the financial management of catastrophic risks. She has recently completed a study of flood risk on the Tisza River in Hungary that combined catastrophe modelling with stakeholder participation for the design of a national flood-insurance pool. In collaboration with the World Bank and InterAmerican Development Bank, she is leading research in close collaboration with developing country policy makers to improve the financial capacity of disaster-prone countries to respond to extreme events. In collaboration with Kyoto University, she organizes an annual conference on Integrated Disaster Risk Management. Dr Linnerooth-Bayer has over 100 publications in the area of risk, and she is on the editorial board of three international journals on this topic.

Michael Thompson (b. UK), originally a professional soldier, studied anthropology (first degree and PhD at University College London, BLitt at Oxford), while also following a career as a Himalayan mountaineer (Annapurna South Face 1970, Everest Southwest Face 1975). His early research on how something second-hand becomes an antique (*Rubbish Theory*, 1979, Oxford University Press) led to work on the 'energy tribes' (in various Western think tanks), on risk, on Himalayan deforestation and sustainable development, on household-product development (in Unilever), on global climate change, on technology and development, and on what might be called 'the even newer institutionalism' (e.g. *Cultural Theory*, co-authored with Richard Ellis and Aaron Wildvasky, 1990, West View). Dr Thompson is a Fellow at the Institute for Science, Innovation and Society, at the Said Business School of Oxford. At the IIASA he is affiliated with the Risk and Vulnerability Program.

Harro van Asselt (b. Netherlands), LL.M. (International Law) is a researcher at the department of Environmental Policy Analysis at the Institute for Environmental Studies (IVM) of the VU University Amsterdam. He has been a research fellow with the Multiple Options, Solutions and Approaches: (Institutional) Interplay and Conflict ('MOSAIC') group of the international Global Governance Project (www.glogov.org) since July 2005. His main expertise lies in international and European climate change policy and law and international environmental governance. He has published extensively on issues related to global climate governance, focusing on the climate and trade interplay, and on the Kyoto Protocol's flexibility mechanisms. He is managing editor of the peer-reviewed journal *International Environmental Agreements* and associate editor of the *Carbon and Climate Law Review*.

Foreword

Climate change is not merely a serious and urgent environmental issue, it also has serious adverse developmental impacts. UN Secretary General Ban Ki-moon labelled it 'a defining issue of our era'. Human activities have contributed significantly to climate change, and still do: much scientific evidence suggests that the changes taking place may be far more rapid and dangerous than is reflected in the latest (2007) IPCC assessment.

While climate change results from activities all over the globe, actual contributions to it have been, and are, rather unevenly spread, with most contributions coming from the industrialized economies. There is little correlation between causing climate change and being exposed to its consequences: it seems clear now that the worst impacts will fall on developing countries. Climate change is likely to undermine the sustainability of livelihoods as well as resource bases for development.

One response to climate change is to cope with its impacts and suffer from the associated damages. Another one is to alter behaviour, institutions, structures and even development paths in such a way as to reduce and curb damage ('adaptation'). A more fundamental response would be for the world economy to reduce its emissions of greenhouse gases and alter its patterns of land use in such a way as to prevent and curb warming itself, and to enhance sinks for greenhouse gases ('mitigation'). There is a need to consider the links and feedbacks between climate change (and policies to address it) and development. On the one hand, development paths vary in the ways in which they affect climate; on the other, different climate policies will have different impacts on development trajectories.

The policy challenge in this is obvious. In 1992, the majority of countries worldwide agreed on the UN Framework Convention on Climate Change (FCCC) aiming at a stabilization of concentrations of greenhouse gases at a level that would prevent 'dangerous anthropogenic interference with the climate system'. The subsequent 1997 Kyoto Protocol was intended to elaborate and implement this

Convention in the context of a framework including objectives for emissions reductions for developed countries by 2008–12. In 2007, the FCCC's Bali Action Plan was established, aiming at a new agreement by the end of 2009 on cooperative action on climate issues beyond 2012. According to this plan, developed countries are to accept commitments regarding mitigation, technology transfer and facilitating adaptation and mitigation efforts in developing countries beyond what these countries consider 'appropriate' in terms of their domestic mitigative actions. The details and extent of these commitments and actions were being negotiated at the time of the editing of this volume.

Many hold that, from an insurance perspective and based on a precautionary strategy, it would be wise to curb global warming at or below an increase of 2 °C above pre-industrial levels. This would appear to be technically feasible. Attempts to weigh the desirability of such action in terms of societal costs and welfare benefits against the implications of inaction vis-à-vis climate change were made by a team led by Sir Nicholas Stern (published in 2006) with a strikingly positive bottom line for going towards that target (to be precise: towards a somewhat less stringent one) if the calculations were based on a reasoning giving serious weight to future consequences. Accepting these different approaches and their outcomes, we are left with the following question: who is expected to do what in order to arrive at and ensure positions within the carbon space? That, to a large degree, is the substance of the current negotiations towards a new global deal.

Any global compact must provide a credible approach that is in the interest of the South. Some of the factors accounting for the present lack of progress are rooted in a deep deficit of trust between negotiating parties. Some crucial questions are the following.

- How and how much will the developed countries contribute to the unavoidable adaptation resulting from the past energy-intensive economic growth in the North?
- How will mitigative actions by developed countries affect development in developing and emerging economies?
- Why should developing countries engage in mitigation whilst industrialized countries do not meet their Kyoto targets (at least, not in their own territory)?
- Why should developing countries be involved at all in mitigation, when it was not these countries but the industrialized ones that primarily caused climate-related problems?
- Why would the developing countries trust industrialized ones when they speak of cooperation and assistance for mitigation undertaken by developing countries, while the developed ones in general have not lived up to their official development assistance commitments?

Developed countries are not only expected to take the lead in finding adequate and appropriate technological answers to the climate challenge as outlined above, but also called upon to support mitigation and adaptation actions in developing countries. Transfer of technologies might facilitate developing countries' getting involved in mitigation, if and to the extent that the trust deficit is overcome through such cooperation. That implies rapid overtures along all these avenues in the decade to come, for countries and regions such as the USA, Canada, Japan and the European Union, as preconditions to bring on board the major developing economies that impact on emission levels, say by 2020. The volume I am introducing here deals with one key set of issues that fall under this general umbrella: the relationship(s) between development policy and climate change policies if the world is to stay on the desirable side of the warming cap. In particular, it deals with how climate change policies could be integrated or even mainstreamed into development cooperation policies of one major player, the European Union, and to what extent that should be done. Among many other things, the study argues that, while climate change should be a central element in development policies, it might be undesirable to lock international climate funding into development cooperation. In doing so, it appears to side with the UN (ECOSOC) Committee on Development Policy in its most recent report (UN Document E/2009/33). In fact, this book provides much argumentation in support of that position. It does so not merely on the basis of academic armchair reflection on the issues and on policies on development and climate as put in place by the European Union thus far, but also by bringing in results of case studies of how the links between development and climate have been shaped and are evolving in a number of important European Union Member States and of a survey of types of assistance needed by developing countries (a total of 10, including Brazil, China, Malawi and Nepal) as manifest in a range of sector studies (including energy, forestry, biodiversity and agriculture).

This book has more to offer, particularly on the best and second-best ways in which the European Union could develop its climate and development policies; the lessons drawn and suggestions made could be of relevance to other parts of the developed world. Since this book provides its analysis in a historical perspective and extrapolates into the future, I am sure that its contents will remain relevant and pertinent in the years to follow.

Professor Dr J. B. (Hans) Opschoor

(Dr J. B. (Hans) Opschoor is Emeritus Professor of the Economics of Sustainable Development at the International Institute of Social Studies, The Hague, and of Environmental Economics at the VU University Amsterdam. He was involved in the work of the Intergovernmental Panel on Climate Change and is currently a member of the UN Committee for Development Policy.)

Acknowledgements

This book is the result of research undertaken within the context of the European Commission-financed project Adaptation and Mitigation Strategies: Supporting European Climate Policy (ADAM) (contract number 98476). We would like to thank our partners in the ADAM project, Joanne Linnerooth-Bayer, Anne Jerneck, Richard Klein, Anthony Patt, Åsa Persson, Michael Thompson and Lennart Olsson, for engaging in discussions on various elements of the project with us.

We would like to thank all those we interviewed, including Aart van der Horst, Michael Linddal, Maria Arce Moreira, Eleanor Briers, Bertrand Loiseau, Ulf Moslener, Imme Scholz, Michael Scholze, Matthias Seiche and Mike Speirs.

We would also like to thank all those who reviewed our individual chapters and those who gave us feedback on our various presentations of the ideas in this book. They include Hans Opschoor, Frans Oosterhuis, Onno Kuik, Jill Jaeger, Mike Hulme, Meine Pieter van Dijk, Marc Pallemaerts, Leo Meyer, Bart Strengers, Eileen Harloff, Marcel Kok, P. J. I. M. de Waart, Philipp Pattberg, Ton Bresser, Eric Massey, Jeltje Kemerink, Marloes Mul, Yunus Mohamed, Frank Jaspers, Amaury Tilmant, Willy Douma, Annelieke Douma, Harrie Oppenoorth, Eco Matser, Harrie Clemens, Daniëlle Hirsch, Ton Dietz, Karin Arts, Wybe Douma, Peter Brinn, Wiert Wiertsema and Pieter van der Zaag.

Furthermore, we are grateful for the research assistance we received from a number of ERM students at the VU University Amsterdam, including Pravesh Baboeram, Milena Garita, Caro Lorika, Matthew Smith, Hsin-Ping Wu, Corinne Cornelisse, Grace Lamminar, Marilen Espinoza, Marit Heinen, Roy Porat, Ruben Zondervan, Belinda McFadgen, Remon Dolevo, Charles Owusu, Laura Meuleman, Ieva Oskolokaite, Emilie Hugenholtz, Hassan El Yaquine, Olwen Davies, Andrej Wout, Chad Rieben, Wouter Wester, Francesca Feller, Brenda Schuurkamp, Anna Harnmeijer, Jens Stellinga, Pieter Pauw, Yvette Osinga, Nguyen Thi Khanh Van, Joao Fontes, Sarianne Palmula, Laybelin Ogano Bichara, Viviana Gutierrez Tobon, Eline van Haastrecht, Coby Leemans, Efrath Silver, Michelle Beaudin and Jorge Triana.

Abbreviations

ACP	African, Caribbean and Pacific (countries)
ALA	Asian and Latin American (countries)
AOSIS	Alliance of Small Island States
BMZ	Federal Ministry for Economic Cooperation and Development (Germany)
CDM	Clean Development Mechanism
CEP	Country Environment Profile
CER	Certified Emission Reduction
CRISP	Climate Risk Impacts on Sectors and Programmes
CSP	Country Strategy Paper
DC	Developing country
DCI	Development Cooperation Instrument
DFID	Department of International Development (United Kingdom)
DG	Directorate General
DKK	Danish Kroner
EBRD	European Bank for Reconstruction and Development
EC	European Community
ECA	European Court of Auditors
EDF	European Development Fund
EIA	Environmental Impact Assessment
EIB	European Investment Bank
ENRTP	Thematic Strategy for the Environment and Sustainable Management of Natural Resources
EU	European Union
FCCC	Framework Convention on Climate Change
FDI	Foreign Direct Investment
G8	Group of 8
G-77	Group of 77

GCCA	Global Climate Change Alliance
GDP	Gross Domestic Product
GEF	Global Environment Facility
GHG	Greenhouse gas
GNI	Gross National Income
GNP	Gross National Product
GTZ	German Agency for Technical Cooperation
IC	Industrialized country
IDA	International Development Agency
IMF	International Monetary Fund
IPCC	Intergovernmental Panel on Climate Change
KP	Kyoto Protocol
LDCs	Least Developed Countries
MDGs	Millennium Development Goals
NAPA	National Adaptation Programme of Action
NC	National Communication
NGO	Non-governmental organization
NIEO	New International Economic Order
ODA	Official development assistance
OECD	Organization of Economic Cooperation and Development
OECD DAC	OECD Development Assistance Committee
ORCHID	Opportunities and Risks of Climate Change
PRSP	Poverty Reduction Strategy Paper
REP	Regional Environment Profile
RSP	Regional Strategy Paper
SEA	Strategic Environment Assessment
TNA	Technology Needs Assessment
UN	United Nations
UNDP	United Nations Development Programme
UNEP	United Nations Environment Programme
UNGA	United Nations General Assembly
USD	US Dollar

Part I

Introduction

1

Climate change, development and development cooperation

JOYEETA GUPTA AND NICOLIEN VAN DER GRIJP

1.1 Introduction

Climate change, development and development cooperation are three complex issues. The political arenas that deal with these issues partially overlap but are not integrated. In recent years, however, there has been a growing trend towards incorporating climate change concerns into the fields of development and development cooperation. Evidently, this is a challenging process. The nature of the challenge is three-fold. Theoretically, the links among the three issues are vast and cover practically all human activities and endeavours. Politically, the nature of North–South relations in all three fields is highly sensitive. Practically, there are limited resources available for global cooperation and it makes sense to use these resources wisely to improve the results for all three fields. But does this practical argument compensate adequately for the other challenges?

This book tackles these issues by combining theoretical, political and practical perspectives. While it focuses on the relationship between climate change and development cooperation, this is undertaken against the broader background of the fundamental links between climate change and development. This book is part of the research project 'Adaptation and Mitigation Strategies (ADAM): supporting European climate policy', the aims of which are to assess the extent to which EU mitigation and adaptation policies can achieve a transition to a world in which the global mean temperature does not rise beyond 2 °C above pre-industrial levels, and to develop strategic policy options to help the EU achieve these goals. This book has emerged from a sub-project intended to 'provide strategic options to the EU and its member countries for mainstreaming and restructuring development assistance, such that it promotes climate mitigation and adaptation in ways that are acceptable to the donor and recipient communities'.

This chapter discusses climate change science and the nature of climate change as a North–South issue to provide the context for this book (see Section 1.2), discusses

Mainstreaming Climate Change in Development Cooperation: Theory, Practice and Implications for the European Union, ed. Joyeeta Gupta and Nicolien van der Grijp. Published by Cambridge University Press. © Cambridge University Press 2010.

Table 1.1 *Glossary of critical terms*

Term	Explanation
Adaptation	Coping with the physical impacts of climate change.
Climate change	A change of climate attributed directly or indirectly to human activity that alters the composition of the global atmosphere and that is in addition to natural climate variability observed over comparable time periods (FCCC, 1992).
Climate change cooperation	Covers financial assistance, technology transfer and market mechanisms between rich and poor on climate change. See Chapter 5.
Climate change regime	Covers the norms, principles, responsibilities and instruments developed in international climate policy. See Chapter 5.
Developed, industrialized, rich, North	These terms are used synonymously to refer to the 40 or so countries listed in Annex I of the Climate Convention 1992. Differentiation between Northern countries is discussed in Chapter 10.
Developing, poor, South	These terms are used synonymously to refer to the 150 or so countries that do not belong to Annex I of the Climate Convention 1992. Differentiation between Southern countries is discussed in Chapter 10.
Development	Development challenges both in rich and in poor countries.
Development cooperation/aid regime	Assistance to developing countries. Covers the activities of donor countries within the OECD in the area of official development assistance and relevant policy decisions and declarations within the UN system. See Chapter 4.
Development regime	Covers the activities of UN agencies, banks, trade institutions and the OECD in relation to promoting development. See Chapter 4.
Incorporation	A loose term to include all kinds of approaches to including climate change perspectives in development and/or development cooperation. See Chapter 3.
Integration	Integration of climate change into development policy implies (a) that existing development and/or development cooperation policies and projects take climate change mitigation aspects into account; (b) possibly that this is done through the use of methods and instruments like check-lists and climate proofing; and (c) that hence the climate change component is more or less an add-on component. See Chapter 3.
International cooperation	Covers both climate change cooperation and development cooperation.
Mainstreaming	Mainstreaming of climate change into development and/or development cooperation is the process by which development policies, programmes and projects are (re)designed, (re)organized, and evaluated from the perspective of climate change mitigation and adaptation. Mainstreaming implies involving all social actors – governments, civil society, industry and local communities – in the process. Mainstreaming calls for changes in policy as far upstream as possible. See Chapter 3.

Table 1.1 *(cont.)*

Term	Explanation
Mitigation	Reducing the emissions of greenhouse gases and enhancing sinks.
Official aid	Assistance provided to recipients as listed in Part II of the DAC list.
Official climate assistance	Refers to official government assistance by industrialized countries to developing countries on climate change.
Official development assistance	Assistance to promote development at concessionary financial terms by the official sector. See Chapter 4.

the evolving political framing of the problem (see Section 1.3), and elaborates on the substantive links between climate change and (sustainable) development and on climate change cooperation (see Section 1.4). It introduces the issue of development cooperation (see Section 1.5), and points to the current political trends towards mainstreaming climate change into development cooperation (see Section 1.6). Finally, it discusses the aims and structure of this book (see Section 1.7). Although issues are defined where they are explored in detail, Table 1.1 provides a preview of the crucial terms used in this book, since it uses terminology that may be unfamiliar to readers.

1.2 Climate change: a serious North–South issue

1.2.1 A brief history

Human influence on the global climate was first signalled at the end of the nineteenth century (Fleming, 2005). In 1979, the first World Climate Conference was convened to discuss its scientific dimensions (WCC, 1979). In 1988, scientists and politicians agreed that this problem had major political ramifications (Toronto Declaration, 1988). In 1989, global heads of state met to put this issue on the top of their agendas (Hague Declaration, 1989); this meeting was followed rapidly by a meeting of environmental ministers (Noordwijk Declaration, 1989). A formalized intergovernmental scientific programme, the Intergovernmental Panel on Climate Change (IPCC), was launched in 1988 to prepare five-yearly reports on climate science, following closely in the footsteps of the Advisory Group on Greenhouse Gases established in 1986. In 1989, the UN established a negotiation process that led to the adoption of the United Nations Framework Convention on Climate Change in 1992 (FCCC, 1992), a follow-up Kyoto Protocol in 1997 (KP, 1997) and annual meetings of the Conference of the Parties. This section elaborates on the current state of climate science and argues that climate change is a typical North–South issue.

1.2.2 The state of the art of climate science

Drawing on the work of the IPCC as reflecting a global scientific consensus, the observed trends of global climate change show unequivocally that the Earth is warming (IPCC-1, 2007). The last decade (1995–2006) has been the warmest since 1850. In fact, the 100-year trend of 0.74 °C is higher than the 0.6 °C indicated by the previous IPCC report (IPCC-1, 2001). Observations reveal that the sea level has been rising since 1993 at an average rate of 3.1 mm per year. Glaciers are melting and snow cover is declining. Precipitation has increased in the northern hemisphere, but has declined in most of Africa, where increasing areas are suffering from drought. Although data on extreme events are less clear, there appears to be an increase in intense tropical weather events. Hydrological systems are influenced by higher and earlier run-off from snow cover, and ecosystems are being affected. Oceans are becoming acidic as carbon dioxide concentrations increase.

IPCC-1 (2007) and the IPCC Synthesis Report (2007) argue that many of these changes are consistent with model predictions and are 'very likely' to be due to increased concentrations of greenhouse gases, since anthropogenic emissions have increased by 70% in the last three and a half decades (1970–2004). The annual emissions of carbon dioxide increased by 80% during the same period, and its concentrations, as well as those of methane, are higher than those calculated for the last 650 000 years. Methane and nitrous oxide emissions have also increased similarly and are mostly from agricultural and fossil-fuel use. Although there has been some periodic cooling as a result of solar and volcanic forcings, this has not compensated for the rise in greenhouse gas emissions.

The IPCC Synthesis Report (2007) expects that these emissions will grow by 25%–90% (CO_2-equivalent) between 2000 and 2030, and will mostly be emitted by fossil-fuel use. This may lead to greater global warming, and these changes may be larger than those experienced in the last century. This could imply a warming of 0.2 °C per decade in the coming two decades, although studies since IPCC-1 (2007) have shown that the increase in temperature might not be so high because of the internal natural variability in the system.

Although serious climate change impacts are expected in the industrialized countries (ICs), these impacts will further exacerbate the existing problems for most developing countries (DCs). Many of the world's poor are already living in circumstances that can be referred to as 'dangerous' (UNDP, 2007: 7). Regional impacts from climate change are expected to vary. While the impacts will be severe in all countries, the IPCC (2007) predicts that in Africa between 75 and 250 million people will be affected by climate-change-related water stress by 2020. In the same period, agricultural produce from rain-fed land could decrease by 50% and the rising sea level will affect coastal regions, leading to a possible adaptation cost of about 5%–10% of GDP. Declining availability of fresh water will affect agriculture in Asia, and rising sea levels will affect densely populated coastal deltas. Temperature rises in Latin America may lead to gradual replacement of tropical forests by grasslands and, hence, species extinction and lower agricultural productivity.

The IPCC Synthesis Report (2007: 13) submits that 'Many of these changes could lead to some impacts that are abrupt or irreversible, depending on the rate and magnitude of the climate change.' The Climate Convention (FCCC, 1992: Art. 2) calls on countries to prevent 'dangerous anthropogenic interference with the climate system'. While many see this as calling for 'value' judgements, others are arguing that atmospheric concentrations of greenhouse gases need to be stabilized (Gupta and van Asselt, 2006). Current concentrations are at 380 ppm CO_2, whereas concentration levels were at 280 ppm prior to the industrial revolution. If the global community wishes to have a fair chance of keeping the temperature rise below 2 °C (compared with pre-industrial levels), emissions need to peak by 2015 or soon thereafter and then rapidly decline. This means that ICs and DCs need to take action. However, this 2° level is disputed. The USA announced finally at the G8 summit in 2009 that it would accept this level; for others (e.g. the Maldives) the target might not be stringent enough.

1.2.3 Climate change as a structural North–South issue

Climate change is a typical North–South issue. The North refers to the approximately 40 rich ICs listed in Annex I of the Climate Convention.[1] The South refers to the remaining 150 or so countries. There are clear problems with this classification (Gupta, 2007); see Chapter 10. However, there are also clear differences between the average rich and the average poor country.

Although in recent years there has been an effort to divert attention and focus on the emissions of China and India, their carbon footprint per capita is 1/5th to 1/15th that of the USA (UNDP, 2007: 7). The ICs have primarily emitted the greenhouse gases of the past. In historic and cumulative terms, British or American emissions are at 1100 tonnes compared with 66 for China and 23 for India. Although China and India are amongst the top gross emitters of CO_2, most emissions are from the ICs, and most DCs are making a very low contribution to the problem. However,

---

[1] Technically speaking, 36 countries are listed in Annex I; but some countries have split up since then (Czech Republic and Slovakia), and new countries have joined in Annex B of the Kyoto Protocol, hence the number of approximately 40.

if the climate change problem is to be kept within 'safe' limits, all countries will have to contribute to emission reductions, and fairly rapidly.[2]

While climate change affects all people, 'the world's poorest people are on the front line' (UNDP, 2007: 27). The countries of the South will be seriously affected by the impacts: they are geographically located where the impacts will be severest; and they are already facing serious development challenges, which will be exacerbated by the climate change problem, implying that their coping capacity is low. Past emissions are more important than future emissions since they determine current warming impacts and they limit the scope for other countries to increase their emissions if the global community collectively wants to stabilize greenhouse gas concentrations (UNDP, 2007: 41). The inertia of the climate system is such that, even if the global community takes serious action to reduce its emissions now, the global impacts will be felt at least for the next 50 years and will seriously affect the world's poor (UNDP, 2007: 4). Furthermore, should the world community wish to attain a 'safe' concentration level, the world budget of energy-related carbon emissions for the twenty-first century is likely to be exhausted as early as in 2032 (UNDP, 2007: 7), meaning that there is little room for Southern emissions to grow.

Politically, the climate issue is a classic North–South issue. The climate change negotiations are designed along Annex I (ICs) and non-Annex I (DCs) lines. The debate typically focuses on the role of the rich versus that of the poor. Although the rich countries created this role division in the 1990s, the poor stick to this configuration, because there is power in numbers. This distinction is increasingly harder to maintain, since some so-called poor countries are quite rich; and there are rich people in poor countries (see Chapter 10). The preamble (namely with reference to the need to develop), the principles (including the recognition of the common but differentiated responsibilities and respective capabilities of countries) and elements (provisions on technology transfer and assistance) of the Climate Convention (FCCC, 1992) build on these differences (see Chapter 5).

1.3 Climate change: the paradigm shift from a technocratic to a development issue

1.3.1 The paradigm shift

This section argues that the political framing of the climate change problem has changed over time (Gupta, 2009). In the 1990s, it was seen as an abstract, global, technological and economic challenge. By the decade's end, the problem had been reframed as a development problem. This paradigm shift has led to the political

---

[2] This does not, of course, mean that DCs should not take action to reduce their emissions (see Gupta, 2008). Box 10.1 discusses this.

discourse on the need to mainstream climate change into development and development cooperation (see Section 1.6).

1.3.2 Climate change: an abstract, technocratic, sectoral, mitigation issue

Climate change was initially defined as an abstract, global, future problem with a technocratic nature. The discourse was framed as a global issue in the academic documentation (e.g. IPCC-3, 1990) and the political declarations (e.g. the Noordwijk Declaration, 1989). The Climate Convention (FCCC, 1992) also qualifies the issue of climate change as a global issue several times, but scarcely uses the term 'local'.

Climate measures were to be 'cost-effective so as to ensure global benefits at the lowest possible costs' (FCCC, 1992: Art. 3(e)). The operating entity for the Convention's financial mechanism is the Global Environment Facility that aims to finance incremental costs or the costs of achieving only the global benefits, not the local benefits. The ICs appeared willing to pay for global benefits. Moreover, the discourses emphasized the present and the future in such a way that climate change was often seen more in terms of its long-term effects rather than as an urgent and current issue. This made the problem appear abstract in terms of the daily priorities of countries and peoples (Gupta, 1997; Gupta and Hisschemöller, 1997). The ICs initially did not link climate change with other development issues – that would imply that other development priorities would have to be addressed before the challenge of climate change could be addressed – and they avoided incorporating the local priorities of DCs, including adaptation (Bodansky, 1993) and desertification.

Furthermore, the problem was broken down into sectoral causes and impacts. The solution was presented in terms of technologies, technology transfer and cost-effective approaches that would encourage the use of such technologies. This technocratic framing possibly resulted from the dominant role of natural scientists in framing the issue; the assumption that environmental aspects can be internalized into existing processes; the need to avoid politicizing the issue and framing it in terms of polluters and liability; and the need to keep the problem focused and small so that it could be addressed effectively, without necessarily bringing in the whole range of development challenges. Hence, DCs saw the climate problem as a 'Western' problem, and one that was caused primarily by the ICs. They felt that the ICs should compensate them for the 'disproportionate or abnormal burden' on them (FCCC, 1992: Art. 3).

1.3.3 Climate change: an urgent, development, political, adaptation issue

However, there has been a subsequent shift in problem framing to focus on development issues. This has happened because different actors (including social scientists and development specialists) are increasingly joining the climate discussions;

scientific emphasis has shifted from the proximate causes (sectors and gases) of emissions to a more systemic analysis; the political consensus has moved towards the achievement of the Millennium Development Goals adopted by the UN in 2000 (see Table 4.1); the potential impacts and the related harm are beginning to manifest themselves (IPCC-1, 2007; Stern, 2007; Parry *et al.*, 2008: 1); and the geopolitics of the contemporary world is also now radically different from 20 years ago.

This shift in problem framing is evident in the recent literature (e.g. IPCC-3, 2007; Halsnæs *et al.*, 2008; Metz and Kok, 2008; Kok *et al.*, 2008) and political rhetoric (see Chapters 4 and 6), and is manifested by legal instruments emphasizing this (e.g. the Clean Development Mechanism, FCCC, 1992: Art. 12; Barera and Schwarze, 2004; Sutter and Parreño, 2007).

The ICs increasingly accept the need to internalize environmental costs into production processes and incorporate climate change into the development process (see e.g. Chapter 6). However, the internalization of such costs, i.e. the implementation of the 'polluter pays' principle, often makes related products and services less competitive internationally. By 2000, diverse driving forces (see Section 3.2) had, nevertheless, led the ICs to revisit their own position on climate change and international cooperation, leading to the mainstreaming discussion at international level (see Section 1.6).

This paradigm shift occurred in response to the DCs' claim that their top priority is development (Gupta, 1997) and has also influenced the DCs that now explicitly link climate change and development (e.g. GOC, 2007). 'Moreover, when one considers that three-quarters of the world's poorest citizens, those living on less than $2 per day, are dependent on the environment for a significant part of their daily livelihoods, climate change presents a serious multifaceted development challenge' (US AID, 2008: 1). 'The poorest countries and communities are likely to suffer the most because of their geographical location, low incomes, and low institutional capacity, as well as their greater reliance on climate-sensitive sectors like agriculture' (Mani *et al.*, 2008: ix). As the UNDP (2007: 1) sums it up: 'Climate change is the defining development issue of our generation.'

1.4 Climate change: the linkage with sustainable development

1.4.1 Introduction

Climate change 'involves complex interactions between climatic, environmental, economic, political, institutional, social, and technological processes' (IPCC-3, 2001: 78) and is thus closely linked to development and sustainable development. Development refers to the process by which society enhances its social, economic and natural resource capital and the development challenges refer to issues ranging from meeting the basic necessities of people to enhancing their national income. Sustainable development redefines development so that it meets the needs of the present without compromising the ability of future generations to meet their own needs (WCED 1987); see Chapters 2 and 4.

Greenhouse gas emissions are generally associated with the process of development in the energy, transport, agriculture, household and other sectors. The use of fossil fuels in industry and transport is a critical source of emissions. Agriculture and land use sectors use inputs produced through the use of energy and also emit greenhouse gases (e.g. wet rice production emits methane and deforestation emits carbon dioxide). Sectoral uses are embedded in the structures of society.

This section elaborates on how the legal agreements link climate change and sustainable development, the theory of the environmental Kuznets curve and the implications for technology-transfer discussions, the substantive links between climate change mitigation and sustainable development, and those between adaptation to climate change and sustainable development.

1.4.2 The Climate Convention: climate change and sustainable development

The relationship between climate change and development is ambiguous in the Climate Convention (Arts and Gupta, 2004). It refers to sustainable development as both a right and a goal (FCCC, 1992: Art. 3(4)). Sometimes, economic development is seen as a pre-condition to taking mitigation and adaptation measures for all countries (FCCC, 1992: Preamble para. 22; FCCC, 1992: Arts. 4(2)(a), 3(4) and 3(5)). However, the ultimate objective of the Convention calls for the problem of climate change to be addressed within a time-frame that allows, inter alia, economic development to proceed in a sustainable manner (FCCC, 1992: Art. 2). This implies that sustainable development can also lead to sustained economic growth.

The Kyoto Protocol is less ambiguous. It calls on all countries to promote sustainable development in their climate-relevant policies (KP, 1997: Art. 2). The article establishing the Clean Development Mechanism – an instrument of cooperation between rich and poor countries – calls on countries to ensure that related projects meet the criteria of sustainable development (KP, 1997: Art. 12); see Chapter 5. At the Conference of the Parties in 2001 at Marrakesh, the Ministerial Declaration (Marrakesh Accords, 2001: Dec. 1/CP7) emphasized the link between climate change and sustainable development, including the developmental issues of poverty, land degradation, access to water and food, and human health. It stated that sustainable development should be included in the reports to the Convention

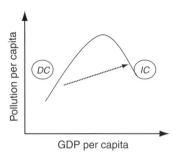

Figure 1.1 The environmental Kuznets curve.

secretariat, including the National Communications[3] and the National Adaptation Programmes of Action (NAPA).[4]

1.4.3 The environmental Kuznets curve

The above confusion about whether countries need to develop first before they can invest in sustainable development is not easily resolved. The environmental Kuznets curve[5] showed, on the basis of empirical evidence, that as societies become richer on a per capita basis they pollute more, see e.g. Malenbaum (1978) and Jänicke *et al.* (1989), for material use, and Grossman (1995) and Selden and Song (1996), for pollutants. Subsequently, as they continue to become richer they can invest more in pollution-control technology and this leads to less pollution per capita (see Figure 1.1). This implied that DCs would first have to invest in becoming richer and only then could they invest in pollution control. It also implied that ICs could more easily decouple their pollution from their economic growth. This was a fairly dominant assumption in the early 1990s.

This implies that investing in pollution control is unlikely in the DCs until they have passed a certain threshold. To address this impasse that DCs would have to develop first before they could invest in climate-friendly technology, the concept of technology transfer and capacity building was adopted (see Section 1.4.4).

This idea was adopted in the Climate Convention (FCCC, 1992: Art. 4(5)) and it referred to the need to develop of the DCs (FCCC, 1992: Preamble paras. 3, 21 and

---

[3] Reports to be provided by all parties to the Convention secretariat demonstrating activities undertaken to fulfil the obligations under the Convention (FCCC, 1992: Art. 12).

[4] Reports on adaptation to be prepared by the least-developed countries using financial support made available under the Convention process (Decision 28/CP.7, para. 7).

[5] The term environmental Kuznets curve is based on its similarity to the time-series pattern of income inequality described by the development economist Simon Kuznets in 1955. A 1992 World Bank Development Report made the notion of an environmental Kuznets curve popular by suggesting that environmental degradation can be slowed down by policies that protect the environment and promote economic development.

22, and Art. 3) and the fact that effective environmental legislation may be too expensive for them (FCCC, 1992: Preamble para. 10; cf. KP, 1997: Art. 10). These views are consistent with the notion of the right to development that DCs have insisted on for a long time (see Chapter 4). Furthermore, policies in DCs were seen as dependent on the assistance provided (FCCC, 1992: Art. 4(7)). A corollary to the implication of the environmental Kuznets curve is the need for ICs to demonstrate leadership by decoupling pollution from economic growth and the need to provide assistance to DCs to help them leapfrog their way to development (see Chapter 5).

However, this approach is problematic. The *mistaken optimism argument* (Gupta, 1997) is that DCs can learn from the past mistakes of the ICs; the problem is defined in terms of technology and is, theoretically speaking, easy to correct incrementally; it gives a direction to the development process; and it reconfirms the leadership of the North and in doing so also allows for unchanged lifestyles for the rich. This is problematic insofar as it assumes that (a) richer societies have the solutions needed, while one could argue that environmental problems including climate change are a symptom of problems associated with the developmental model adopted by the ICs (Nyerere *et al.*, 1990; Khor, 2001, Agarwal *et al.*, 1992); (b) growth-oriented capitalism is inherently ecologically destructive (Gorz, 1994) and compatible with the existence of poverty; and (c) adopting modern technologies is expensive, is often incompatible with local circumstances, and may be compensated for by the growth of consumption. Such ideas come not only from Southern scholars but also from Northern economists such as Sir Nicholas Stern (2007: 1), who states that 'Climate change presents a unique challenge for economics: it is the greatest and widest-ranging market failure ever seen.'

Second, since the 1990s, scientific papers in the area of the environmental Kuznets curve suggest that, while the hypothesis may hold for individual local pollutants, there are serious questions applicable to the theory in general. Copeland and Taylor (2004) argue that, although growth leads to increased pollution, there is also evidence that it leads to more investment in environmental improvement and so there is no simple relationship and the net impacts may be difficult to predict. Further, basing this relationship simply on per capita income is a little simplistic since other structural national features, such as natural resources and capital, and issues related to trade might have a greater impact.

Several scholars have argued that, although the environmental Kuznets curve may apply for some individual local pollutants involving short-term costs, empirical evidence does not show that it holds for other global pollutants such as carbon dioxide, which have long-term and dispersed costs (see e.g. Opschoor, 1995; De Bruyn and Opschoor, 1997; Dinda, 2004). There may be an N-shaped curve instead (see Figure 1.2). Further, at aggregate levels of local pollution, the environmental

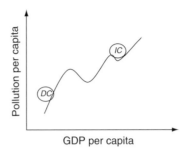

Figure 1.2 The N-shaped curve.

Kuznets curve might not hold. Finally, the curve might not be valid for certain types of pollution such as land use change and biodiversity loss.[6] The authors of a more recent study (Caviglia-Harris *et al.*, 2009) suggest that there is no inverted-U-curve relationship between environmental impacts and economic growth – and, hence, argue that growth alone cannot lead to sustainable development.

On the other hand, there are also more positive expectations. Stern (2004) argues that there is evidence that DCs are adopting relevant policies. 'However, emissions may be declining simultaneously in low- and high-income countries over time, *ceteris paribus*, though the particular innovations typically adopted at any one time could be different in different countries' (Stern, 2004: 1435).

This implies that, if there is no inverted-U curve, the likelihood that the ICs can lead in decoupling is much less evident; although, of course, if the ICs have the political will, this can be done. At the same time, the notion that development precedes sustainable development becomes more questionable – touching at the heart of many of the arguments used during the negotiation process.

The N-shaped curve perhaps explains why representatives both from ICs and from DCs often talk about development first, before committing to far-reaching measures on climate change, and can explain why such countries as Australia, Iceland, Norway and Spain asked to increase their emissions in the Kyoto Protocol and why the USA has, at the time of writing, not ratified the Kyoto Protocol. But this does not imply that there are no technological solutions available (von Weiszäcker *et al.*, 1997; Metz *et al.*, 2000). It also does not imply that the rich do not have more options for decoupling and that the poor cannot learn from the rich. However, technologies without institutional support and behavioural change may be inadequate; they may also be irrelevant and unaffordable for the DCs, so the DCs may need to find their own context-relevant solutions.

---

[6] Dinda also refers to the critique of methods, the over-reliance on single sectors, the assumptions regarding the development trajectories, the focus on income rather than human development index and so on.

1.4.4 Leapfrog technology

In 1990, on the basis of early insights from the environmental Kuznets curve, scientists at the Second World Climate Conference recommended that DCs should be assisted in using modern technologies: 'It is clear that developing countries must not go through the evolutionary process of previous industrialization but rather, must "leapfrog" ahead directly from a status of under-development through to efficient, environmentally benign, technologies presently found only in the most advanced industrial economies' (SWCC, 1990: para. 2.2).

The Convention adopted this idea (FCCC, 1992: Art. 4(5)) and there were hopes that this would help to decouple economic growth from environmental pollution. Such decoupling could be undertaken through decarbonization and dematerialization processes (the former aims to reduce the associated carbon content of production and consumption processes while the latter reduces the energy and mass throughput in the system). However, technology transfer is complicated because, first, most technologies are owned privately and available only on commercial terms, and, where such technologies are public, there are no easy processes to make them available (Andersen *et al.*, 2000a: 19). This means that there need to be financial provisions to ensure such transfers (Henikoff, 1997), and this has been partly arranged through the financial mechanism (FCCC, 1992: Art. 21) and the Clean Development Mechanism (KP, 1997: Art. 12), which promote technology transfer in return for emission credits. Second, the history of technology transfer suggests that it works only under the condition that the focus is on leapfrog technology transfer. This would imply focusing on adapting leapfrog technologies to fit into the circumstances of DCs and avoiding selling older technologies to DCs as the pollution-haven hypothesis suggests.

The IPCC Special Report on Technology Transfer (Metz *et al.*, 2000) takes a technologically optimistic look at the role of technology transfer in addressing climate change. It calls for rapid technological innovation to develop environmentally sound technologies and their transfer to DCs. 'Development with modern knowledge offers many opportunities to avoid past unsustainable practices and move more rapidly towards better technologies, techniques and associated institutions' (Metz *et al.*, 2000: 3). It argues that technology transfer can be made more effective through effective capacity building (in terms of human skills, organizational skills and information-assessment and monitoring-capacity skills), the creation of an enabling environment and developing useful mechanisms for technology transfer.

The technology-transfer debate is hopeful because capacity building can be used to make leapfrog technologies compatible with local situations in DCs. The DCs also provide greenfield situations where countries are not yet in an infrastructural, industrial and technological lock-in situation. This means that change is relatively easier than in countries with one dominant techno-economic paradigm characterized by interlocking social and industrial infrastructure that resists change for as long as possible (Mansley *et al.*, 2000). On the other hand, the N-shaped curve cited above suggests that technology transfer may also exacerbate the situation by selling Northern lifestyles and patterns to the South.

The IPCC defines capacity building as follows:

Capacity building is required at all stages in the process of technology transfer. Social structures and personal values evolve with a society's physical infrastructure, institutions and the technologies embodied within them. New technological trajectories for an economy therefore imply new social challenges. This requires a capacity of people and organisations to continuously adapt to new circumstances and to acquire new skills.

(Andersen et al.*, 2000b: 4–5)*

Unlike technology transfer, through which markets are created, there are no immediate returns on capacity building for the ICs, except where technical assistance is involved.

1.4.5 Climate change mitigation and sustainable development

Technology transfer aims to enhance opportunities for mitigation and adaptation. Mitigation is 'an anthropogenic intervention to reduce anthropogenic forcing of the climate system; it includes strategies to reduce greenhouse gas sources and emissions and enhancing greenhouse gas sinks' (IPCC-2, 2007: 878 [italics removed]). The IPCC Third Assessment Synthesis Report (2001) argued that climate change and sustainable development are linked and that policies could be made more effective if the deliberative process took short- to long-term thinking into account, used an expanded list of tools for deliberation, took a broader set of policies and criteria into account, and used a portfolio approach to developing policies. The Fourth Assessment Report argued that mitigation can have ancillary benefits for sustainable development, and that development that has many sustainable features can provide fruitful conditions for promoting mitigation (IPCC-3, 2007: 693). Sustainable development is not about choosing a pre-mapped path but about 'navigating through an unchartered and evolving landscape' (IPCC-3, 2007: 693). 'The more climate change issues are mainstreamed as part of the planning perspective at the appropriate level of implementation, and the more all the relevant parties are involved in the decisionmaking process in a meaningful way, the more likely they are to achieve the desired goals' (IPCC-3, 2007: 693).

Arguments regarding the relationship between development and mitigation include the following.

• Development, for the poor, implies increased access to basic needs including food, water, clothing and shelter, but also to education and health services and enhanced opportunities to participate in the labour force, all of which could increase greenhouse gas emissions but could also enhance resilience with respect to the impacts of climate change. This is where the distinction between survival and luxury emissions is relevant (Agarwal and Narain, 1991; Opschoor, 2009).

- Development, for the rich, involves more production and consumption, which implies using more energy, transport, greater amounts of waste, greater degrees of agricultural production and investment in infrastructure and housing, all of which leads to greater greenhouse gas emissions. Although economic growth may lead to increased greenhouse gas emissions, policy choices could affect the level of emissions and it is very possible for countries to adopt good policies (IPCC-3, 2007). Sectors that have not yet reached their maximum production level have win–win opportunities to reduce their greenhouse gas emissions per unit of production at the same time as maximizing production (IPCC-3, 2007). Beyond that point there may be trade-offs.
- Policy choices need to be sustained and interconnected with decisions in other fields to ensure long-term sustained emission reduction (IPCC-3, 2007).
- Energy efficiency, increasing the productivity of resource use, better waste management, efficient and sustainable transport systems and combating deforestation are areas in which win–win opportunities for reducing emissions and promoting sustainable development exist (IPCC-3, 2007). For example, in the energy-supply sector, mitigation options include fuel-switching from GHG-intensive sources to others, improved efficiency in production, distribution and use, and conservation practices at the user end. In the transport sector, this could mean shifting to less energy-intensive modes of transport (e.g. bicycling, but also hybrid vehicles), and land use and transport planning. In the building sector, this could have implications for building design through improving access to natural light and insulation, but also more efficient delivery of services in the buildings and better use by users.

The key question is does the global community have the political will, social willingness, technologies, resources and institutions to promote such fundamental change in society?

1.4.6 Climate change adaptation and sustainable development

In order to cope with the impacts of climate change, societies will have to adapt. Adaptation implies 'adjustment in natural or *human systems* in response to actual or expected climatic stimuli or their effects, which moderates harm or exploits beneficial opportunities' (IPCC-2, 2007: 869). Adaptation can be anticipatory (prior to an impact), autonomous (triggered by ecological, market or welfare changes) or planned (resulting from a deliberate policy decision).

In general, the vulnerability to climatic impacts will be higher where there are other stresses on individuals, such as 'poverty, unequal access to resources, food security, environmental degradation, and risks from natural hazards' (IPCC-2, 2007: 813). Efforts to reduce vulnerability and to promote sustainable development 'share common goals and determinants including access to resources (including information and technology), equity in the distribution of resources, stocks of human and social capital, access to risk-sharing mechanisms and abilities of decision-support mechanisms to cope with uncertainty' (IPCC-2, 2007: 813). However, some development activities exacerbate vulnerabilities. Development can exacerbate adaptation through (mal)development – where societies develop without taking the potential impacts of climate change into account (Tearfund, 2006). Finally, adaptation activities can have ancillary benefits for development.

1.4.7 Climate change cooperation

Climate change cooperation between ICs and DCs refers to the cooperative mechanisms set up under the climate regime. It was clear from the start of the negotiations that substantial resources would be necessary to deal with the climate change problem and that DCs would be particularly vulnerable. However, who should pay for these costs? This depends on criteria for dividing responsibilities between countries. The pre-1992 political declarations discussed the need for the ICs to provide 'new and additional' resources (see Section 5.3.2) to the DCs to help them deal with climate change. Such new and additional resources were justified on the grounds that the ICs, the primary emitters of greenhouse gases, would need to compensate the DCs for the abnormal burden they would have to bear and because the ICs had the ability to pay. Hence, the Climate Convention states that developed country Parties 'shall also provide such financial resources, including for the transfer of technology, needed by the developing country parties to meet the agreed full incremental costs of implementing measures' (FCCC, 1992: Art. 4(3)). This would include the making of national inventories of emissions, national programmes, the promotion of technology transfer and sustainable management of natural resources, cooperation on adaptation, integration of climate change considerations into national policies, cooperation in research, information and education, and the preparation of National Communications.

Article 11 of the Kyoto Protocol states that the ICs shall 'provide new and additional financial resources to meet the agreed full costs' incurred by DC Parties in implementing certain activities such as emission inventories and should provide financial resources, including for the transfer of technology, needed by the DC Parties to meet the agreed full incremental costs of advancing the implementation of existing commitments in Article 4.

In follow-up processes the climate regime established three types of cooperative mechanisms: technology transfer under the Convention and Protocol, financial mechanisms under the regime and a market mechanism – the Clean Development Mechanism.

1.5 Development cooperation: concepts and figures

1.5.1 Introduction

International cooperation includes, inter alia, development cooperation and climate change cooperation (see Section 1.4.7). Development aid/cooperation/partnership is a post-World War II phenomenon whereby some rich countries took on the responsibility to provide financial resources to DCs to help them develop. Chapter 2 discusses the literature with respect to development cooperation. The process of aid provision and related policymaking ran parallel to the process during which the DCs demanded the New International Economic Order (see Section 4.2.2) and the right to development (see Section 4.4.2). It also ran parallel to other UN processes promoting development in various parts of the world. Chapter 4 explores these different processes and discusses the Millennium Development Goals to provide the context for the current discussions on aid.

This section provides a definition of official development assistance (ODA) and puts it into the context of other international flows. It shows how ODA flows have evolved over the years and what ODA resources are spent on.

1.5.2 Conceptual clarification

International aid consists of resources transferred primarily by OECD donors to recipients to promote human welfare and enhanced development of the recipients. The financial flows include official development assistance (ODA), other official development finance (ODF), international bank lending, foreign direct investment and other private flows, including remittances from migrant workers. On a global scale, financial flows (both private and official, or governmental) originate mainly from the members of the OECD Development Assistance Committee (DAC).[7] There are also flows from non-DAC countries (see Box 1.1). Official development assistance or official aid is a distinct part of the cooperation with DCs (see Chapter 4).

ODA includes transactions undertaken by the official sector to promote the economic development and welfare of DCs; is concessional in character; and includes a grant element of at least 25%, calculated at a discount rate of 10% (i.e. a grant or a 'soft loan') (OECD, 2007). ODA can be provided bilaterally to countries listed as 'ODA recipients' on the DAC list (low- or middle-income

---

[7] The 23 DAC countries are Australia, Austria, Belgium, Canada, Denmark, Finland, France, Germany, Greece, Ireland, Italy, Japan, Luxembourg, the Netherlands, New Zealand, Norway, Portugal, Spain, Sweden, Switzerland, the UK, the USA and the European Community (EC). The World Bank, IMF and UNDP are observers. South Korea became a DAC member in 2010, raising the total to 24.

Box 1.1 Official assistance by non-OECD DAC members

Traditionally, the discussion on aid centres on the aid provided by OECD DAC
countries to DCs. At the same time, some DCs also provide aid to other DCs.

Assistance provided by DCs is increasing, quadrupling from a stable annual
average of USD 1.2 billion between 1997 and 2001, to USD 5.5 billion by 2007
(OECD, 2009b). Most of this assistance comes from the Arab states. The
Coordination Group of Arab National and Regional Development Institutions has
coordinated Arab aid to 140 DCs since 1975. Cumulatively they have given
unconditional and highly concessional aid of USD 76 billion excluding state-to-state
assistance (Al-Hamad, 2003). They provide 0.85% of their income as aid (Al Saud,
2003): mostly to Asia (energy, industry and mining) and Africa (New Partnership for
Africa's Development) (Querishi, 2003; Jatta, 2003). Saudi Arabia and Kuwait are
the two largest donors. Their aid projects have reduced costs, enhanced services and
improved social welfare (ESCWA, 2007), but are mostly motivated by religious and
commercial goals, although these are probably supply driven (Villanger, 2007).
Environmental issues have not appeared to play a critical role.

China is another important donor, providing aid since 1950, initially to support other
communist countries and sometimes to secure support in UN meetings (Lancaster,
2007). The Department of Foreign Aid falls under the Ministry of Commerce. It
sponsors infrastructure and development projects in Africa. Lancaster (2007: 1) argues
that the 'Chinese provide their aid largely without the conditions that typically
accompany Western aid – a good human rights performance, strong economic
management, environmentally responsible policies and political openness on the part of
recipient governments. We know that Chinese aid emphasizes infrastructure, something
many poor countries need and want but often find traditional Western aid donors
reluctant to fund.' Van Dijk (2009) argues that Chinese aid is often driven by conflicting
objectives: the aid is not transparent and open to discussion; and Chinese assistance can
bring opportunities and also threats to Africa.

---

countries), or through multilateral institutions for redistribution to DCs. The 1969
definition was modified in 1972, to distinguish aid for economic and social development and welfare in DCs from 'other official flows' including, for example,
military assistance and trade-related flows such as reduced tariffs and concessions
on imports from DCs. Bilateral aid includes grants and grant-like flows (e.g. for
technical cooperation, developmental food aid, humanitarian aid, debt forgiveness
and administrative costs), and loans.

Table 1.2 shows that the total net financial flows from DAC donor states to DCs
since 1995 have increased and peaked in 2005–6, reflecting an increase in financial
flows due to exceptional Paris Club debt-relief operations (in particular for Iraq and

Table 1.2 *Total net flows from DAC countries by type of flow in USD million*

	1995–9 average	2000–4 average	2005–6 average	2007
Official development assistance	54 191	62 614	105 749	103 491
Other official flows	10 162	−2 381	−4 172	−6 438
Private flows at market terms	113 751	49 047	187 169	325 350
Net grants by NGOs	5 758	8 907	14 680	18 508
Total net flows	183 862	118 057	303 426	440 912

Source: Based on OECD DAC statistical databases, at 2007 prices and currency rates.

Nigeria). Private flows, transactions at commercial terms, have trebled in this period and are three times as large as ODA. However, the 2007 figures possibly reflect not a trend, but a sudden peaking in investment. Grants from NGOs have also trebled but are about a fifth of ODA. Grants from NGOs often include resources provided by the state, so there may be some double counting here.

1.5.3 Data on flows, donors and sectors

ODA is provided by the OECD DAC countries (75%) and multilateral organizations (20%) and amounted to about USD 100 billion annually in the period 2005–7, almost double the annual average of USD 55 billion for the two previous decades. This increase includes exceptional debt-relief operations and special-purpose grants, rather than programmable aid. Excluding debt relief, the amount of ODA disbursements by DAC countries rose by 2.4% in 2007 compared with 2006. With the end of the three-year Paris Club debt-relief operations in 2007, the total ODA flow and the share of debt relief are expected to diminish and normalize from 2008 onwards. Figure 1.3 on the trends in total net ODA flows from DAC donors throughout the past five decades[8] shows a slowly rising trend in ODA, decreasing briefly in the late 1990s and increasing since 2005.

In absolute terms, in 2007, the USA was the largest donor, followed by Germany and France. However, in relative terms, when ODA is seen as a percentage of GNI, these countries score much lower, with the USA as one of the least generous donors globally (see Section 4.4.3 on the 0.7% commitment). The 15 EU countries that are DAC members together provided 59.5% of total net ODA. The share provided by

---

[8] Not all current DAC countries have been members since the DAC's establishment in 1960. The latest country to join was South Korea, in 2010.

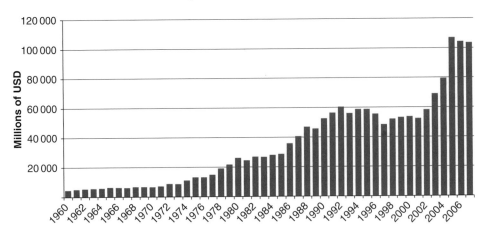

Figure 1.3 Net ODA by all DAC countries: 1960–2007. Source: Based on OECD DAC statistical databases.

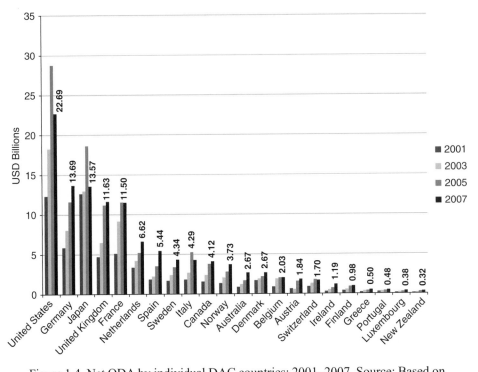

Figure 1.4 Net ODA by individual DAC countries: 2001–2007. Source: Based on OECD DAC statistical databases.

the EU and its Member States is projected to increase to some 65% by 2010, in contrast with the projected further decline in US ODA disbursements (OECD, 2007). Figure 1.4 provides an overview of individual DAC countries' total net aid disbursements in 2007, with comparative statistics from 2005, 2003 and 2001,

Table 1.3 *Major aid uses: ODA by sector in percentages*

	1984–5 (%)	2004–5 (%)	2006–7 (%)
Social and administrative infrastructure	26.5	33.4	36.7
Economic infrastructure	18.4	13.3	12
Agriculture, forestry and fishing	11.5	3.4	3.7
Industry and other production	5.9	2.3	1.7
Commodity aid and programme assistance	18.4	2.8	4.1
Humanitarian aid	1.9	10.0	7.2
Other	17.4	34.8	34.7

Source: Based on OECD DAC (2009b).

to correct for incidental increases or decreases in aid, such as for exceptional debt relief.

ODA disbursements are spent on development programmes to improve the social, administrative and economic infrastructure in DCs. In 2007, nearly 60% consisted of programmable aid for these sectors. According to the OECD DAC (2009b), this ratio of 'country programmable aid' is fairly typical in annual ODA disbursements. Country programmable aid refers to the amount of aid that can be programmed by DCs in their budgets. Such aid excludes forms of aid such as emergency relief, debt relief and other aid that is not programmable by donors. During 2007, debt relief made up 9.4% of the ODA allocations, half of which alleviated Iraqi debt. In the preceding two years, Iraq and Nigeria alternated as top recipients of exceptional debt relief at respectively 21% (2006) and 24.3% (2005) of the total ODA disbursements. Emergency and humanitarian aid accounted for 6.75% of the total aid spent in 2007.

Major aid uses by individual DAC donors include social and administrative infrastructure; economic infrastructure; the production sectors, including agriculture, forestry and fishing, industry, mining, construction, trade and tourism; commodity aid and general programme assistance; emergency and humanitarian aid; and other expenses. Table 1.3 shows that most aid is used for improvement of social and administrative infrastructure. The OECD DAC data provide sparse details on aid disbursements for environmental protection projects. The only mention of the environment in the breakdown of annual ODA allocations is to environmental protection as a multi-sector/cross-cutting category. During 2007, USD 2.3 billion was spent under this heading, which is double the 10-year average.

In 2007, nearly 19% of the total net ODA disbursements went to the Least Developed Countries (LDCs), and Sub-Saharan Africa received 20.7% of the total aid flows.

1.6 The political trend towards incorporating climate change cooperation into development cooperation

Section 1.3 argued that there has been a paradigm shift towards linking climate change and development. Together with this shift, there is a recent trend towards 'mainstreaming' climate change in development cooperation. This section previews this trend and the substantive links between climate change mitigation, adaptation and development cooperation. Both are elaborated upon in subsequent chapters.

There is a paradigm shift from seeing climate change as a sectoral, technocratic issue to giving it a human, local, development face (see Section 1.3). In this decade, diverse social actors with vastly different motivations are all arguing that climate change should be incorporated and 'mainstreamed' into development and development cooperation (see Table 3.2). This trend is visible within the EU, where mainstreaming of climate change into development cooperation is seen as consistent with its existing policy to integrate environmental issues into development issues (see Chapter 6). This trend is also visible in UN agencies, where various bodies are promoting mainstreaming. International banks are also linking climate change to development activities (see Chapter 4).

Although this trend exists, there is considerable confusion regarding the difference between mainstreaming and integration. The authors of this book see mainstreaming as a political concept that moves a marginal idea to the centre of discussions in order to redesign other existing discourses. In contrast, integration is considered a policy approach and tool to ensure coherence between sectoral activities and activities at centralized and decentralized levels (see Chapter 3; Table 3.3).

Behind this trend towards incorporating climate change into development cooperation are several substantive arguments. First, climate change and development are intimately linked, so climate change and development cooperation should be linked too (see Section 1.3.3). Second, waiting for DCs to develop before addressing climate change may be inappropriate given what is known about the environmental Kuznets curve (see Section 1.4.3). Third, both ICs and DCs are increasing their investments in fossil fuels, despite the consequent greenhouse gas emissions. Some of the investment in fossil fuels in DCs is financed through international development cooperation, bank loans and export credits, as well as foreign direct investment. As poorer societies follow the Western model of growth, their greenhouse gas emissions also increase accordingly. Thus, as China liberalizes, its emissions continue to grow exponentially, in line with its national income, although there may be no clear one-to-one relationship between the two. Hence, these transfers may be counter-productive and might lock DC societies into the use of specific technologies and processes. At the same time, on the adaptation front, the reasons for taking action are that the achievement of the Millennium Development

Goals may be negatively affected by the impacts of climate change (Agrawala and van Aalst, 2008; UNDP, 2007); and development programmes and projects may increase the vulnerability of local people by promoting inappropriate development (Tearfund, 2006). However, development cooperation activities may be designed to promote sustainable development and enhance the resilience of local people with respect to the impacts of climate change.

The mainstreaming and integration of climate change into development cooperation is *politically sensitive* in the context of the past international negotiations both on development issues and on climate change, and these sensitivities are discussed in future chapters (see Chapters 4 and 5). This does not imply that there are no *practical reasons* for mainstreaming and integration in order to enhance the synergies between the different systems of cooperation and to minimize the contradictions. But there are very real fears. Will these resources be diverted to meet what ICs see as critical goals, or will they also be used for the needs of the DCs themselves? Will they be diverted for environmental goals at the cost of social goals?

A key challenge is to understand how DCs can be persuaded to participate more effectively in the climate change regime because, without their effective participation, the motivation for the ICs to take action is reduced, and because ultimately it is in the best interest of the DCs themselves to mitigate emissions and reduce the need for adaptation.

At the same time, there are also major institutional challenges in the ICs that affect their ability to act in accordance with their rhetoric and to design cooperative mechanisms that are truly effective. Sometimes, it is also simply the complications of aid delivery and the number of aid providers that make the process cumbersome.

Any effort to link climate change with development cooperation, and DCs with ICs, will have to take the complex chain of events into account in determining how aid policy can best be improved. It will have to take into account the legal basis in the climate change negotiations on the need for assistance; the political sensitivities in mainstreaming climate change in development cooperation from a DC point of view; the political and practical perspective of combining climate change with development cooperation from an IC point of view; the role of development cooperation in a broader range of cooperative instruments between ICs and DCs; and how the differences of perspectives can be reconciled.

1.7　The aims and structure of this book

As this chapter has shown, the trend of linking climate change and development cooperation raises a number of critical questions. What is the relationship between climate change cooperation and development cooperation? Is mainstreaming climate change into development cooperation a good idea? To what extent is climate change being mainstreamed into development cooperation in practice and in theory? To what extent should it be mainstreamed? To address these questions, this book combines literature analysis on climate change, development and development cooperation, policy overviews at international and EU level and case studies reviewing the needs of ten selected DCs and the supply of assistance from the EU.

The book begins by framing the issue in terms of the literature on development cooperation and mainstreaming. It highlights the goals of development cooperation and the lessons learned from development cooperation (see Chapter 2). It then defines the ways in which climate change can be incorporated into development and development cooperation and discusses the concept of mainstreaming (see Chapter 3). It then examines the evolving policy on global cooperation and climate change cooperation. It looks at the evolution of UN policy on development and development cooperation and OECD policy on development cooperation, and highlights issues such as the right to development, the financial commitment to development cooperation of ICs and the need for resources (see Chapters 4 and 5). It examines the position of the EU and its Member States on development cooperation and the role of climate change therein (see Chapters 6 and 7). It assesses the needs of the DCs, as reflected in national reports to international agencies, e.g. the National Communications and the National Adaptation Programmes of Action, and the supply of assistance as reflected in development cooperation documents, e.g. the Country Strategy Papers (CSPs) for selected DCs (see Chapters 8 and 9). Chapter 10 then draws conclusions.

By filling a gap in the literature on linking climate change, development and development cooperation, and articulating the views of ICs and DCs, this book aims to contribute to the ongoing discussions on this important issue.

Acknowledgements

The authors would like to thank Mike Hulme, Onno Kuik, Meine Pieter van Dijk, Leo Meyer, Bart Strengers, Eileen Harloff and Marcel Kok for their comments on (parts of) this chapter.

References

Agarwal A. and Narain, S. (1991). *Global Warming in an Unequal World: A Case of Environmental Colonialism*. New Delhi: Centre for Science and Environment.

Agarwal, A., Carabias, J., Peng, M. K. K. *et al.* (1992). *For Earth's Sake: A Report from the Commission on Developing Countries and Global Change*. Ottawa: International Development Research Centre.

Agrawala, S. and van Aalst, M. (2008). Adapting development cooperation to adapt to climate change. *Climate Policy*, **8**(2), 183–93.

Al Saud, T. b. A. A. (2003). Arab aid: a durable and steadfast source of development finance, in *Arab Aid: Past, Present and Future*. Dubai: The OPEC Fund for International Development, pp. 15–25.

Al-Hamad, A. Y. (2003). The coordination group of Arab national and regional development institutions, in *Arab Aid: Past, Present and Future*. Dubai: The OPEC Fund for International Development, pp. 9–14.

Andersen, S. O., Buckley, E. N., Chandler, W. *et al.* (2000a) Technical summary, in *Methodological and Technological Issues in Technology Transfer*, ed. B. Metz, O. R. Davidson, J.-W. Martens, S. N. M. van Rooijen and L. v. W. McGrory. Cambridge: Cambridge University Press, pp. 5–43.

Andersen S. O., Chandler, W., Christ, R. *et al.* (2000b). Summary for policy makers: methodological and technological issues in technology transfer, in *Methodological and Technological Issues in Technology Transfer*, ed. B. Metz, O. R. Davidson, J.-W. Martens, S. N. M. van Rooijen and L. v. W. McGrory. Cambridge: Cambridge University Press, pp. 1–9.

Arts, K. and Gupta, J. (2004). Sustainable development in climate change and hazardous waste law: implications for the progressive development of international law, in *The Law of Sustainable Development*, ed. N. Schrijver and F. Weiss. Dordrecht: Kluwer Academic Publishers, pp. 519–51.

Barrera, J. and Schwarze, R. (2004). Does the CDM contribute to sustainable development? Evidence from the AIJ pilot phase. *International Journal of Sustainable Development*, **7**(4), 353–68.

Bodansky, D. (1993). The United Nations Framework Convention on Climate Change: a commentary. *Yale Journal of International Law*, **18**, 451–588.

Caviglia-Harris, J. L., Chambers, D. and Kahn, J. R. (2009). Taking the "U" out of Kuznets: a comprehensive analysis of the EKC and environmental degradation. *Ecological Economics*, **68**(4), 1149–59.

Copeland, B. R. and Taylor, M. S. (2004). Trade, growth, and the environment. *Journal of Economic Literature*, **42**(1), 7–71.

De Bruyn, S. M. and Opschoor, J. B. (1997). Developments in the throughput–income relationship: theoretical and empirical observations. *Ecological Economics*, **20**(3), 255–69.

Dinda, S. (2004). Environmental Kuznets curve hypothesis: a survey. *Ecological Economics* **49**(4), 431–55.

ESCWA (2007). *Economic Trends and Impacts: Foreign Aid and Development in the Arab Region*. UN Document E/ESCWA/EAD/2007/1. Beirut: UN Economic and Social Commission for Western Asia.

FCCC (1992). United Nations Framework Convention on Climate Change. Signed 9 May 1992, in New York, NY; entered into force 21 March 1994. Reprinted in (1992). *International Legal Materials*, **31**(4), 849.

Fleming, J. R. (2005). *Historical Perspective on Climate Change*. Oxford: Oxford University Press.

GOC (2007). *China's National Climate Change Programme: 2007*. Beijing: National Development and Reform Commission of the People's Republic of China.

Gorz, A. (1994). *Capitalism, Socialism, Ecology*. London: Verso Books.

Grossman, G. M. (1995). Pollution and growth: what do we know?, in *The Economics of Sustainable Development*, ed. I. Goldin and L. A. Winters. Cambridge: Cambridge University Press, pp. 19–42.

Gupta, J. (1997). *The Climate Change Convention and Developing Countries: From Conflict to Consensus?* Dordrecht: Kluwer Academic Publishers.

(2007). International law and climate change: the challenges facing developing countries, in *Yearbook of International Environmental Law*, Vol. 16, ed. O. K. Fauchald and J. Werksman. Oxford: Oxford University Press, pp. 114–53.

(2008) *Engaging Developing Countries in Climate Change Negotiations*. Study for the European Parliament's Temporary Committee on Climate Change (CLIM), IP/A/CLIM/IC/2007-111. Institute for European Environmental Policy (IEEP) and Ecologic, Briefing number 631–715. Brussels/London/Berlin: Institute for European Environmental Policy and Ecologic.

(2009). Climate change and development (cooperation), in *Sustainable Development: New Challenges for Poverty Reduction*, ed. M. Salih. Cheltenham: Edward Elgar, pp. 94–108.

Gupta, J. and Hisschemöller, M. (1997). Issue-linkages: a global strategy towards sustainable development. *International Environmental Affairs*, **9**(4), 289–308.

Gupta, J. and van Asselt, H. (2006). Helping operationalise Article 2: a transdisciplinary methodological tool for evaluating when climate change is dangerous. *Global Environmental Change*, **16**(1), 83–94.

Hague Declaration (1989). *Declaration of the Hague*, Meeting of Heads of State, 11 March 1989.

Halsnæs, K., Shukla, P. R. and Garg, A. (2008). Sustainable development and climate change: lessons from country studies. *Climate Policy*, **8**(2), 202–19.

Henikoff, J. (1997). Bridging the intellectual property debate: methods for facilitating technology transfer in environmental treaties, in *Innovations in International Environmental Negotiation*, ed. L. E. Susskind, W. M. Moomaw and T. L. Hill. Cambridge, MA: Pon Books, pp. 48–59.

IPCC Synthesis Report (2001). *Climate Change 2001*. Cambridge: Cambridge University Press.

(2007). *Climate Change 2007*. Cambridge: Cambridge University Press.

IPCC-1 (2001). *Climate Change 2001: The Scientific Basis. Contribution of Working Group I to the Third Assessment Report of the Intergovernmental Panel on Climate Change*. Cambridge: Cambridge University Press.

(2007). *Climate Change 2007: The Physical Science Basis. Contribution of Working Group 1 to the Fourth Assessment Report of the Intergovernmental Panel on Climate Change*. Cambridge: Cambridge University Press.

IPCC-2 (2007). *Climate Change 2007: Impacts, Adaptation and Vulnerability. Contribution of Working Group II to the Fourth Assessment Report of the Intergovernmental Panel on Climate Change*. Cambridge: Cambridge University Press.

IPCC-3 (1990). *Climate Change: The IPCC Response Strategies*. Geneva: IPCC secretariat, UNEP and WMO.

(2001). *Climate Change 2001: Mitigation*. Cambridge: Cambridge University Press.

(2007). *Climate Change 2007: Mitigation of Climate Change. Contribution of Working Group III to the Fourth Assessment Report of the Intergovernmental Panel on Climate Change*. Cambridge: Cambridge University Press.

Jänicke, M., Monch, H., Ranneberg, T. and Simonis, U. E. (1989). Economic structure and environmental impacts: East–West comparisons. *The Environmentalist*, **9**(3), 171–82.

Jatta, F. (2003). Arab aid in Africa: building the foundations for social and economic development, in *Arab Aid: Past, Present and Future*. Dubai: The OPEC Fund for International Development, pp. 45–50.

Khor, M. (2001). *Rethinking Globalization: Critical Issues and Policy Choices*. London/New York, NY: Zed Books.

Kok, M., Metz, B., Verhagen, J. and van Rooijen, S. (2008). Integrating development and climate policies: national and international benefits. *Climate Policy*, **8**(2), 103–18.

KP (1997). Kyoto Protocol to the United Nations Framework Convention on Climate Change. Signed 10 December 1997, in Kyoto; entered into force 16 February 2005. Reprinted in (1998). *International Legal Materials* **37**(1), 22.

Lancaster, C. (2007). *The Chinese Aid System*. Washington, D.C.: Centre for Global Development.

Malenbaum, W. (1978). *World Demand for Raw Materials in 1985 and 2000*. New York, NY: McGraw-Hill.

Mani, M., Markandaya, A. and Ipe, V. (2008). *Climate Change: Adaptation and Mitigation in Development Programs: A Practical Guide*. Washington, D.C.: World Bank.

Mansley, M., Martinot, E., Onchan, T. *et al.* (2000). Financing and partnerships for technology transfer, in *Methodological and Technological Issues in Technology Transfer*, ed. B. Metz, O. R. Davidson, J.-W. Martens, S. N. M. van Rooijen and L. v. W. McGrory. Cambridge: Cambridge University Press, pp. 143–74.

Marrakesh Accords (2001). Report on the Conference of the Parties on its Seventh Session, held at Marrakesh from 29 October to 10 November 2001, FCCC/CP/2002/13.

Metz, B. and Kok, M. (2008). Integrating development and climate policies. *Climate Policy*, **8**(2), 99–102.

Metz, B., Davidson, O. R., Martens, J.-W., van Rooijen, S. N. M. and McGrory, L. v. W., eds. (2000). *Methodological and Technological Issues in Technology Transfer*. Cambridge: Cambridge University Press.

Noordwijk Declaration (1990). Noordwijk Declaration on Climate Change, in *Noordwijk Conference Report: Volume I*, ed. P. Vellinga, P. Kendall and J. Gupta. The Hague: Netherlands Ministry of Housing, Physical Planning and Environment.

Nyerere, J. K., Singh, M., Abdalla, I. S. *et al.* (1990). *The Challenge to the South: The Report of the South Commission*. Oxford: Oxford University Press.

OECD DAC (2007). *Statistical Reporting Directives*, DCD/DAC(2007)34. Paris: Organization for Economic Co-operation and Development.

    (2009a). *Development Co-operation Report 2008*. Paris: Organization for Economic Co-operation and Development.

    (2009b). *Geographical Distribution of Financial Flows to Developing Countries: Disbursements, Commitments, Country Indicators 2003–2007: 2009 Edition*. Paris: Organization for Economic Co-operation and Development.

Opschoor, H. (2009). *Sustainable Development and a Dwindling Carbon Space*, Public Lecture Series 2009, No. 1. The Hague: Institute of Social Studies.

Opschoor J. B. (1995). Ecospace and the fall and rise of throughput intensity. *Ecological Economics*, **15**(2), 137– 41.

Parry, M., Canziani, O. and Palutikof, J. (2008). Key IPCC conclusions on climate change impacts and adaptations. *WMO Bulletin*, **57**(1), 1–8.

Querishi, M. (2003). Arab aid in Asia: a case for building public sector investments with private initiatives, in *Arab Aid: Past, Present and Future*. Dubai: The OPEC Fund for International Development, pp. 27–33.

Selden, T. M. and Song, D. (1996). Environmental quality and development: is there a Kuznets curve for air pollution emissions? *Journal of Environmental Economics and Management*, **27**(2), 147–62.

Stern, D. I. (2004). The rise and fall of the environmental Kuznets curve. *World Development*, **32**(8), 1419–39.

Stern, N. (2007). *The Economics of Climate Change: The Stern Review*. Cambridge: Cambridge University Press.

Sutter, C. and Parreño, J. C. (2007). Does the current Clean Development Mechanism (CDM) deliver its sustainable development claim? An analysis of officially registered CDM projects. *Climatic Change*, **84**(1), 75–90.

SWCC Scientific Declaration (1990). *Scientific Declaration of the Second World Climate Conference*. Geneva: World Meteorological Organization.

Tearfund (2006). *Overcoming the Barriers: Mainstreaming Climate Change Adaptation in Developing Countries*, Tearfund Climate Change Briefing Paper 1. Teddington: Tearfund.

Toronto Declaration (1988). Conference statement of the conference *The Changing Atmosphere: Implications for Global Security*, organized by the Government of Canada, Toronto, 27–30 June 1988.

UNDP (2007). *Fighting Climate Change: Human Solidarity in a Divided World. UNDP Human Development Report 2007–2008*. New York: Palgrave Macmillan.

US AID (2008). *Integrating Climate Change into Development*. Washington, D.C.: United States Agency for International Development.

van Dijk, M. P., ed. (2009). *The New Presence of China in Africa*. Amsterdam: Amsterdam University Press.

von Weiszäcker, E., Lovins, A. and Lovins, H. (1997). *Factor Four, Doubling Wealth and Halving Resource Use*. London: Earthscan.

Villanger, E. (2007). *Arab Foreign Aid: Disbursement Patterns, Aid Policies and Motives*. Bergen: Chr. Michaelson Institute.

WCC (1979). *Proceedings of the World Climate Conference: A Conference of Experts on Climate and Mankind*. Geneva: World Meteorological Organization.

WCED (1987). *Our Common Future: Brundtland Report of the World Commission on Environment and Development*. Oxford: Oxford University Press.

Part II

Theoretical Exploration

2

Development and development cooperation theory

JOYEETA GUPTA AND MICHAEL THOMPSON

2.1 Introduction

Since this book is about development, sustainable development and development cooperation, this chapter examines these complex issues and their interlinkages. While some see development as measurable by gross national product (GNP), others argue that development is much more complex. Whereas sustainable development has been embraced by the scientific community (e.g. the IPCC), the political community (e.g. the Climate Convention and the Rio Declaration of 1992) and the private sector (e.g. as part of corporate social responsibility), its content is often not clear. Although development cooperation appears to be entrenched in the psyche of countries and in international law (Cárdenas et al., 1995), others focus on aid fatigue and call for its ending (Ruyter, 2005).

This chapter examines how the development literature has evolved over time, the evolution of the sustainable-development literature and why most DCs are still developing after 50 years of theory and practice (see Section 2.2). It examines the evolution of the development cooperation literature, the motives for providing assistance, the debate on the effectiveness of aid, the modalities of aid and the problems with aid (see Section 2.3). Finally, it makes recommendations (see Section 2.4) and draws conclusions (see Section 2.5).

2.2 Development theory

2.2.1 Introduction

This section discusses the evolution of development theory, the evolution of sustainable development and why popular images about how countries should develop are fading in the light of the reality. It shows that in the development economics world there is frustration regarding whether singular elegant formulae can deliver the goals of development and sustainable development. In this context, it

Mainstreaming Climate Change in Development Cooperation: Theory, Practice and Implications for the European Union, ed. Joyeeta Gupta and Nicolien van der Grijp. Published by Cambridge University Press. © Cambridge University Press 2010.

argues that development is not an economic issue and that a much more comprehensive approach to understanding development processes needs to be taken. Second, transferring ideas to other contexts works only if those contexts are similar to the context from which the idea comes. Otherwise projects are likely to fail. A key way to pre-empt project and programme failure is to allow for active debate among the three constituencies in society – government, civil society and market – about how local problems can best be defined and, hence, addressed, to allow the development of 'clumsy' solutions.

2.2.2 The evolution of development theory

Since the end of World War II, the development studies literature has examined how nations develop and the critical factors in promoting such development. This section, which is based on Meier (2001), Thorbecke (2006) and Easterly (2007), sums up the evolution of development theory (see Table 2.1).

In the 1950s, development aimed to increase gross domestic product (GDP) through investments in large-scale industry and infrastructure and exploitation of the notion of economies of scale. This was expected to have spill-over effects through reducing social inequality and was possibly influenced by the emerging successful Russian model of growth. The goal was based on theories of a 'big push' (Rosenstein-Rodan, 1943; Rosenstein-Rodan, 1976), 'balanced growth' (Nurkse, 1953) and 'take-off into sustained growth' (Rostow, 1956). Rostow, for example, emphasized the need for national income to grow faster than population, calling for a minimum investment-to-GNP ratio. Rosenstein-Rodan focused on promoting massive aid to help countries deal with the constraints on their development.

In the 1960s, it became clear that a singular focus on industry would not lead to balanced growth, and would not help agricultural labour to shift to industries. South Korea and Taiwan were able to use agricultural gains in the industrial sector, and this may have influenced theorists. Thorbecke (2006) lists the influential theories of this time – namely economic dualism, balanced versus unbalanced growth, inter-sectoral linkages, effective protection, human capital, shadow prices, choice of technique, patterns of growth and the role of agriculture. The policy recommendations that flowed from these theories included price policy, balanced growth between agriculture and industry, export promotion, regional integration, fiscal reforms and sectoral plans.

In the 1970s, growing underemployment and unemployment, wide gaps between the rich and poor, and increasing debt led to a greater focus on employment, income redistribution, poverty and meeting basic needs. Development focused on enhancing real per capita GDP since increases in income were often lost through the rapid

Table 2.1 *Development theories over time*

Period	Goal	The problem	Theories	Policy recommendations
1950s	Increase GDP	Poor government and infrastructure; market failures	Big push; balanced growth; take-off	Government intervention; develop infrastructure and large-scale projects; economies of scale; import substitution
1960s	Increase GDP and employment; promote balance-of-payment equilibrium	Employment did not shift easily from agriculture to industry; balance-of-payment problems	Economic dualism; balanced versus unbalanced growth; inter-sectoral linkages; human capital; shadow prices; role of agriculture	Pricing policy; 'balanced growth' (industry and agriculture); export promotion; foreign aid; regional integration; fiscal reforms; sectoral plans
1970s	Increase per-capita GDP	Poor entrepreneurship; unemployment and underemployment; too much government intervention; inequalities; high debts	Role of informal sector and villages; migration to cities; appropriate technology; relationship among output, employment, income distribution and poverty; socio-economic investment criteria; underdevelopment and dependency theories	Improve entrepreneurship; minimize role of governments; integrated rural development; employment strategies; redistribution of wealth; government and markets complementary

Table 2.1 (cont.)

Period	Goal	The problem	Theories	Policy recommendations
1980s	Increase macro-economic stability and fiscal discipline	Need to pay back debts	Endogenous growth; links between trade and growth, human capital and technology transfer; new institutional economics and role of institutions; focus on markets; policy analysis	Improve policies and instruments; liberalization and privatization
1990s	Enhance human development and reduce poverty; increase entitlements and capability; enhance freedom	Need to increase knowledge; need to enhance social capital; crisis of governance; institutional failure	Role of institutions; endogenous policies; roles of markets and government; corruption; social and human capital	Enhance social capital; improve institutions (property rights, contract enforcement, reduce corruption); promote good governance; stakeholder participation; improve markets; deregulation and liberalization
Since 2002	Promote sustainable development and human dignity	Bankruptcy of grand ideas	Need for accountability and effectiveness	Generate ideas; no simple solutions

Source: Building on Meier (2001), Thorbecke (2006) and Easterly (2007).

rate of population growth and inflation. GDP could be increased by the promotion of entrepreneurship and minimizing government intervention. Subsequently, emphasis was put on non-monetary indicators, leading to the Human Development Index, and on the need to invest in markets and governments as complementary institutions. Influential theories focused on the role of the informal sector; rural–urban migration; appropriate technology; relationships among output, employment, income distribution and poverty; underdevelopment theory; and dependency theory (Thorbecke, 2006).

In the 1980s, the debt problem overwhelmed many DCs and development now focused on promoting endogenous growth and efficiency. Theories on trade and growth, human capital and technology transfer, institutional economics, markets and policy analysis were influential. The policy implications were that one should support, inter alia, privatization and liberalization. These changes were influenced by the stagflation in the ICs which led to reducing the size of the public sector and encouraging private-sector participation (Riddell, 2007).

Subsequently, attention shifted once more to mitigation of poverty. It was argued that investing in the poor would be a key factor in promoting development. This was to be achieved by investing in new and different policies and encouraging good governance. However, mitigating poverty in itself was not enough, since it often increased consumption but not production. Subsequently, the emphasis shifted to a focus on entitlements and capabilities and enhancing freedom (Sen, 1999). The idea was that investing in social capital and freedom would release the creative genius in humans and promote development. Important theories here focused on the role of institutions, the complementary role of markets and government, the negative influence of corruption and the significance of social capital. The focus in recent years has shifted towards promoting sustainable development (WCED, 1987; see Section 1.4), and today there is a focus on the need for ideas (cf. Easterly, 2007; Meier, 2001). Probably the next generation of ideas will focus on enhancing human dignity. But, as ideas keep changing, Lindauer and Pritchett (2002: 2) state that

Any push toward deepening market reforms will be seen as a continuation of the failed strategies of the present, while any strategy that calls for government intervention and leadership … will be seen as a reversion to the failed strategies of the past. What is of even deeper concern than the lack of an obvious dominant set of big ideas that command (near) universal acclaim is the scarcity of theory and evidence based research on which to draw.

Thus, each decade has had its own fixation about how DCs could become richer and about development-related prescriptions. However, these prescriptions often created new problems, such as exacerbating income inequality (Vos, 2002), or creating 'non-viable national economies' (De Rivero, 2001). Much of the literature assumed that underdevelopment could be related to a single cause and could be solved through tinkering with single instruments (Adelman, 2001). De Rivero (2001) argues that DCs should forget about dreams of wealth and focus on meeting basic needs first. Others argue that this will only encourage consumption and that without focusing on productive processes development will not take place. At the same time, experiences from implementation challenged perceptions of development (see Box 2.1).

2.2.3 The evolution of the concept of sustainable development

The concept of sustainable development was introduced in the 1980s (IUCN, 1980; Brown, 1981). The World Commission on Environment and Development defined it as progress 'that meets the needs of the present without compromising the ability of future generations to meet their own needs' (WCED, 1987: 43) and that '[i]n essence, sustainable development is a process of change in which the exploitation of resources, the direction of investments, the orientation of technological development, and institutional change are all in harmony and enhance both current and future potential to meet human needs and aspirations' (WCED, 1987: 46). The notion of inter- and intra-generational equity was elaborated by Brown-Weiss (1989).

Initial work focused on sustainability indicators that continue to be used today (e.g. Kuik and Verbruggen, 1991; Markandya and Halsnæs, 2002). Initially, DC actors perceived sustainable development as a means to prevent their own development and argued that development should come first, flowing out of a reference to the environmental Kuznets curve (see Section 1.4.3), or that sustainable development was merely interpreted as more of the same (Chatterjee and Finger, 1994). More recent literature (Beg et al., 2002; Cohen et al., 1998; Schneider et al., 2000; Banuri et al., 2001; Markandya and Halsnæs, 2002; Metz et al., 2002; Morita et al., 2001; Munasinghe and Swart, 2000; Najam et al., 2003; Smit et al., 2001; Swart et al., 2003; Wilbanks, 2003; Opschoor, 2009) shows a convergence in ideas emphasizing both inter- and intra-generational aspects as well as the need to balance social, economic and environmental aspects in hard sustainability or trading off among those in soft sustainability (Barnett, 2001; Lehtonen, 2004; Robinson, 2004). Sustainable development is seen as an open-ended concept like democracy (Lafferty, 1996) and as combining substantive and procedural elements (Dovers and Handmer, 1993; Mebratu, 1998; Sachs, 1999; Dasgupta, 1993; Sen, 1999). In 2002, the International Law Association, a non-governmental association of legal scholars, adopted the New Delhi Declaration on Principles of International Law relating to Sustainable Development at its 70th Conference (ILA, 2002), which defines the law of sustainable development as incorporating seven principles. These include the duty of states to ensure sustainable use of natural resources and the principles of equity and the eradication of poverty; common but differentiated responsibilities; a

Box 2.1 Towards a new development paradigm: lessons from the Nepal case study

Development is not just an economic story

Throughout the Age of Aid, development has been seen as very much an economic process. The interactions of state, market and civil-society actors have been put into a 'black box', into which aid is then fed in some way and the result – development if you are lucky – is read off in terms of *per capita GDP*. Increasingly, however, it has become evident that development is economic only in its consequences; it is something else – entitlements, democratization, social capital – that makes development possible (e.g. Sen, 1981; Dasgupta, 1993; Putnam, 1993). A 'new paradigm' is therefore called for: one that goes inside the black box.

Dharma gone wrong: the Nepal experience

The Finnish-funded Bara Forest Management Plan set out to privatize a large tract of forest, which, unbeknown to the aid-providers, was already a complex mosaic of property rights: private, public and common-pool (often rather informally established rights, and at small social-scale levels, but rights nonetheless). The idea was that a Finnish private company, in partnership with some Nepali business houses, would be given responsibility for the regeneration of the entire forest, together with near-monopoly rights to its commercial exploitation. The aim, in line with the now-defunct Washington consensus, was to introduce radical changes in a sector that had not hitherto been oriented to market-led approaches. From that point on, it was all downhill, with the ambitious project eventually turning into a disaster so unmitigated and so universally acknowledged that the Finns swore never again to get involved in forestry in Nepal.

To their great credit, however, they commissioned a post-mortem (carried out by three Nepalis and a Finn), which arrived at the following conclusion:

As a villager in Nepal is apt to say, it is the *dharma* of the bureaucracy to regulate, of the markets to innovate and of activist groups to advise caution. The case of Bara was one in which *dharma* had gone wrong – a situation characterized by the bureaucracy assuming the role of the market and vice versa. It further showed that the hierarchic order that has broken down is no instrument for the implementation of radical reforms, without first mending that order and restoring its legitimacy, or as the villager might say, restoring the *dharma*
(Sharma *et al.*, 2004: 241–2).

Dharma, especially in the West, is often equated with fate, but it is much more than that. *Dharma* is 'the law' or, more properly, 'the righteousness that underlies the law'. Hence the emphasis on legitimacy, and the accompanying consequence – breakdown – when that legitimacy is eroded. Legitimacy, moreover, cannot spring from the interaction of just markets and hierarchies: the only sort of interaction that is entertained by the development paradigm (for instance in 'the commanding heights' versus 'the

Washington consensus'). Legitimacy requires a third form of solidarity (egalitarianism) and it is manifested here as the 'activist groups' whose *dharma* it is to advise caution. Fatalism, however, *is* involved, in the sense that the more the *dharma* goes wrong the more likely it is that people will find themselves labouring under the 'double burden': increasingly impoverished and increasingly subject to social exclusion.

Dharma-restoration: the new paradigm

So here, in this deceptively simple diagnosis of the Bara fiasco, is the black box opened up: four forms of solidarity – hierarchy (i.e. state or bureaucracy), individualism (markets from the global to the local), egalitarianism (activist groups or civil society) and fatalism (the double burden-carriers) – each of which is (or, rather, should be) in contentious engagement with the other three: something that you are most unlikely to get if you are operating with a development paradigm that, at the most, allows for just markets and hierarchies. In other words, the development paradigm does not encompass the requisite variety.

On the rare occasions when the policy process has encompassed the requisite variety (Nepal's community forests, for instance, which stand in such contrast to the Bara fiasco; see Ives, 2004) we find that each of the three contending voices – from the state, the market and civil society – is able (a) to make itself heard and (b) is then responded to by the others (fatalists don't really have a voice; if they did they would not be fatalistic). In this situation – *dharma* restored – none of the solidarities undermines its own morality. State actors behave like Edmund Burke's (1790) 'trustees' focusing on the long-term general interest rather than on opportunistic and narrow claims; market actors are guided by Adam Smith's (1776) 'hidden hand', and do well only when others also benefit; and civil-society actors, like Edmund Burke's 'small platoons', are genuinely of the grassroots. But that, thanks to the over-elegant and voice-silencing way in which aid has been fed in, is not how things are in Nepal.

In '*dharma* gone wrong' – typified by Bara – government actors, forgetting all about Edmund Burke, direct their energies to what is euphemistically called 'rent-seeking'; market actors, thanks to what has been dubbed 'licence raj' (or 'crony capitalism'), increasingly deal in club goods that only look like private goods (in other words, they do well even when others do not benefit); and NGOs, though they may walk and quack like genuinely egalitarian actors, turn out on closer inspection to be BONGOs, GONGOs, DONGOs and PONGOs (business-organized NGOs, government-organized NGOs, donor-organized NGOs and party-organized NGOs, respectively). Historians (e.g. Schama, 1997) can show us that these distortions were largely absent in all those countries that we now label 'developed'.

The normative implications that flow from this opening-up of the old paradigm's black box are profound, simple and (for those who have bought into the old paradigm) challengingly counterintuitive. We can, for instance, distinguish between Good Aid (aid that is actually aiding development) and Bad Aid (aid that, because of all this morality-undermining, is working in the opposite direction). Indeed, there now exists a whole array of methods, schemas and indicators by which this key distinction can be made operational. They include the case-study method (e.g. Verweij and Thompson, 2006), systemic constructivist-based discourse analysis (e.g. Thompson *et al.*, 1998), similarly theory-based scenario planning (Ney and Thompson, 2000), indicators of technological inflexibility (Collingridge, 1980) and a template – the refurbished theory of pluralist democracy (Ney, 2009; Thompson, 2008) – for assessing policy sub-systems in terms of the extent to which they encompass the requisite variety.

By *Michael Thompson*

precautionary approach to human health, natural resources and ecosystems; public participation and access to information and justice; and good governance, integration and interrelationship, in particular in relation to human rights and social, economic and environmental objectives. The International Law Association and experts associated with it are developing this further (Schrijver, 2008).

Although the concept of sustainable development has been widely adopted in the policy world (United Nations Conference on Environment and Development, 1992; Commission on Sustainable Development, World Summit on Sustainable Development, 2002) and in treaties (e.g. the Climate Convention of 1992 and the Biodiversity Convention of 1992), courts remain sceptical[1] (see Chapter 4). A major problem is that the open-ended character of sustainable development allows room for both interpretation and manipulation.

2.2.4 DCs and development: the elusive transition

Popular images about how DCs can develop include those relating to the demographic-transition theory, Rostow's take-off theory and more recently the environmental Kuznets curve. All these images convey the message of hope and the potential for a linear progression towards development.

Many argue that the driving force behind hunger and poverty is population growth (Malthus, 1994; see also Ehrlich, 1968: 1; Ehrlich and Ehrlich, 1990; Pimental and Pimental, 1979: 142–3; Brown, 1996; Snow, 1969: 25, 30 and 39–40). If population is the cause of the problem, the solution requires 'curbing' population growth according to the neo-Malthusians. But when will population growth stabilize? The demographic-transition theory (van der Tak *et al.*, 1979) predicts that, as societies become developed, the death rate falls first and then the birth rate, leading eventually to falling or stable population numbers. There is adequate

---

[1] Case concerning the Gabčíkovo–Nagymaros project (Hungary/Slovakia), para. 140, Judgment of 25 September 1997, ICJ: Reports of Judgments, Advisory Opinions and Orders.

empirical proof to support this hypothesis. In fact, global growth rates have peaked already (Population Commission, 1996: 11), but UN predictions indicate that global populations will continue to grow well into the twenty-first century and the tapering of population growth depends on a number of interrelated issues (e.g. poverty, access to food, health care, the impacts of climate change).

This brings us to the following question: how do societies develop? Rostow's theory of economic take-off suggests that societies pass through phases of traditional society based on subsistence economies, through a transitional stage when specialization creates surpluses for trading, to take-off when industrialization and investment increase in a society, accompanied by the development of political and social institutions, culminating in a drive to maturity when the economy diversifies to a stage of high mass consumption in which the service sector becomes more and more dominant. Since then, others have argued that there are different routes to development.

But as societies develop they tend to over-exploit their natural resources and sinks, thereby damaging the very resource base that nourishes them. The environmental Kuznets-curve theory (see Section 1.4.3) argues that, as societies pass beyond a critical point in income per capita, they will pollute less in relation to their income, see Malenbaum (1978) and Jänicke *et al.* (1989), for material use, and Grossman (1995) and Selden and Song (1996), for pollutants. Thus, as DCs become richer they will invest more in pollution-control technology and be less inclined to exhaust their natural resources.

The above three theories provide hope. They emphasize that, as societies develop, their population stabilizes, poverty decreases, basic needs are met, education and health care increase, and local environmental pollution decreases. The expectation is that development will be accompanied by greater equity within societies, economic stability could lead to more political stability, and people would demand and get accountable systems of governance at national and international level.

Yet, empirical evidence over the last 50 years suggests that over time there is an increasing divergence in income between rich and poor, often also in absolute terms. The world's richest 50 people earn more than 416 million poor people. The poorest 40% of the world's population receives only 5% of global income, while the richest 10% receives 54% of the global income (UNDP, 2005: 3). Within countries, income inequality is also growing, e.g. within the OECD (OECD, 2008). The current financial crises may exacerbate this.

If this is the case, then there is no confidence that countries will evolve along the paths mentioned above. Instead, countries with high population and hunger are perhaps trapped in a population–hunger cycle while countries with low income are trapped in the poverty cycle (Moran, 1986), and local degradation may begin to affect the resource base of these countries. However, there are always countries that can escape the trap of poverty and underdevelopment.

This may mean that poverty and underdevelopment constitute a structural problem of developing countries. At the same time, there is also no evidence that the environmental Kuznets curve holds for global environmental pollutants (see Section 1.4.3); hence, as countries become richer, they increase their wealth and reduce their pollution by, inter alia, sending their pollution abroad (e.g. by transporting waste) and acquiring resources as far away as possible (e.g. fishing in foreign waters) (Gupta, 2004). Their greenhouse gas emissions also continue to increase. This, in turn, affects the ability of the poor countries to support themselves and climb out of the poverty trap, calling for a major injection of aid (Sachs *et al.*, 2004).

2.3 Development cooperation practice and theory

2.3.1 Introduction

Unlike development theory, which has a highly academic tone, development cooperation practice has strong political undertones, especially insofar as it has to be justified to the electorates of donor countries. This section first elaborates on the concept and discusses the evolution of development cooperation (see Section 2.3.2). It then discusses the motives for providing aid (see Section 2.3.3), the challenges in aid provision and its effectiveness (see Section 2.3.4).

2.3.2 The history of aid

Although the genesis of aid can be traced back to philosophers such as Hume and to early missionary programmes (e.g. the first Swedish missionaries went to Ethiopia in 1860), development cooperation, in its current incarnation, is a post-World War II, post-colonial, phenomenon. Early non-state actors of the twentieth century were already working in the poorer parts of the world to address key development issues. Nevertheless, during the era of colonization, colonial powers did not promote development cooperation (Thorbecke, 2006).

Following the establishment of the UN (UN Charter, 1945), President Truman emphasized that the proposed aid to Greece and Turkey would help these countries become self-supporting and independent, leading to enhanced economic stability in the region and thereby supporting the principles promoted by the UN: 'If we falter in our leadership, we may endanger the peace of the world – and thereby endanger the welfare of this nation' (Truman, 1949). Post-colonial powers developed foreign policies designed to support their former colonies and the Marshall Plan of 1947 was developed to promote the reconstruction of Europe after the war and disbursed USD 13 billion. The establishment of the International Bank for Reconstruction and

Development following the Bretton Woods Conference of 1944, the aid agencies in some industrialized countries (ICs) since the 1950s and the Development Assistance Committee (DAC) of the OECD in 1961 further institutionalized the process of aid provision to the DCs (see Chapter 5).

This sub-section provides a history of development cooperation (Pronk, 2001; Thérien, 2002; Wuyts, 2002). In the 1940s, aid was born under a social-democratic left-wing culture (Thérien, 2002), focusing on post-war reconstruction and technical assistance. However, by the 1950s, aid was defined more by a right-wing approach and was used to finance investment in infrastructure and promote trade with DCs (Wuyts, 2002). It was influenced by the development of the welfare state in the ICs. In this period 90% of aid came from the USA, UK and France. 'The USA used it as an instrument of containment; the UK and France turned it into a substitute for colonial domination' (Thérien, 2002: 454). Assistance from the USA was focused primarily on non-Communist countries, namely South Korea, Taiwan and Latin America.

In response to the call by DCs for the creation of a financial institution to support development, the left supported the establishment of the United Nations Development Programme (UNDP) in 1965, and the right the International Development Agency (IDA) in 1960 and the OECD DAC in 1961 (Thérien, 2002). The latter two bodies kept the control of ODA in the hands of the ICs and, although the DAC has since debated on many development-aid-related issues, this has stayed a club of donor countries (see Chapter 4). In the 1960s, aid was limited to the use of public resources for development purposes, and aid for military purposes was excluded from the definition of aid – a result of the victory of the left in defining aid (Thérien, 2002).

In the 1970s, the failure of the approach that resources from the rich would gradually trickle down to the poor, and growing disparities between rich and poor, led to a new focus on meeting basic needs (Pronk, 2001) and left-wing ideas prevailed (Thérien, 2002). At UN level, the adoption of the New International Economic Order (NIEO) gave a new forum for North–South dialogue, although the NIEO was itself a stillborn regime. Despite the use of resources for basic needs and subsistence agriculture, the results were not spectacular.

In the 1980s, the right wing was once more dominant and DC debt increased heavily. Aid was used to redesign national policies through structural development programmes and debt relief (Pronk, 2001; Thérien, 2002). Government was given less emphasis, and the private sector and markets acquired more importance. Public expenditures were reduced, while trade was liberalized and currencies were devalued. This implied that the previous emphasis on basic needs evaporated, and the decade is seen as a lost decade for development. At the same time, images of affluence accompanied by extreme poverty in DCs led the public in ICs to demand greater accountability in aid schemes and this led to conditional policy to meet donor expectations and ensure institutional survival (Sobhan, 2002).

By the 1990s, aid was being used once more to emphasize basic needs, humanitarian assistance and good governance, and thus reflected a more left-wing approach (Thérien, 2002). In 1996, the OECD emphasized the reduction of poverty, promotion of social development and integration of environmental values in policymaking. It included goals such as halving the number of people living in poverty by 2015, and capacity building for democratic and accountable governance and the rule of law. At the end of the 1990s there was still poverty in the DCs and there was a feeling that aid agencies had once more failed to catalyse change (Sobhan, 2002). In the late 1990s, the goal was to provide aid only to those countries that had good governance in place and five principles were adopted (Paul, 2006). These principles, inspired by the World Bank's Comprehensive Development Framework and included in OECD DAC (2003), are promotion of developing country (government and stakeholder) ownership of policies; accountability of government for development policies; donor coordination and policy coherence; the need to develop a strategy on poverty; and promotion of partnerships involving donors, recipients and civil society (Paul, 2006). The idea was that investing resources in countries with good governance where stakeholders could be actively engaged in policy processes would lead to improved welfare for the poorest (Sobhan, 2002). Sobhan (2002: 544) states that

It is arguable that the original development design failed to address the sources of poverty which originate in the structural injustices which underwrite the political economy of most DCs. In such an institutional context reform merely served to perpetuate poverty, accentuate inequalities and empower a small elite who have used their wealth to monopolize state power. In such an unjust social order, a few so-called poverty centred projects will not ensure a sustainable assault on poverty or the empowerment of the poor.

In the 1990s, the definition of aid was modified: military aid was included in the 1990s; both grants and loans were included; and administration costs, help to refugees in their first year in the host country and education of DC students in donor countries could be counted as aid (Thérien, 2002).

With the adoption in 2000 of the Millennium Development Goals, which emphasize meeting basic needs (see Table 4.1), aid efforts are increasingly being focused on meeting these goals. The idea of country ownership was also institutionalized in the writing of Poverty Reduction Strategy Papers (PRSPs) for the World Bank and Country Strategy Papers (CSPs) for the EU. The history of development cooperation is linked to the theory of development and is summed up in Table 2.2.

2.3.3 Motives for providing aid

The literature (Schraeder *et al.*, 1998) identifies motives for providing aid (Table 2.3), which can be clustered to include altruistic (charitable, humanitarian, solidarity) goals such as poverty alleviation and disaster relief (Pronk, 2001; FitzGerald, 2002; Sobhan,

Table 2.2 *The history of aid summarized*

Period	Political context	Goals	Solutions	Development cooperation trends
1940s	Establishment of UN and Bretton Woods bodies; Truman's Four Point Programme; Marshall plan; US aid programme	Increase GDP	Government important; develop infrastructure and large-scale projects; economies of scale; import substitution	Reconstruction; support for self-sufficiency; economic stability; political security; community development
1950s	Post-war and post-colonial reconstruction; aid programmes in UK, USA and France	Increase GDP and employment; promote balance-of-payment equilibrium	Pricing policy; balanced growth; export promotion; regional integration; sectoral plans	Technical assistance; diffusion of innovation; community development; support to newly independent governments
1960s	Establishment of UNDP, IDA and OECD DAC	Increase per-capita GDP	Improve entrepreneurship; minimize governments; integrated rural development; employment strategies; redistribution of wealth; governments and markets	Filling of trade and investment gaps
1970s	Adoption of NIEO and North–South dialogue.	Increase macro-economic stability and fiscal discipline	Improve policies and instruments; privatization and liberalization	Basic human needs; bridge savings investment gap; bridge balance-of-payment gap
1980s	Debt crises; development of Asian tigers	Enhance human development and reduce poverty; increase entitlements and capability; enhance freedom	Enhance social capital; improve institutions; promote good governance; stakeholder participation; improve markets; deregulate	Conditional aid; structural adjustment and debt relief; 'Washington consensus'; gender and empowerment issues; rural development

Table 2.2 (*cont.*)

Period	Political context	Goals	Solutions	Development cooperation trends
1990s	Fall of Berlin wall; end of communism; aid in EU Treaty; OECD policy of 1996; adoption of the right to development in UN; UNDG; environmental crises	Promote sustainable development	Generate ideas; no simple solutions	Humanitarian assistance; democratic governance; good governance; greening of aid; neo-liberal agenda; sustainable development; rural development phased out
2000s	MDGs; Monterrey Conference; Paris Declaration; EU Consensus; financial crises			Country ownership of policies emphasized; PRSPs; aid only to countries with good governance; aid for development and environment; issue of remittances

2002); enlightened self-interest such as the promotion of democracy and peace (cf. FitzGerald, 2002); political and strategic interests such as maintaining a sphere of influence and links with former colonies and rewarding 'good' policy (Pronk, 2001; Alesina and Weder, 2002; Sobhan, 2002); security interests such as preventing the expansion of communism (Cárdenas *et al.*, 1995) and terrorism, especially since 9/11; economic (promoting open economies, creating markets for products and services) (Pronk, 2001); and environmental (promoting solutions to environmental challenges) motives, which may be altruistic, or enlightened self-interest, and may also be fuelled by political, strategic and economic interests in terms of creating markets for technologies. These political, commercial and other criteria have often negatively impacted on aid effectiveness – and '[t]he politics of aid remains central to any discussion of whether and how aid works' (Riddell, 2007: 106).

Donor behaviour differs (see also Chapter 7). Scandinavian donors and Australians, probably because they did not have colonies and do not nurture geopolitical ambitions, tend to provide aid as a reward for less-corrupt governments, while the USA supports democracies and aims to protect itself, but does not always

Table 2.3 *Motives for providing aid*

Goals	Why	Content
Altruistic	Solidarity; based on ability to pay; humanitarian reasons; charity	Promotion of poverty alleviation programmes; disaster relief for genocide, famine or other disasters
Enlightened self-interest	To promote stability and peace and thereby a flourishing global economy; peace, stability, democracy	Promotion of peace keeping; political stability; transition to democracy; helping countries retain independence
Political and strategic interest	To maintain political relations and protect geopolitical interests	Promotion of good relations with former colonies; creation of satellite countries
Security interests	Containment of communism; to suppress the rise of terrorism	Promotion of relations with countries neighbouring communist states and with countries that involuntarily or voluntarily provide space for terrorists
Economic	To create markets for goods and services, sometimes through tied aid	Promotion of open economies, foreign investment, transfer of technologies
Environmental	To address global environmental challenges	Creation of the conditions for effective environmental problem solving

take corruption into account (Alesina and Weder, 2002; Riddell, 2007). The French tend to provide aid more consistently to past colonies. Japan tends to give aid to Asian countries and countries that support its position in UN negotiations (Alesina and Weder, 2002). Apart from Norway, other European countries label debt cancellation as aid; Denmark, France, Sweden, the Netherlands and the Czech Republic include the costs of hosting refugees as aid; while Germany, France and Portugal also try to include the costs of hosting foreign students as aid (Joint European NGO Report, 2006). For the period 1980–99, Berthélemy (2006) finds that altruism is high in Switzerland, Ireland, Denmark, Norway, Austria, New Zealand, the USA (because it provides assistance to regions in conflict) and Finland, while the UK, Italy, France and Australia are relatively selfish. Other authors using different criteria find that the USA primarily promotes its geopolitical interests (Sobhan, 2002) and its own commercial interests since 70% of US ODA is tied (Riddell, 2007). McGillivray (2004) found that, during the 1970s and 1980s, bilateral donors promoted primarily their own interests. France and Japan do not focus on poverty reduction (Nunnenkamp and Thiele, 2006). Japanese aid focuses on promoting peace and development in order to secure Japan's own security and prosperity (Riddell, 2007). Evaluation of country behaviour is dependent on the criteria used, but shows that donors are neither always nor consistently altruistic and that there is a gap between the rhetoric and implementation of policy goals (Nunnenkamp and Thiele, 2006).

2.3.4 *The debate about the effectiveness of aid*

Themes on aid

The history of aid shows that there are three core themes concerning the role of aid with respect to development. The first theme is that aid should be seen as the key, but temporary, factor promoting economic development in the DCs. Emmerij (2002) explains the history. There is thus a constant search to demonstrate either that aid has been effective and therefore it should be stopped, or that aid has been ineffective and therefore it should be stopped. However, Pronk (2001) argues that there is no reason to think that international development assistance should be temporary; there will always be a need for the rich to support the poor. Of course, the definition of who is rich and who is poor may change over time.

A second, related, theme is the focus on the effectiveness of aid. While many scholars indicate that aid has scarcely led to improved GDP in partner countries, others argue that the effectiveness of aid should be measured in terms of other indicators.

A third theme in the literature is that aid should be seen more modestly as merely a catalyst for change – a catalyst that through strategic intervention can lead to substantial improvement in the circumstances of DCs (Pronk, 2001). For, where aid works to promote consumption or provide disaster relief, it has little effect on productive efforts and hence on growth. Easterly (2007: 331) states that

In sum, we don't know what actions achieve development, our advice and aid do not make those actions happen even if we knew what they were, and we are not even sure who this 'we' is that is supposed to achieve development. I take away from this that development assistance was a mistake. Yet it doesn't necessarily follow that foreign aid should be eliminated. Once freed from the delusion that it can accomplish development, foreign aid could finance piecemeal steps aimed at accomplishing particular tasks for which there is clearly a huge demand – to reduce malaria deaths, to provide more clean water, to build and maintain roads, to provide scholarships for talented but poor students, and so on. It could seek to create more opportunities for poor individuals, rather than try to transform poor societies.

Simple correlations

Much of the debate on aid effectiveness has focused on econometric correlations between aid and various indicators of effectiveness. Some argue that no statistically sound correlation can be found between aid and economic development in the partner country. Boone (1994; 1996) argued that aid has had no effect on investment or growth in specific case-study countries. Using cross-country analysis, Rajan and Subramanian (2005) have argued that there is no statistical correlation, whether negative or positive, to show that there is a relationship between aid flows and economic growth in the partner country even if one takes into account what aid is used for, who gives it, who it is given to, and whether there are short-term or long-term impacts (Rajan and Subramanian, 2005). Dollar and Easterly (1999) have come to a similar conclusion, especially with respect to Africa. Gomanee *et al.* (2005), however, argue that statistical correlations obtained using different approaches can show that, for each 1% increase in the aid/GNP ratio, there is an increase of 0.25% in Africa's growth rate.

Riddell's review (2007) shows, however, that, although much of the literature has argued that aid was ineffective, others have shown that aid has been successful. He argues that evaluating the quality of data and deciding whether success can be attributable to aid are key methodological problems in making national and international assessments. The use of sample analysis being generalized across all aid processes is a major problem. However, most project-related assessments show that these projects are successful, though the results might not be sustainable either because local actors lose interest or because contextual conditions change.

Part of the problem with aid may have been that governments hoped that applying simple formulae for promoting growth would be effective. Each decade had a different focus and different formulae regarding aid provision. Each decade led to disappointment, since there was a perception of failure. Following the focus on aid to promote technical change and infrastructural development, which did not lead to closing the gap between rich and poor and achieving a balance of payment equilibrium, there was a major discussion concerning whether aid should be used to change policies in partner countries. This was done by linking aid to conditionalities, such as through the Structural Adjustment Programmes.

Conditionality

Donors insisted on conditionality when they wished to ensure that (a) the partner country adopted a specific policy; (b) aid resources were spent in specific policy environments; (c) partner countries purchased specific goods or services from the donor; (d) partner countries committed to a specific set of policies under threat of aid cancellation; and (e) private-sector costs were reduced, by sending a message regarding partner-government behaviour (Collier *et al.*, 1997). Conditionality was intended to ensure that the goals of the donor were met, and to reduce the costs for the donor while ensuring that aid was effective from the donor's viewpoint. However, more often than not, the prescriptions prevented real reform in partner governments, had perverse effects, and lacked credibility (Collier *et al.*, 1997); reduced government expenditure in the health and education sectors; led to a shift of agricultural production from food for subsistence to export-oriented cash crops; and tended to reduce welfare in the partner countries. The DCs resisted conditionality, often seeing it as alien to their interests, and donors were also often not consistent in terms of what they wanted (Paul, 2006). Sometimes, up to 100 conditions have been imposed on partners, which is 'administratively burdensome' and distorts the existing policy processes (Joint European NGO Report, 2006). Donor conditionality does not substitute for DC ownership and commitment to ideas (Goldin *et al.*, 2002; Thérien, 2002). It can undermine existing policy ideas and lead to loss of institutional capacity while delegitimizing the state in the view of its own populace (Sobhan, 2002). Conditionality also did not work because of a fundamental ignorance on the part of donors about what works (Riddell, 2007: 243). The Bretton Woods Institutions continue to support a package of macroeconomic policies, even though there is evidence that this does not work and that 'fiscal austerity, pursued blindly, in the wrong circumstances, can lead to high unemployment and a shredding of the social contract' (Stiglitz, 2002: 84). Conditionality may work where partner countries believe in the conditions, and this led subsequently to a shift to providing conditional aid to countries with 'good governance' (or an acceptance of the conditions).

An element of conditionality is aid tied[2] to economic interests in the donor countries. This helps donors more than partners, thus providing support for the transfer-paradox hypotheses (Ohlin, 1966; Pronk, 2003). About 41.7% of current OECD aid is not tied (OECD, 2006: 31). The use of tied aid can increase the costs of a project by 20%–30% (Jepma, 1991) or 15%–40% (Joint European NGO Report, 2006; Cassen *et al.*, 1994; Krueger *et al.*, 1989; La Chimia, 2004). The UK International Development Act (effective since 2002) has declared tied aid illegal.

A part of such tied aid is reliance placed on technical assistance as an effective tool of cooperation. This amounts at present to 36%–40% of total aid (Riddell, 2007: 202). Such assistance works in terms of its short-term goals of providing training; however, technical assistance is often seen as too expensive, inappropriate and fostering dependence (Joint European NGO Report, 2006; Riddell, 2007). Nevertheless, despite the rising critique this activity continues. Increasingly, there is also a focus on capacity building, but this too suffers from the challenge that donors often do not know how to improve capacity in the poorest countries (Riddell, 2007: 211).

Another element of conditionality is accountability, calling for the preparation of reports on how aid has been used. However, increasing requirements to ensure

---

[2] 'Tied aid credits are official or officially supported loans, credits or associated financing packages where procurement of the goods or services involved is limited to the donor country or to a group of countries which does not include substantially all developing countries ...' (Glossary of Key Terms and Concepts. From the Development Co-operation Report: Efforts and Policies of Members of the Development Assistance Committee, http://www.oecd.org/glossary/0,2586,en_2649_33721_1965693_1_1_1_1,00.html).

accountability have led to a high administrative burden on DCs. An average African country has to prepare 10 000 quarterly reports to meet the differing requirements of the various donors (Joint European NGO Report, 2006).

Good governance

In the 1990s, aid was concentrated on promoting good governance in partner countries because there was an increasing perception that aid money was spent better where the governance system was effective; but this too often had unexpected perverse effects (Anders, 2005). Furthermore, it was expected that good governance would lead to good policies, but this is not always clear from the empirical evidence (Acemoglu *et al.*, 2005). There is little evidence that aid has significantly changed policies (Burnside and Dollar, 1997), since domestic politics will have a more definitive impact (Dollar and Easterly, 1999), and partner-country 'ownership' of reforms is more important (Dollar and Easterly, 1999: 574; Bourguignon and Sundberg, 2007). Hence, donors should identify reformers, not create them (Dollar and Svensson, 2000). This meant that aid should be concentrated in places with good governance from the donor's point of view. This idea was adopted in the Monterrey Consensus (2002), but research shows that it was adopted more by the multilateral than by the bilateral organizations (Dollar and Levin, 2006). As a result, in 1999, countries with good policies received twice as much aid as countries with poor policies. This contrasted with the practices in 1990 and led to higher poverty reduction effectiveness per dollar of ODA (Goldin *et al.*, 2002), although the poor in poor-policy environments were neglected. However, there is also evidence that aid can work in contexts where the policies are not always supportive (Addison *et al.*, 2005). Increasingly, the focus has been on ensuring partner-country ownership of policies; however, it was soon realized that ownership is often not genuine and can become another sort of conditionality (Thérien, 2002: 459).

Some donors have sought to give aid on the basis of the performance of partner countries in terms of good governance, decentralization, macro-economic policies and investment support (Bourguignon and Sundberg, 2007). This allows partners to pick their own policies, and effectiveness implies giving aid to those countries where measurable progress is being made (Bourguignon and Sundberg, 2007). However, this is also complicated. Over what time period should the measurable progress have been made, especially given that different kinds of expenditures and policies have different payback times (e.g. education, health care, disaster relief and war relief) and may occur in unstable political contexts, and a credible negative result is difficult to establish (Fitzpatrick *et al.*, 2007)? Furthermore, should aid be given only to countries that perform well, or to all countries that need aid? Some authors recommend that a balance is needed and that, where aid is provided to 'fragile states', the emphasis should be on bypassing the government (Bourguignon and Sundberg, 2007), while others argue that it is of vital importance to support governments in these fragile political entities.

On the basis of economic modelling that linked foreign aid with policies on fiscal surplus, inflation and trade openness, it has been argued that aid works best when partners have good policies (Burnside and Dollar, 2000). Some also argue that aid works better when external environmental features (terms of trade, climate shocks, the real value of exports) are negative. This is because the need for aid is higher and hence the productivity of aid will be greater (Guillaumont and Chauvet, 2001). Here aid is seen as compensating for and cushioning against the negative effects of policies at the international level.

Aid could be seen as a transfer. In such a situation, the donor loses in welfare and the partner gains in welfare. However, there may be transfer paradoxes such that the donor gains and the partner country loses (Leontief, 1936). Transfer paradoxes can occur in a number of different situations; see Brahman *et al.* (2006), who cite the literature on the issue. Sometimes aid is self-serving, insofar as there appears to be a correlation between foreign aid supporting political reform and foreign direct investment (FDI) supporting economic reform (Alesina and Dollar, 2000). However, Brahman *et al.* (2006) conclude through the use of new economic-geography models that such paradoxes cannot occur and that aid in general has a temporary effect.

Furthermore, the ineffectiveness of aid can perhaps also be attributed to the motives of the donors (see Section 2.3.3): the support of specific countries for political reasons, the use of conditionalities and the transfer of Western practices and solutions have not been successful. The ineffectiveness of aid can be attributed to poor understanding in aid agencies of other cultures (Matz, 2008), processes and local motivations (e.g. the aid processes become captives of local power politics (Theesfeld, 2008; Herrfahrdt-Pähle, 2008)), and the policy instruments (e.g. public–private partnerships (Kluge and Scheele, 2008; Krause, 2008)).

The top 2% of aid partners, receiving about 17% of their GDP through aid over four decades, have experienced zero per capita growth; and the success stories of South Korea, Ghana and Botswana cannot be attributed merely to aid (Easterly, 2007). While some argue that aid to Africa is ineffective (Easterly, 2006), others state that, since the start of aid, the standard of living in the poor countries has increased by about 20% (Doucouliagos and Paldam, 2008; cf. Collier, 2006). Addison *et al.* (2005) submit, on the basis of an assessment of 30 papers, that in a situation without aid most partner countries would have been much worse off. That there have been a few bad projects does not imply that aid does not work. Donors are continually learning and aid delivery is improving (Burnside and Dollar, 1997; Collier and Dollar, 2004; McGillivray, 2004). Pronk (2001) argues that at times aid has 'been a spectacular success'.

Most aid works in terms of its own objectives (Cassen *et al.*, 1994). Individual development cooperation projects have often been successful, especially in the agricultural and health fields. For example, development cooperation projects in soybean production have meant that Brazil has become a major producer and exporter, and has economic returns equivalent to USD 6 billion per year (Hungria *et al.*, 2005). Similar success stories in using soybeans have been reported in Zimbabwe and Nigeria. The World Health Organization's programme to eradicate small-pox has been successful; and the programme to end river blindness has been effective (Levine *et al.*, 2004).

Part of the assessment problem is that there has been no systematic evaluation of aid programmes and projects. There have been few meta-studies using a holistic analysis that evaluates the effectiveness of aid (Riddell, 2007: 4). In the 1980s, aid effectiveness was being measured through the introduction of logical frameworks, incremental learning and process-based approaches. In the 1990s the emphasis was on understanding baseline situations prior to the interventions, undertaking annual reports to measure progress and longitudinal analysis. A key issue is how to attribute success in a project – is it because of the intervention or the combination of the intervention and local contextual features, or is it more attributable to the local features? However, the neglect of existing power structures at partner-country level is a critical factor associated with failure.

Challenges for donors and partner countries, and their relationship

One can also look at the challenges that aid poses by deconstructing the aid-delivery process and examining the challenges in the donor and partner countries, and in their relationship. Seven challenges exist in the donor countries: (a) the choice of partner countries, which is often dictated by centralized strategic decisions, inadvertently supporting corrupt and non-democratic regimes (e.g. Joint European NGO Report, 2006; Burnside and Dollar, 2000; Svensson, 1998; Tornell and Lane, 1999; Riddell, 2007); (b) the way policy is made, often a centralized decision made at ministerial or cabinet level and reflecting donor interests (see Table 3.3), while the money is disbursed through decentralized processes (Svensson, 2003: 383; Riddell, 2007); (c) the difficulty of aggregating the results of diverse projects, programmes and sectoral support by the donor countries (Goldin *et al.*, 2002); (d) the use of context- and history-neutral, theory-based, single-formula approaches (Pronk, 2001; Goldin *et al.*, 2002); (e) the continuing use of conditionality and tied aid as a way to ensure that donor goals are met (Riddell, 2007); (f) the volatility of aid and the use of statistical manoeuvring ('saving face, not lives') in order to include controversial statistics in aid data (e.g. debt cancellation, hosting refugees in Western countries for one year and student scholarships in donor countries) and the mismatch with needs

(Riddell, 2007); and (g) changes in donor policy following changes of government. Other challenges include poor coordination between donors and institutional challenges in donor agencies (Riddell, 2007).

Three major challenges exist in the partner countries: (a) poor systems of governance and low capacity; (b) competition for assistance in countries which may exacerbate local tensions (Paul, 2006, cf. Theesfeld, 2008; Herrfahrdt-Pähle, 2008); (c) poor aid design may distort the domestic context (e.g. displace domestic savings and create dependency, displacement of local expertise and a policy-substitution effect (Ndulu, 2002; cf. Pronk, 2001; Svensson, 2000)).

Challenges in the donor–partner relationship include (a) mismatches at the level of priorities being addressed through conditionality and tied aid; (b) mismatch in expectations; (c) mismatch in the use of tools; (d) mismatch between the demand for accountability and the resulting administrative burden; (e) the lack of discussion between donors and partner governments (Easterly, 2006); and (g) the difficulty in establishing good inter-personal chemistry between agents from donors and partners. While Verweij *et al.* (2006) see the major challenge as a lack of understanding of how the market, civil society and government relate to each other in specific contexts over specific time-frames and hence recommend clumsy solutions, Riddell (2007) describes the situation differently, calling for a more active engagement with politics both in the donor and in the partner country.

Policies and modalities of development cooperation

Aid is generally provided through four different modalities. Technical assistance and project-based help are the oldest forms of assistance. Programmatic forms, including sector-wide approaches and general budget support, are of more recent origin. Other instruments of cooperation include microfinance, co-financing, venture capital and risk guarantees (Bouab, 2004).

The literature cited above is quite negative about both technical and project-based assistance, arguing that these are relatively ineffective forms of assistance in terms of making a difference to society. However, they can make a significant difference to the individuals for whom these projects are designed. Arguments in favour of sector-wide approaches and general budget support include that these forms of aid (a) are untied; (b) give partner governments a free hand in deciding how to formulate nationwide policies based on domestic priorities and where to focus their limited resources, thus ensuring ownership; (c) provide resources for strengthening institutions; (d) help in developing accounting tools to guarantee accountability; (e) allow a coherent national policy; and (f) allow donor harmonization (Bird and Cabral, 2007; Lawson and Booth, 2004). Budget support is generally measured by its impact on macro-economic indicators. Sectoral support, or budget support for specific sectors, focuses on helping individual sectors and is generally measured in terms of improvements in sectoral indicators. However, such support requires a minimum level of credibility in the ability of the receiving government to manage the resources and be accountable. Prior to such assistance, countries are often asked to prepare their own strategies for poverty reduction, including through preparation of Poverty Reduction Strategy Papers (PRSPs), and Country Strategy Papers (CSPs).

Sector-wide and general budget approaches are relatively young, and assessments are few. While sector-wide approaches have been effective in the health, education and road-building sectors, the benefits for the poorest are not always obvious, the long-term effects are difficult to predict, and the costs of such aid might not be lower than with project-based approaches (Riddell, 2007). General budget support is more difficult to evaluate, because causal links and other data are not easy to ascertain; it is not necessarily cheaper than project-based aid, and the process of aid provision is complex (Riddell, 2007).

In the area of environmental management, general budget support can be helpful to support the environmental area through policies and institutional support; provide relevant ministries with additional funding that can be spent on a discretionary basis; help develop budgetary discipline, since the finance ministry will have control; help to mainstream environment in national policy; create a role for sector working groups to coordinate in the area of multi-sector policy dialogues; and enhance the transparency of policy processes (Bird and Cabral, 2007).

The primary differences among aid modalities are in terms of whether any conditions are placed on the resources, whether there is some need for using domestic management procedures, and the point of entry and level of interactions. Scholars suggest that avoiding aid conditionality and moving towards partnership with a focus on endogenously designed policy change based on dialogue with relevant stakeholders is useful.

2.4 Lessons learned: clumsy BASICS

This brings us to the next question – how should aid be designed and can it, in fact, be designed? Some design principles emerge in the literature. These can be summed up in the acronym BASICS (van den Berg and Feinstein, 2009).

- Broader assessments: development cooperation needs to be effective in order to justify its continuation. There have been few, if any, meta-transdisciplinary assessments of development cooperation as a whole. Authors of many macro-economic assessments have argued either that aid has not helped (e.g. Boone, 1996) or that countries would not have been as well off without aid (e.g. Gomanee *et al.*, 2005). Small-scale projects and technical support are easier to measure in terms of intermediate goals; while programme, sector and budget-support assessments are more difficult to measure. Projects and technical support provide examples both of success stories and of challenges. Examining the micro-indicators such as impacts on the lives of people that aid programmes are meant to improve (cf. Pronk, 2003), on the local ecosystems and in relation to the specific goals of each approach is more likely to yield positive results than examinations of large-scale macro-indicators. The jury is still out on how effective sector-wide and budget support is.

- Align type of aid to type of partner: the nature of development cooperation should be consistent with the 'type' of partner country. The literature indicates that long-term budget (Paul, 2006), programme and sectoral assistance (Harford and Klein, 2004; cf. Ndulu, 2002) may be effective in countries with good governance. Assistance to states with good governance may often result in activities that are continued long after the aid has ceased. Where governance systems are weak, it may be more useful to identify 'reformers' or talented individuals in NGOs or local government (cf. Ndulu, 2002) and support short-term projects (Harford and Klein, 2004; Paul, 2006). However, there is a risk that the projects end when support ends. Sometimes, providing resources to NGOs exacerbates their relationship with the government, by changing power relations between these two actors. Where the state is very weak, the focus should be on meeting the basic needs of the people and providing support to the sectors that the government itself tries to support (Paul, 2006). At the same time, there are also some who argue that it is precisely in weaker states that it is important to invest in social harmony, political stability and peace, since these may eventually lead to development (Pronk, 2001: 628).

- Simplistic solutions do not work: the use of simple formulae in the past to promote development has not always been successful (see Table 2.1). The key message is that design principles should not be simplistic. For example, poverty alleviation is not simply done through providing employment. It must also focus on the root causes of poverty – the structural features that lead to marginalization of the poor – for long-term results. This implies providing the poor with access to markets, credit, and restructuring resources and services in favour of the poor (Sobhan, 2002), dealing with local power politics, empowering the poor and investing in poor people's health, education and infrastructure (Goldin *et al.*, 2002).

- Imbalances in, or distortions to, the local economy should be avoided: aid should not create a distortion in local salaries or lead to a brain drain (Pronk, 2003); and should not distort local economies by subsidizing activities that should be phased out. Furthermore, it should not inadvertently lead to a policy-substitution effect (Pronk, 2003), which may happen in countries that become dependent on aid. Providing cash may help partner countries use this resource more effectively and is less tied to donor preferences! Others argue that aid should be in the form of loans so that it is used for productive purposes and promotes discipline (Djankov *et al.*, 2006). Aid is most effective below 15% of GDP of the partner countries and, beyond 15%–45% of GDP in partner countries, there may be diminishing returns (Collier and Dollar, 2002; Addison *et al.*, 2005).

- Conditionalities and tied aid are problematic. Conditionalities have been introduced to ensure the effectiveness of aid use but have often been counter-productive. Conditionalities may be effective where there is a domestic constituency in the partner country that supports such a reform (Pronk, 2003) and if the foreign donor is consistent and persistent (Svensson, 2003).

But, for each partner, a careful balance should be achieved between ownership and reform, using domestic governance systems and managing risks (Paul, 2006). Tied aid is a specific type of condition when assistance is linked to specific commodities and services from the donor country. Tied aid, including food aid and technical assistance, has high economic costs (UNDP, 2005:102; Commission for Africa, 2005: 92; IMF and World Bank, 2006: 7, 83; United Nations Millennium Project, 2005:197; Easterly and Pfutze, 2008). However, tied aid and technical assistance create a constituency in donor countries that support aid (Wolfenson cited in La Chimia, 2004); removing tied aid removes this constituency. Conditionalities and tied aid tend to be 'catch 22' situations.

- Stakeholders have to be included and mobilized. The literature unanimously supports the increasing engagement and mobilization of stakeholders in the scoping, design and implementation of aid projects (e.g. Bird and Cabral, 2007).

Many of these ideas have been adopted in the Paris Declaration on Aid Effectiveness of 2005 (see Section 4.3.4). Although the word 'design' is used, this may be a misnomer. Development cooperation is a 'clumsy' process (coined by Shapiro, 1988) through which partner countries and stakeholders collectively engage in the preparation of context relevant ideas that may have a likelihood of success. But there is a 'method in the madness' as the popular saying goes. The clumsiness of this process is rooted in the idea that all people, societies and epochs are inherently different and the interdependence between these actors is equally so. Clumsy solutions allow for contestation among hierarchical, egalitarian and individualistic notions about how problems can be solved (Verweij et al., 2006).

2.5 Conclusions

This chapter argues that the evolution of theories on development and what helps countries develop is in a state of flux. Each decade, insights are replaced by newer insights. It comes to three main conclusions.

First, the idea that aid is a short-term concept is a fallacy. As long as there are poor and rich, there will be a need for aid, although the composition of these groups may keep changing.

Second, the idea that aid should contribute to national income is long outdated. The purpose of aid is to contribute to improving the welfare of the poorest; and such welfare can be measured only by examining the multiple benefits that may emerge in terms of improved health conditions, education and employment opportunities, improved ecosystems and ecosystem services and so on.

Third, history shows us that aid can no longer be conceived as a donor-driven, pre-planned, context-neutral approach to development that is based on simple theoretical formulae. Instead, clumsy approaches that take into consideration in a balanced manner the BASICS design principles (broader assessments, aligning aid type to context, avoiding simplistic approaches, avoiding the creation of imbalances in the local economy, minimizing conditionalities and tied aid, and engaging stakeholders in the clumsy processes) may be necessary.

Acknowledgements

This chapter has benefited from the critical and constructive comments of Jill Jaeger and Eileen Harloff.

References

Acemoglu, D., Johnson, S. and Robinson, J. A. (2005). Institutions as a fundamental cause of long-run growth, in *Handbook of Economic Growth*, Volume 1A, ed. P. Aghion and S. N. Durlauf. Amsterdam: Elsevier, pp. 385–472.

Addison, T., Mavrotas, G. and McGillivray, M. (2005). *Aid, Debt Relief and New Sources of Finance for Meeting the Millennium Development Goals*. Tokyo: UNU World Institute for Development Economics Research.

Adelman, I. (2001). Fallacies in development theory and their implications for policy, in *Frontiers of Development Economics: The Future in Perspective*, ed. G. M. Meier and J. E. Stiglitz. Oxford/Washington, D.C.: Oxford University Press and World Bank, pp. 103–34.

Alesina, A. and Dollar, D. (2000). Who gives foreign aid to whom and why? *Journal of Economic Growth*, **5**(1), 33–63.

Alesina, A. and Weder, B. (2002). Do corrupt governments receive less foreign aid? *American Economic Review*, **92**, 1126–37.

Anders, G. (2005). *Civil Servants in Malawi: Moonlighting, Kinship and Corruption in the Shadow of Good Governance*. PhD thesis, Law Faculty, Erasmus University Rotterdam.

Banuri, T., Weyant, J., Akumu, G. *et al.* (2001). Setting the stage: climate change and sustainable development, in *Climate Change 2001: Mitigation. Contribution of Working Group III to the Third Assessment Report of the Intergovernmental Panel on Climate Change*, ed. B. Metz, O. Davidson, R. Swart and J. Pan. Cambridge: Cambridge University Press, pp. 73–114.

Barnett, J. (2001). *The Meaning of Environmental Security, Ecological Politics and Policy in the New Security Era*. London: Zed Books.

Beg, N., Morlot, J. C., Davidson O. *et al.* (2002). Linkages between climate change and sustainable development. *Climate Policy*, **2**(3), 129–44.

Berthélemy, J.-C. (2006). Bilateral donors' interest vs. recipients' development motives in aid allocation: do all donors behave the same? *Review of Development Economics*, **10** (2), 179–94.

Bird, N. and Cabral, L. (2007). *Changing Aid Delivery and the Environment: Can General Budget Support Be Used to Meet Environmental Objectives?* London: Overseas Development Institute.

Boone, P. (1994). *The Impacts of Foreign Aid on Savings and Growth*. Mimeo. London: London School of Economics.

(1996). Politics and the effectiveness of foreign aid. *European Economic Review*, **40**, 289–329.

Bouab, A. H. (2004). Financing for development, the Monterrey consensus: achievements and prospects. *Michigan Journal of International Law*, **26**, 359–69.

Bourguignon, F. and Sundberg, M. (2007). Aid effectiveness: Opening the black box. *American Economic Review*, **97**(2), 316–21.

Brahman, S., Gerretsen, H. and van Marrewijk, C. (2006). *Agglomeration and Aid*. CESifo Working Paper No. 1750. Munich: CESifo Group.

Brown, L. R. (1981). *Building a Sustainable Society*. Washington, D.C.: Worldwatch Institute.

    (1996). *Tough Choices: Facing the Challenge of Food Security*. New York, NY: W. W. Norton & Co.

Brown-Weiss, E. (1989). *In Fairness to Future Generations: International Law, Common Patrimony, and Intergenerational Equity*. Tokyo: United Nations University Press.

Burke, E. (1790). *Reflections on the Revolution in France*. Republished as Clark, J. C. D., ed. (2001). *Reflections on the Revolution in France*. Palo Alto, CA: Stanford University Press.

Burnside, C. and Dollar, D. (1997). *Aid, Policies and Growth*. World Bank Policy Research Working Paper 1777. Washington, D.C.: World Bank.

    (2000). Aid, policies, and growth. *American Economic Review*, **90**(4), 847–68.

Cárdenas, E. J., Di Cerisano, C. S. and Avalle, O. (1995). The changing aid environment: perspectives on the official development assistance debate. *ILSA Journal of International & Comparative Law*, **2**, 189–201.

Cassen, R. and associates (1994). *Does Aid Work? Report to an Intergovernmental Task Force*. Oxford: Clarendon Press

Chatterjee, P. and Finger, M. (1994). *The Earth Brokers*. London: Routledge.

Cohen, S., Demeritt, D., Robinson, J. and Rothman, D. (1998). Climate change and sustainable development: towards dialogue. *Global Environmental Change*, **8**(4), 341–71.

Collier, P. (2006). *What Can We Expect from More Aid to Africa?* Unpublished manuscript. Centre for the Study of African Economies, Oxford University.

Collier, P. and Dollar, D. (2002). Aid allocation and poverty reduction. *European Economic Review*, **26**(8), 1475–500.

    (2004). Development effectiveness: what have we learnt? *Economic Journal*, **114**(496), F244–71.

Collier, S., Guillamont, P., Guillamont, S. and Gunning, J. W. (1997). Redesigning conditionality. *World Development*, **25**(9), 1399–407.

Collingridge, D. (1980). *The Social Control of Technology*. Milton Keynes: Open University Press.

Commission for Africa (2005). *Our Common Interest*. London: Commission for Africa.

Dasgupta, P. (1993). *An Inquiry into Well-being and Destitution*. Oxford: Clarendon Press.

De Rivero, O. (2001). *The Myth of Development: The Non-Viable Economies of the 21st Century*. London: Zed Books.

Djankov, S., Montalvo, J. G. and Reynal-Querol, M. (2006). Does foreign aid help? *Cato Journal*, **26**(1), 1–28.

Dollar, D. and Easterly, W. (1999). The search for the key: aid, investment and policies in Africa. *Journal of African Economies*, **8**(4), 546–77.

Dollar, D. and Levin, V. (2006). The increasing selectivity of foreign aid: 1984–2003. *World Development*, **34**(12), 2034–46.

Dollar, D. and Svensson, J. (2000). What explains the success or failure of structural adjustment programmes? *Economic Journal*, **110**(466), 894–917.

Doucouliagos, H. and Paldam, M. (2008). Aid effectiveness on growth: a meta-study. *European Journal of Political Economy*, **24**(1), 1–24.

Dovers, S. and Handmer, J. (1993). Contradictions in sustainability. *Environmental Conservation*, **20**(3), 217–22.

Easterly, W. (2006). *The White Man's Burden: Why the West's Efforts to Aid the Rest Have Done So Much Ill and So Little Good*. Oxford: Oxford University Press.

(2007). Was development assistance a mistake? *American Economic Review*, **97**(2), 328–32.

Easterly, W. and Pfutze, T. (2008). Where does the money go? Best and worst practices in foreign aid. *Journal of Economic Perspectives*, **22**(2), 29–52.

Ehrlich, P. R. (1968). *The Population Bomb*. New York, NY: Ballantine.

Ehrlich, P. R. and Ehrlich, A. H. (1990). *The Population Explosion*. New York, NY: Simon and Schuster.

Emmerij, L. (2002). Aid as a catalyst: comments and debate. Aid as a flight forward. *Development and Change*, **33**(2), 247–60.

FitzGerald, E. V. K. (2002). Rethinking development assistance: the implications of social citizenship in a global economy, in *Social Institutions and Economic Development: A Tribute to Kurt Martin*, ed. E. V. K. FitzGerald. Dordrecht/Boston, MA: Kluwer Academic Publishers, pp. 125–42.

Fitzpatrick, B., Gelb, A. and Sundberg, M. (2007). *Aid to Sub-Saharan Africa: Whither $650 Billion?* Washington, D.C.: World Bank.

Goldin, I., Rogers, H. and Stern, N. (2002). *The Role and Effectiveness of Development Assistance: Lessons from World Bank Experience*. Washington, D.C.: World Bank.

Gomanee, K., Girma, S. and Morrissey, O. (2005). Aid and growth in Sub-Saharan Africa: accounting for transmission mechanisms. *Journal of International Development*, **17**, 1055–75.

Grossman, G. M. (1995). Pollution and growth: what do we know?, in *The Economics of Sustainable Development*, ed. I. Goldin and L. A. Winters. Cambridge: Cambridge University Press, pp. 19–42.

Guillaumont, P. and Chauvet, L. (2001). Aid and performance: a reassessment. *Journal of Development Studies*, **37**(6), 66–92.

Gupta, J. (2004). Global sustainable food governance and hunger: traps and tragedies. *British Food Journal*, **106**(5), 406–16.

Harford, T. and Klein, M. (2004). *Donor Performance: What Do We Know, and What Should We Know?* Public Policy for the Private Sector, Note Number 278, October 2004. Washington, D.C.: World Bank.

Herrfahrdt-Pähle, E. (2008). Two steps forward, one step back: institutional change in Krygyz water governance, in *Water Politics and Development Cooperation: Local Power Plays and Global Governance*, ed. W. Scheumann, S. Neubert and M. Kipping. Berlin: Springer, pp. 277–300.

Hungria, M., Franchini, J. C., Campo, R. J. and Graham, P. H. (2005). The importance of nitrogen fixation to soybean cropping in South America, in *Nitrogen Fixation in Agriculture, Forestry, Ecology and Environment*, ed. D. Werner and W. E. Newton. Berlin: Springer, pp. 25–42.

ILA (2002). *Resolution 3/2002 of the International Law Association: The New Delhi Declaration of Principles of International Law Relating to Sustainable Development*.

IMF and World Bank (2006). *Global Monitoring Report 2006: Strengthening Mutual Accountability: Aid, Trade and Governance*. Washington, D.C.: World Bank/ International Monetary Fund.

IUCN (1980). *World Conservation Strategy*. Geneva: IUCN, UNEP and WWF.

(1991). *Caring for the Earth*. Geneva: IUCN, UNEP and WWF.

Ives, J. D. (2004) *Himalayan Perception: Environmental Change and the Well-being of Mountain Peoples*. Oxford: Routledge.

Jänicke, M., Monch, H., Ranneberg, T. and Simonis, U. E. (1989). Economic structure and environmental impacts: East–West comparisons. *The Environmentalist*, **9**(3), 171–82.

Jepma, C. K. (1991). *The Tying of Aid*. Paris: Organisation for Economic Co-operation and Development.

Joint European NGO report (2006). *EU Aid: Genuine Leadership or Misleading Figures? An Independent Analysis of European Governments' Aid Levels*. Brussels: Concord.

Kluge, T. and Scheele, U. (2008). Private sector participation in water supply and sanitation, in *Water Politics and Development Cooperation: Local Power Plays and Global Governance*, ed. W. Scheumann, S. Neubert and M. Kipping. Berlin: Springer, pp. 205–26.

Krause, M. (2008). The political economy of water and sanitation services in Colombia, in *Water Politics and Development Cooperation: Local Power Plays and Global Governance*, ed. W. Scheumann, S. Neubert and M. Kipping. Berlin: Springer, pp. 237–58.

Krueger, A., Michalopoulos, O. C. and Ruttan, V. W. (1989). *Aid and Development*. Baltimore, MD: Johns Hopkins University Press.

Kuik, O. J. and Verbruggen, H., eds. (1991). *In Search of Indicators of Sustainable Development*. Dordrecht: Kluwer Academic Publishers.

La Chimia, A. (2004). International steps to untie aid: the DAC/OECD recommendation on untying official development assistance to the least developed countries. *Public Procurement Law Review*, **13**(1), 1–29.

Lafferty, W. M. (1996). The politics of sustainable development: global norms for national development. *Environmental Politics*, **5**(2), 185–208.

Lawson, A. and Booth, D. (2004). *Evaluation of General Budget Support: Evaluation Framework*. London: Overseas Development Institute.

Lehtonen, M. (2004). The environmental–social interface of sustainable development: capabilities, social capital, institutions. *Ecological Economics*, **49**(2), 199–214.

Leontief, W. (1936). A note on the pure theory of transfer, in *Explorations in Economics: Notes and Essays Contributed in Honour of F. W. Taussig*. New York, NY: McGraw-Hill, pp. 84–92.

Levine, R. and What Works Working Group with Kinder, M. (2004) *Lives Saved: Proven Successes in Global Health*. Washington, D.C.: Center for Global Development.

Lindauer, D. L. and Pritchett, L. (2002). What's the big idea? The third generation of policies for economic growth. *Economia*, **3**(1), 1–39.

Malenbaum, W. (1978). *World Demand for Raw Materials in 1985 and 2000*. New York, NY: McGraw-Hill.

Malthus, T. R., reprint ed. Winch, D. (1994). *An Essay on the Principle of Population*. Cambridge: Cambridge University Press.

Markandya, A. and Halsnæs, K., eds. (2002). *Climate Change and Sustainable Development: Prospects for Developing Countries*. London: Earthscan.

Matz, M. (2008). Rethinking IWRM under cultural considerations, in *Water Politics and Development Cooperation: Local Power Plays and Global Governance*, ed. W. Scheumann, S. Neubert and M. Kipping. Berlin: Springer, pp. 177–204.

McGillivray, M. (2004). Descriptive and prescriptive analyses of aid allocation: approaches, issues, and consequences. *International Review of Economics and Finance*, **13**, 275–92.

Mebratu, D. (1998). Sustainability and sustainable development: historical and conceptual review. *Environmental Impact Assessment Review*, **18**, 493–520.

Meier, G. M. (2001). Ideas for development, in *Frontiers of Development Economics: The Future in Perspective*, ed. G. M. Meier and J. E. Stiglitz. Oxford/Washington, D.C.: Oxford University Press and World Bank, pp. 1–12.

Metz, B., Berk, M., den Elzen, M., de Vries, B. and van Vuuren, D. (2002). Towards an equitable climate change regime: compatibility with Article 2 of the climate change convention and the link with sustainable development. *Climate Policy*, **2**(2/3), 211–30.

Monterrey Consensus (2002). *Monterrey Consensus of the International Conference on Financing for Development*, UN Document A/Conf/198. Available online at http:// www.un.org/esa/ffd/monterrey.

Moran, T. H. (1986). Overview: the future of foreign direct investment in the third world, in *Investing in Development: New Roles for Private Capital*, ed. T. H. Moran. New Brunswick, NJ: Transaction, pp. 3–34.

Morita, T., Robinson, J., Adegbulugbe, A. *et al.* (2001). Greenhouse gas emission mitigation scenarios and implications, in *Climate Change 2001: Mitigation, Contribution of Working Group III to the Third Assessment Report of the Intergovernmental Panel on Climate Change*, ed. B. Metz, O. Davidson, R. Swart and J. Pan. Cambridge: Cambridge University Press, pp. 115–66.

Munasinghe, M. and Swart, R., eds. (2000). *Climate Change and Its Linkages with Development, Equity and Sustainability*. Geneva: Intergovernmental Panel on Climate Change.

Najam, A., Rahman, A. A., Huq, S. and Sokona, Y. (2003). Integrating sustainable development into the Fourth Assessment Report of the Intergovernmental Panel on Climate Change. *Climate Policy*, **3**(S1), S9–17.

Ndulu, B. J. (2002). Partnerships, inclusiveness and aid effectiveness in Africa, in *Social Institutions and Economic Development: A Tribute to Kurt Martin*, ed. E. V. K. FitzGerald. Dordrecht/Boston, MA: Kluwer Academic Publishers, pp. 143–68.

Ney, S. (2009). *Resolving Messy Policy Problems: Handling Conflict in Environmental, Transport, Health and Ageing Policy*. London: Earthscan.

Ney, S. and Thompson, M. (2000). Cultural discourses in the global climate change debate, in *Society, Behaviour and Climate Change Mitigation*, ed. E. Jochem, J. Sathaye and D. Bouille. Dordrecht/Boston, MA: Kluwer Academic Publishers, pp. 65–92.

Nunnenkamp, P. and Thiele, R. (2006). Targeting aid to the needy and deserving: nothing but promises? *World Economy*, **29**(9), 1177–201.

Nurkse, R. (1953). *Problems of Capital Formation in Developing Countries*. Oxford: Blackwell.

OECD (2006). *Development Co-operation Report 2005*. Paris: Organisation for Economic Co-operation and Development.

(2008). *Growing Unequal? Income Distribution and Poverty in OECD Countries*. Paris: Organisation for Economic Co-operation and Development.

OECD DAC (2003). *Harmonizing Donor Practices for Effective Aid Delivery*. Paris: Organisation for Economic Co-operation and Development.

Ohlin, G. (1966). *Foreign Aid Policies Reconsidered*. Paris: Organisation for Economic Co-operation and Development.

Opschoor, J. B. (2009). Sustainability, in *Handbook on Economics and Ethics*, ed. I. van Staveren and J. Peil. Cheltenham: Edward Elgar, pp. 531–8.

Paul, E. (2006). A survey of the theoretical economic literature on foreign aid. *Asian–Pacific Economic Literature*, **20**(1), 1–17.

Pimental, D. and Pimental, M. (1979). *Food, Energy and Society*. London: Edward Arnold.

Population Commission (1996). *Caring for the Future: Report of the Independent Commission on Population and Quality of Life – A Radical Agenda for Positive Change*. Oxford: Oxford University Press.

Pronk, J. P. (2001). Aid as a catalyst. *Development and Change*, **32**(4), 611–29.

(2003). Aid as a catalyst: a rejoinder. *Development and Change*, **34**(3), 383–400.

Putnam, R. D. (1993). *Making Democracy Work: Civic Traditions in Modern Italy*. Princeton, NJ: Princeton University Press.

Rajan, R. G. and Subramanian, A. (2005). *Aid and Growth: What Does the Cross-Country Evidence Really Show?* IMF Working Paper WP/05/127. Washington, D.C.: International Monetary Fund.

Riddell, R. (2007). *Does Foreign Aid Really Work?* Oxford: Oxford University Press.

Rio Declaration (1992). *Rio Declaration and Agenda 21*. Report on the UN Conference on Environment and Development, Rio de Janeiro, 3–14 June 1992, UN Document A/CONF.151/26/Rev. 1 (Vols. 1–III). New York, NY: United Nations.

Robinson, J. (2004). Squaring the circle? Some thoughts on the idea of sustainable development. *Ecological Economics*, **48**, 369–84.

Rosenstein-Rodan, P. (1943). Problems of industrialization of Eastern and Southeastern Europe. *Economic Journal*, **53**(210/211), 202–11.

(1976). The theory of the 'big push', in *Leading Issues in Economic Development*, 3rd edn, ed. G. Meier. Oxford: Oxford University Press, pp. 632–6.

Rostow, W. W. (1956). The take-off into self-sustained growth. *The Economic Journal*, **66** (261), 25–48.

Ruyter, T. (2005). *Requiem voor de hulp: de ondergang van een bedrijfstak*. Breda: Papieren Tijger.

Sachs, W. (1999). Sustainable development and the crisis of nature: on the political anatomy of an oxymoron, in *Living with Nature*, ed. F. Fischer and M. Hajer. Oxford: Oxford University Press, pp. 23–42.

Sachs, W., McArthur, J. W., Schmidt-Traub, G. *et al.* (2004). *Ending Africa's Poverty Trap*. Brookings Papers on Economic Activity. Washington, D.C.: Brookings Institution.

Schama, S. (1997). *The Embarrassment of Riches: An Interpretation of Dutch Culture in the Golden Age*. New York, NY: Vintage.

Schneider, S. H., Easterling, W. E. and Mearns, L. O. (2000). Adaptation: sensitivity to natural variability, agent assumptions and dynamic climate changes. *Climatic Change*, **45**(1), 203–21.

Schraeder, P. J. (1998). Clarifying the foreign aid puzzle: a comparison of American, Japanese, French, and Swedish aid flows. *World Politics*, **50**(2), 294–323.

Schrijver, N. J. (2008). *The Evolution of Sustainable Development in International Law: Inception, Meaning and Status*. Leiden: Martinus Nijhoff Publishers.

Selden, T. M. and Song, D. (1996). Environmental quality and development: is there a Kuznets curve for air pollution emissions? *Journal of Environmental Economics and Management*, **27**(2), 147–62.

Sen, A. (1981). *Poverty and Famines: An Essay on Entitlement and Deprivation*. Oxford: Oxford University Press.

(1999). *Development as Freedom*. Oxford: Oxford University Press.

Shapiro, M. (1988). Judicial selection and the design of clumsy institutions. *Southern California Law Review*, **61**, 1555–69.

Sharma, S., Koponen, J., Gyawali, D. and Dixit, A. (2004). *Aid under Stress: Water, Forests and Finnish Support in Nepal*. Lalitpur: Himal Books.

Smit, B., Pilifosova, O., Burton, I. *et al.* (2001). Adaptation to climate change in the context of sustainable development and equity, in *Climate Change 2001: Impacts, Adaptation and Vulnerability. Contribution of Working Group II to the Third Assessment Report of the Intergovernmental Panel on Climate Change*, ed. J. J. McCarthy, O. F. Canziani, N. A. Leary, D. J. Dokken and K. S. White. Cambridge: Cambridge University Press, pp. 877–912.

Smith, A. (1776). *An Inquiry into the Nature and Causes of the Wealth of Nations*. Reprinted as Campbell, A., Skinner, A. S. and Todd, W. B., eds. (1976). *An Inquiry into the Nature and Causes of the Wealth of Nations*. Oxford: Oxford University Press.

Snow, C. P. (1969). *The State of Siege*. New York, NY: Charles Scribner's Sons.

Sobhan, R. (2002). Aid effectiveness and policy ownership. *Development and Change*, **33**(3), 539–48.

Stiglitz, J. (2000). Vers un nouveau paradigme du développement. *L'Economie Politique*, **5**, 6–39.

Svensson, J. (1998). *Foreign Aid and Rent Seeking*. Policy Research Working Paper No. 80. Washington, D.C.: World Bank.

(2000). When is foreign aid policy credible? Aid dependence and conditionality. *Journal of Development Economics*, **61**(1), 61–84.

(2003). Why conditional aid does not work and what can be done about it? *Journal of Development Economics*, **70**, 381–402.

Swart, R., Robinson, J. and Cohen, S. (2003). Climate change and sustainable development: expanding the options. *Climate Policy*, **3**(S1), S19–40.

van der Tak, J., Haub, C. and Murphy, E. (1979). Our population predicament: a new look. *Population Bulletin*, **34**(5), 4.

Theesfeld, I. (2008). Political power play in Bulgaria's irrigation sector reform, in *Water Politics and Development Cooperation: Local Power Plays and Global Governance*, ed. W. Scheumann, S. Neubert and M. Kipping. Berlin: Springer, pp. 259–76.

Thérien, J.-P. (2002). Debating foreign aid: right versus left. *Third World Quarterly*, **23**(3), 449–66.

Thompson, M. (2008). *Organising and Disorganising: A Dynamic and Non-Linear Theory of Institutional Emergence and Its Implications*. Axminster: Triarchy Press.

Thompson, M., Rayner, S. and Ney, S. (1998). Risk and governance: part 2. *Government and Opposition*, **33**(3), 330–54.

Thorbecke, E. (2006). *The Evolution of the Development Doctrine: 1950–2005*. UNU-WIDER Discussion Paper 2006/155. Helsinki: World Institute for Development Economics Research of the United Nations University.

Tornell, A. and Lane, P. R. (1999). The voracity effect. *American Economic Review*, **89**(1), 22–46.

Truman, H. S. (1949). Special Message to the US Congress on Greece and Turkey: The Truman Doctrine. Available online at http://www.trumanlibrary.org.

UN Charter (1945). Charter of the United Nations (San Francisco), 26 June 1945, and amended on 17 December 1963, 20 December 1965 and 20 December 1971. ICJ Acts and Documents No. 4.

UNDP (2005). *Human Development Report 2005: International Cooperation at a Crossroads, Aid, Trade and Security in an Unequal World*. New York, NY: United Nations Development Programme.

United Nations Millennium Project (2005). *Investing in Development: A Practical Guide to Achieving the Millennium Development Goals*. New York, NY: United Nations.

van den Berg, R. and Feinstein, O., eds. (2009). *Evaluating Climate Change and Development*. World Bank Series on Development, Volume 8. New Brunswick/ London: Transaction Publishers.

Verweij, M. and Thompson, M., eds. (2006). *Clumsy Solutions for a Complex World: Governance, Politics and Plural Perceptions*. Basingstoke: Palgrave.

Vos, R. (2002). Economic reforms development and distribution: were the founding fathers of development theory right?, in *Social Institutions and Economic Development: A Tribute to Kurt Martin*, ed. E. V. K. FitzGerald. Dordrecht/Boston, MA: Kluwer Academic Publishers, pp. 85–99.

WCED (1987). *Our Common Future: Brundtland Report of the World Commission on Environment and Development*. Oxford: Oxford University Press.

Wilbanks, T. (2003). Integrating climate change and sustainable development in a place-based context. *Climate Policy*, **3**(S1), S147–54.

Wuyts, M. (2002). Aid, the employment relation and the deserving poor: regaining political economy, in *Social Institutions and Economic Development: A Tribute to Kurt Martin*, ed. E. V. K. FitzGerald. Dordrecht/Boston, MA: Kluwer Academic Publishers, pp. 169–86.

3

Mainstreaming climate change: a theoretical exploration

JOYEETA GUPTA

3.1 Introduction

There is increasing pressure in the policy world to mainstream climate change into development and development cooperation (see Section 1.6). Mainstreaming is a concept that brings marginal, sectoral, issues into the centre of discussions, thereby attracting more political attention, economic resources and intellectual capacities. The term 'mainstreaming' is often used loosely in climate discussions to mean a range of ideas that are referred to here as 'incorporation'. It is also often used interchangeably with integration. However, this chapter argues that it is preferable to give a more specific meaning to the concept of mainstreaming and to make a clear distinction between mainstreaming and integration.

This chapter discusses the policy evolution of, and the driving factors behind, mainstreaming discussions to set the stage for the theoretical analysis (see Section 3.2). It examines what mainstreaming climate change means for development and development cooperation (see Section 3.3). It operationalizes the different elements of such processes (see Section 3.4) before drawing conclusions (see Section 3.5).

This chapter argues, first, that the policy attention to mainstreaming climate change into development and development cooperation arises out of the diverse motivations of the various actors. Second, the current discussions on climate-change mainstreaming follow mainstreaming discussions in other fields (e.g. gender) and its proponents should learn from and link up, where relevant, with them. Third, mainstreaming is the last step in the stages of incorporating climate change into policy processes. Fourth, it examines the concept of mainstreaming climate change *both* in general policy processes *and* within official development assistance. This chapter does not address the issue of whether mainstreaming climate change into development cooperation is a good idea or not (see Chapter 10).

Mainstreaming Climate Change in Development Cooperation: Theory, Practice and Implications for the European Union, ed. Joyeeta Gupta and Nicolien van der Grijp. Published by Cambridge University Press. © Cambridge University Press 2010.

3.2 Climate change mainstreaming: driving forces behind policy

In the initial stages of the climate negotiations, the Climate Convention focused on setting up a specific field of action to deal with climate change (see Chapter 1). However, the initial assessments showed that greenhouse gases were emitted from practically every sector in society, and all sectors would have to take appropriate mitigation measures. Furthermore, most sectors were likely to be affected by the potential impacts of climate change and measures to adapt would need to be integrated into sectoral plans. This section sums up the policy evolution towards mainstreaming and discusses the motivating factors behind the idea of mainstreaming. It shows that there is a convergence of opinion amongst different actors that climate change needs to be incorporated into development and development cooperation processes, and points to the relative lack of discussion on whether this should also happen in other fields of international cooperation.

3.2.1 The policy evolution

The policy evolution on mainstreaming climate change has passed through three phases (see Table 3.1). In the first phase, in the 1990s, climate change was seen more as a stand-alone issue. In the second phase, towards the end of the 1990s, policy-makers and scholars made the link with development issues (see Section 1.4), and aid agencies began exploring the potential of mainstreaming climate change into development cooperation. In the third phase, a policy process was launched to implement this in multilateral (e.g. World Bank and EU) as well as bilateral aid, and mainstreaming increasingly came to be seen as 'good development practice' (World Bank, 2008; US AID, 2007; Danida, 2008; Roberts, 2007; Roberts *et al.*, 2008). The details of the policy processes are provided in the relevant chapters: the UN and the OECD (see Chapter 4); the climate change regime (see Chapter 5); the European Union (see Chapter 6); and specific EU countries (see Chapter 7).

Formally, the UN Development Group (see Chapter 4), the OECD (see Chapter 4) and the European Union (see Chapter 6) have adopted mainstreaming policy frameworks, each with their own interpretation and emphasis. While the OECD focuses on mainstreaming primarily adaptation in development policy, the EU focuses on both adaptation and mitigation. Furthermore, the priority given to mainstreaming and the nature of mainstreaming differ from country to country, and North European countries tend to be more proactive (Gigli and Agrawala, 2007). Although these countries started to incorporate climate change on an ad hoc basis, there is an increasing desire to make this much more structural in their development cooperation policies. Michaelowa and Michaelowa (2008: 18) argue that 'the "development community" will appreciate the stabilization of the overall aid budget and the

Table 3.1 *Mainstreaming climate change: policy evolution*

Year	Actor	Document	Key element
Phase 1			
1992	Negotiators, researchers	Climate Convention and scientific papers	Climate change as distinct policy field; different also from desertification
Phase 2			
1997 onwards	Researchers	Articles	Keeping climate change distinct from development issues creates an artificial distinction between the two issues
1997	EU	Treaty of European Union	Integrate environment in other sectors
Phase 3			
2002	Donor agencies	Poverty and climate change	Link climate change to development cooperation
2003–4	EU Commission and Council	Climate change in the context of development cooperation	Link climate change to development cooperation
2005	G7	Gleneagles Communique and Plan of Action	Focus on using development cooperation to redesign energy systems and adaptation projects
2006	World Bank	Clean Energy and Investment Development Framework	Aims at enhancing access in Sub-Saharan Africa, promoting transition to low-carbon economies and assisting with adaptation
2006	OECD	Declaration on Integrating Climate Change Adaptation into Development Co-operation	Focus on integrating adaptation into development cooperation
2007	UNEP	Global Environment Outlook	Links globalization, ecosystem services and human well-being
2007	UNEP–UNDP	Partnership on climate change	Aims to integrate adaptation into national development plans and facilitate access to carbon finance
2007	EU	Global Climate Change Alliance	Aims at integrating climate change into poverty reduction efforts
2008	World Bank	Strategic Framework for Climate Change and Development	Aims at economic growth, poverty reduction, the MDGs, energy access and adaptive capacity
2009	OECD	Joint High Level Meeting on Environment and Development	Policy guidelines on integrating adaptation into development cooperation

"climate community" will appreciate the effort from an environmental perspective'. In fact, the outcome can be sold in two different policy fields, thus also creating a constituency in the donor countries.

3.2.2 The driving forces and arguments in favour of mainstreaming

The convergence in policy processes towards mainstreaming climate change in development cooperation results from different motivations of different stakeholders.

Scientists realized that climate change was not a sectoral (e.g. energy or transport) or a marginal issue and that it had to be linked to development issues and poverty reduction (IPCC, 2007; UNEP–UNDP, 2007). While earlier research on climate change did not make significant links with development, since the 2000s scientists have increasingly made these links (see Section 1.4).

Actors in the development cooperation world worried about (a) the difficulties in defining and delivering development assistance; (b) whether development cooperation is perceived as successful (see Chapter 2); and (c) the lack of high-quality projects aimed at poverty amelioration. They were particularly receptive to new ideas and help with the design of new adaptation projects from the climate perspective (Michaelowa and Michaelowa, 2008). Second, the environmental critique of past multilateral development cooperation projects (Hicks *et al.*, 2008; Werksman, 1993) led to environmental concerns being taken into account where possible. This resolve was strengthened by the growing realization that climate policies will be rendered ineffective because economic growth in DCs may increase emissions (see the discussion of the environmental Kuznets curve, Chapters 1 and 10). In other words, if development cooperation promotes carbon-intensive development paths in DCs, this will be counter-productive. Past development cooperation has often focused on energy technologies that have negative environmental impacts. Furthermore, if other international flows apart from of ODA (see Table 1.3) promote GHG intensive investments, this could also further aggravate the problem. For example, rich countries have provided export credits to subsidize the export of, inter alia, fossil fuels to DCs (Maurer and Bhandari, 2000; Gupta, 2007). In addition, development aid projects may be seriously affected by the potential impacts of climate change, e.g. up to 50%–65% of the ODA assistance to Nepal between 1998 and 2000 (Agrawala, 2005).

Taking climate change into account may lead to win–win situations (US AID, 2008). For example, Mani *et al.* (2008: ix) state that, '[a]lthough poor developing countries contribute the least to GHG emissions, some development programs – such as expanding access to clean energy (including through regional projects), as well as financing improved land management and forest management programs –

can offer true win–win opportunities in terms of both supporting good development and reducing global GHG emissions'. They (Mani *et al.*, 2008: ix) go on to state that, 'Given the inherent costs involved, adaptation should be pursued not as an end in itself, but as a means to meet the development objectives of countries.' Finally, declining domestic support in donor countries over the years had led to aid fatigue. From within the development aid bureaucracy (e.g. in Germany), a desire to keep development aid intact may have been a key reason for introducing climate change into the discussion (Michaelowa and Michaelowa, 2008). Since greenhouse gases are increasingly emitted by the new emerging economies, mitigation efforts would clearly be focused there and adaptation would be focused on the poorer countries (Michaelowa and Michaelowa, 2008).

Most DCs argue in various fora that they need to prioritize their own development challenges first before they can take on challenges such as climate change (e.g. Gupta, 1997; G-77 and China Summit Declaration, 2005). Engaging DCs constructively thus inevitably implies taking this link on board. Climate projects will not be 'owned' by DCs if the link to sustainable development is not obvious. While this is their perspective with respect to climate change assistance in general, they are also sceptical about the use of existing ODA funds for climate change (see Section 3.5 and Chapter 5) and for any conditionality in assistance (G-77 and China, Summit Declaration, 2005). However, the small island states and the LDCs probably support any effort at raising money. In the Caribbean, adaptation work is already being integrated into development under a 1997 project (Caribbean Planning for Adaptation to Climate Change) which led to the Mainstreaming Adaptation to Climate Change programme (Tearfund, 2006; van Aalst, 2006).

For most IC governments, mainstreaming climate change follows logically from the existing discourses on environmental integration and mainstreaming. In relation to climate change assistance, it is clear that this must be linked to sustainable development if it is to be effective in DCs. The lack of new and additional resources for climate change (see Chapter 4) and the waning enthusiasm for development cooperation justified the argument that whatever resources are available should be used more efficiently and the transaction costs of setting up a new bureaucracy should be minimized. These are also reasons for mainstreaming climate change (Gupta *et al.*, 2010).

For multilateral organizations, mainstreaming climate change makes sense because of the close substantive links between climate change and development and the Millennium Development Goals (MDGs) (UNEP–UNDP, 2007). For development banks, mainstreaming follows logically from the past efforts at environmental integration and from their accession to the Equator principles (see Chapter 4). Furthermore, mainstreaming avoids 'legacy' and 'reputational'

Table 3.2 *Motives of actors, countries and agencies for mainstreaming climate change in development and development cooperation*

Actor	Arguments in support of mainstreaming
At actor level	
Scientists	Climate change is too serious a global problem to be dealt with through a sectoral approach
Aid agencies	Ideas lacking on how best to promote development, thus openness to new ideas
	Need to deal with environmental critique of past assistance and existing aid projects may aggravate climate change; climate change will impact on existing aid projects and needs to be taken into account in project design; aid fatigue, climate change thus providing a new source of funding
Climate lobby groups	Lack of new and additional funding, hence need to use existing aid more effectively using existing aid agencies reduces transaction costs
At country level	
DCs	For middle-income countries: climate change is not an immediate priority of the developing world, poverty alleviation and development are; engaging these countries and ensuring their ownership of climate projects implies making a link
	For AOSIS and LDCs: climate change will impact on poverty and therefore needs to be mainstreamed
ICs	Mainstreaming of adaptation and mitigation is theoretically sound; lack of new and additional resources, and lack of ODA, means using available resources more efficiently; the Country Strategy Papers requested by the EU are based on PRSPs taking climate change impacts into account
At multilateral level	
Multilateral agencies	The Climate Convention calls for NAPAs – which integrate adaptation and poverty
Development banks	The World Bank asks countries to write PRSPs – since climate change affects poverty, these PRSPs should take climate change into account
UNDP/UNEP	Climate change and development/poverty are closely linked and therefore justify mainstreaming

problems (see Table 3.2). Some agencies have already established tools that help with incorporating climate change concerns. For example, the World Bank's Poverty Reduction Strategy Papers (PRSPs) written by DCs must take the impacts of climate change on poverty into account if they are to be effective assessments on the basis of which policies are to be made. The first generation of these documents did not make such references, but increasingly climate change is being incorporated. The climate regime calls on the poorest countries to prepare National Adaptation Programmes of Action. However, increasing realization of the links between the two issues has called for closer coordination (see Chapter 9). Since the preparation of these documents is mostly financed directly or indirectly by development cooperation resources, this is also seen as a driving force for mainstreaming adaptation (Mani *et al.*, 2008). The EU, too, has Country Strategy Papers (CSPs), which build on the PRSPs, and, although the first generation of these papers did not take climate change into account, increasingly these documents are more integrative (see Chapters 6 and 9).

On the face of it, mainstreaming climate change into *development cooperation* appears to make sense because substantively mainstreaming climate change into *development* makes sense (see Chapter 1). But mainstreaming climate change into development is not easy. Most ICs have not been very active or successful in doing so (Tearfund, 2006). In DCs, climate change is too often seen as a separate subject, e.g. in the Philippines (Lasco *et al.*, 2008). Furthermore, mainstreaming climate change into other international cooperative efforts – such as trade and investment – is not being discussed.

3.3 Incorporating climate change

3.3.1 Introduction

Most of the literature and policy documents on climate change and development cooperation uses mainstreaming and integration interchangeably (e.g. ECA, 2006; UNEP–UNDP, 2007; Agrawala and van Aalst, 2008; Persson and Klein, 2009; Bouwer and Aerts, 2006). However, some of the literature on mainstreaming gives the term a much more specific interpretation, distinguishing it from the other terms often used in the field. In this chapter, we argue that the latter approach is preferable from a conceptual point of view.

Generally, climate change mainstreaming is considered a subset of environmental mainstreaming, and incorporating climate change into development cooperation strategies is thus a subset of incorporating environmental issues into development strategies. Environmental incorporation has been on the political agenda since the publication of the Brundtland report (WCED, 1987); see Chapter 4. The EU has promoted this both within its own development policy and with respect to its development cooperation policy through, inter alia, environmental-impact assessments and expansion of budget lines (CEC, 2000; CEC, 2001; EU, 2005; OECD, 1999; OECD, 2001; Persson *et al.*, 2009); see Chapter 6. The development banks have also adopted this idea.

This brings us to the question – what is mainstreaming? 'Mainstreaming' has been used with respect to education for handicapped children, gender, intercultural relations, disaster management and environment, including climate-change-related issues. Three of these fields are relevant for climate change – environment, gender and disaster mainstreaming.

Gender mainstreaming, *en vogue* since the UN Fourth World Conference on Women in Beijing in 1995 (Beijing Declaration, 1995), has subsequently been adopted at UN level, by the EU and its Member States, by some aid agencies and by many countries. Gender mainstreaming reflects a paradigm shift from science and policy processes that took a gender-neutral look at issues reflecting dominant male perspectives to an understanding of how the different sexes are influenced by and influence social and economic interactions and policy processes.

Disaster mainstreaming, which became important in 1990 with the launch of the International Decade for Natural Disaster Reduction,[1] reflected a paradigm shift from the time when people thought disasters were one-off, single, unpredictable events to when disasters were seen as a repetitive element of modern life, the mainstreaming of which into policy processes would help society cope with them (Benson and Twigg, 2007). This paradigm shift resulted from the media presentations of disasters, the increasing costs associated with disasters,[2] the costs associated with disaster relief and the likelihood that disasters will negatively impact on the MDGs. Pre-disaster mitigation measures that integrate disaster risk management into the development processes may cost relatively little compared with disaster impacts. For example, incremental changes to infrastructure to make it disaster-resistant have a relatively lower cost than repairing and reconstructing the infrastructure and repairing the social damage once the disaster has occurred.

This section defines mainstreaming (see Section 3.3.2), elaborates the stages of incorporation of climate change (see Section 3.3.3), explains the differences between mainstreaming and integration (see Section 3.3.4), relates mainstreaming to the levels of governance (see Section 3.3.5), and explores the linkages between climate mainstreaming and other mainstreaming discourses (see Section 3.3.6).

3.3.2 *Climate change mainstreaming: a theoretical exploration*

Climate change mainstreaming should build on the existing literature on mainstreaming. Mainstreaming in gender studies challenges the 'gendered nature of

---

[1] Political commitment was evident in Agenda 21 (1992), the Yokohama 'Action Plan for the 21st Century for a safer world' (1994), the Beijing Platform for Action (1995), the strategy 'A Safer World in the 21st Century: Disaster and Risk Reduction' (1999), the UN General Assembly's International Strategy for Disaster Reduction (2000), discussion at the World Summit on Sustainable Development (2002), the Delhi Declaration (2004) and the Hyogo Framework for Action (2005) with 168 signatories. The World Bank set up a Global Facility for Disaster Reduction and Recovery and integrates disasters in PRSPs.

[2] The member states of the South Asian Association for Regional Cooperation may have lost about 2%–20% of their GDP and about 12%–66% of revenues as a result of disasters annually. This excludes losses to the informal sectors (SAARC, 2008). At least 86 countries face losses from '30–95% of their GDP and/or mortality in areas at risk' (Dilley *et al.*, 2005).

assumptions, processes and outcomes' (Walby, 2005: 321). 'Gender mainstreaming is the (re)organization, improvement, development and evaluation of policy processes, so that a gender equality perspective is incorporated in all policies at all levels at all stages, by the actors normally involved in policymaking' (Council of Europe, 1998: 15). ECOSOC Resolution 2006/36 (ECOSOC, 2006) defines gender mainstreaming as

The process of assessing the implications for women and men of any planned action, including legislation, policies or programmes, in all areas and at all levels. It is a strategy for making women's as well as men's concerns and experiences an integral dimension of the design, implementation, monitoring and evaluation of policies and programmes in all political, economic and societal spheres so that women and men benefit equally and inequality is not perpetuated.

This means that mainstreaming is more than integration. Gender studies show that integration is often achieved through check-lists, but this does not address mainstreaming. Mainstreaming calls for using a gender lens to study and design policy processes and outcomes. As such, it raises the question of whether mainstreaming is an end or a process (Walby, 2005). Furthermore, gender studies show that, even though political intentions may aim at mainstreaming, policy outcomes may often fall back towards integration and invisibility, that mainstreaming often ends up giving a different and faulty image of gender, i.e. application of male standards (Rees, 1998), or dominant neo-liberal approaches regarding women. Hence, in addition to mainstreaming approaches it is strategically wise to have specifically targeted policies. Finally, gender mainstreaming is successful only when the dynamic interlinkages between connected systems are taken into account, when the roles of different actors and their patterns of interaction are understood, and when the balance of power between men and women is accounted for.

Similarly, disaster studies argue that disasters, including weather-related disasters, are unlikely to be one-off events and that it is necessary to ensure that infrastructure and society are resilient with respect to such disasters. Furthermore, disaster mainstreaming questions the development model encouraged by globalization that leads, inter alia, to greater urban concentrations and climate change. It also focuses on managing risks through appropriate governance and innovative approaches (including disaster-resistant construction, compensatory disaster risk management, public–private partnerships, early-warning systems and emergency management, training for disaster management, insurance and reinsurance 'to transfer risk from local to global level', and a capital pool) to promote post-disaster recovery (Delhi Declaration, 2004). Disaster mainstreaming considers 'risks emanating from natural hazards in medium-term strategic frameworks and institutional structures, in country and sectoral strategies and policies and in the design of individual projects in hazard-prone countries. Mainstreaming requires analysis of how potential hazard events could affect the performance of policies, programmes and projects and of the impact of those policies, programmes and projects, in turn, on vulnerability to natural hazards. This analysis should lead to the adoption of related measures to reduce vulnerability, where necessary, treating risk reduction as an integral part of the development process rather than as an end in itself' (Benson and Twigg, 2007: 5).

In the environmental arena, Seymour *et al.* (2005: 2) argue that

> mainstreaming requires a conceptual shift that identifies environmental sustainability as an objective of the development process, rather than focusing on compliance with environmental standards as a side condition to the achievement of other objectives. It thus requires a focus on proactive investment in policies and projects that promote integration of environmental sustainability into development strategies themselves, rather than as an 'add-on' component to policies and projects conceptualized without reference to environmental sustainability.

UNEP–UNDP (2007) defines environmental mainstreaming as follows:

> Environmental mainstreaming is defined as integrating poverty–environment linkages into national development planning processes and their outputs, such as Poverty Reduction Strategy Papers (PRSPs) and Millennium Development Goal (MDG) strategies. It involves establishing the links between poverty and environment – including climate change – and identifying the policies and programmes to bring about better pro-poor environmental management. It is targeted at influencing national plans, budget processes, sector strategies and local level implementation – reflecting the need to integrate the valuable contribution of environmental management to improved livelihoods, increased economic security and income opportunities for the poor. The overall aim is to establish enduring institutional processes within government, from national to local levels, and within the wider stakeholder community, to bring about environmental mainstreaming that is focused on the government bodies responsible for poverty reduction and growth policies, and that strengthens the role of environmental agencies and non-governmental actors.

In the climate change area, Tearfund (2006: 10) defines mainstreaming as follows:

> Awareness of climate impacts and associated measures to address these impacts are integrated into the existing and future policies and plans of developing countries, as well as multilateral institutions, donor agencies and NGOs. At the national level, mainstreaming shifts responsibility for climate change adaptation from single ministries or agencies to all sectors of government, civil society and the private sector. However, to ensure mainstreaming does not lead to adaptation efforts becoming fragmented and the priority given to it being reduced, a coordinating mechanism such as a multi-stakeholder committee is required, which is afforded political power by being attached to a senior political office or powerful ministry of government.

This definition focuses only on adaptation and on integration, and appears to emphasize sectors. However, mainstreaming is much more than that and as a consequence this book uses the following definition:[3]

Mainstreaming of climate change into development and/or development cooperation is the process by which development policies, programmes and projects are (re)designed, (re)organized, and evaluated from the perspective of climate change mitigation and adaptation. It means assessing how they impact on the vulnerability of people (especially the poorest) and the sustainability of development pathways – and taking responsibility to re-address them if necessary. Mainstreaming implies involving all social actors – governments, civil society, industry and local communities – in the process. Mainstreaming calls for changes in policy as far upstream as possible.

3.3.3 The stages of incorporation of climate change in development and development cooperation

A theoretical distinction among five different stages of incorporating an issue into an agenda can be made. These stages go from relatively easy steps to more complex and potentially expensive steps (at least in the short term). In the first stage, there is a search for ad hoc pilot projects and approaches that aim to reduce emissions or enhance adaptation. In the following stage, there is a more systematic search for win–win solutions that simultaneously deal with climate change and development goals. In the third stage, all policies, programmes and projects are subjected to climate proofing to ensure that they are resilient with respect to the impacts of climate change. In the fourth phase, all policies, programmes and projects are subjected to GHG-emission screening to ensure that these emissions are taken into account in project design (mitigation integration). In the final stage, a climate change lens is used to view all policies, programmes and projects, and changes are made upstream. Mainstreaming often implies reorganization and redesign of development policies from a climate change perspective; this is much more difficult to achieve since it calls for a major structural shift in society (Picciotto, 2002: 323). In other words, if there is a policy focusing on the development of fossil-fuel technology, an integration approach may mean that the emissions of greenhouse gases are taken into account and are based on a number of criteria including economic costs. However, in a mainstreaming approach, the focus will be on redesigning society to reduce the growth of emissions (see Figure 3.1). As society moves towards mainstreaming the costs are likely to go up; however, beyond a certain threshold, perhaps the costs for society come down as there are fewer negative environmental impacts and as vested interests lose out to new interests that support the new order. For

---

[3] Inspired by the gender mainstreaming definition of the Council of Europe (1998: 15).

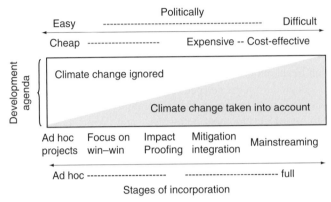

Figure 3.1 Incorporating climate change into development and development cooperation.

example, deforestation in the short term may bring economic benefits; conversely, protecting the forests may bring costs. However, in the long term, protecting the forests brings benefits for current and future generations, and other social actors may benefit. A simple focus only on financial costs may indicate that the more environmental aspects are taken into account, the higher the costs; however, if one takes medium to long term social, environmental and economic benefits into account, mainstreaming could be cost-effective.

3.3.4 Integration and mainstreaming

As mentioned in Section 3.3.1, the terms mainstreaming and integration are often used interchangeably. However, we argue that there are distinct differences. In environmental studies, mainstreaming implies taking the environment into account from the earliest moment of decision-making, i.e. in the scoping stages before even talking about decisions, and throughout the process of programme implementation and/or a project's life cycle. The benefits of mainstreaming include identifying win–win solutions in the long term, catalysing positive outcomes for development and higher cost-effectiveness, enhancing the way in which sequential steps are taken, pre-empting negative lock-in situations or 'legacy issues', and reducing the risks to the reputation and investments of the investor (Seymour et al., 2005). Similar benefits may accrue in climate change mainstreaming.

Mainstreaming is much more than check-lists and includes the design of questions such as the following. Would a general environmental analysis have implications for national priorities? Are the drivers for development influenced by environmental issues? Have the development objectives integrated the environmental aspects? Can environmental goals be promoted through development? What

Table 3.3 *Differences between mainstreaming and integration*

	Integration	Mainstreaming
Discourse level	Policy level	Political or strategic level and possibly as part of the social discourse
Implications	Development policies take climate change into account	Climate change is used to redesign development policies at strategic level and all stages of the planning process
Tools	Check-lists and screening	Debate on winners and losers; making trade-offs
Approach	Reactive	Proactive, innovative
Actors	Sectors	Dialogue with sectors, civil society, NGOs

governance conditions are needed to promote sustainability? What time-frame has to be taken? Have local actors been consulted?

Integration of climate change into development policy implies that existing development policies and projects take climate change aspects into account, possibly through the use of methods and instruments like check-lists and climate proofing, and consequently the climate change component is an add-on component. Mainstreaming goes beyond integration to imply that climate change and development issues are treated equally, attributes and components of climate change are actively used to create a conceptual shift to reframe development starting from the perspective of climate change, and there is a proactive and innovative, rather than a reactive, response to development. Finally, mainstreaming is a political and ideological concept that moves climate change from a marginal discourse and puts it in the centre of discussions to redesign other issues, whereas integration is a policy discourse and tool to ensure coherence between sectoral activities and hierarchical activities at centralized and decentralized levels (Table 3.3). Unlike integration, which calls for greater sectoral cooperation, coordination and coherence, mainstreaming is clearly not focused on sectoral issues, looks at upstreaming, takes short- and long-term impacts into account, is flexible, and calls on all social actors to participate in the process (Tearfund, 2006; Schipper and Pelling, 2006).

Mainstreaming can be seen as a goal or as a process. Mainstreaming as a goal aims to ensure that the key item to be mainstreamed is effectively determining the normative, substantive and procedural aspects of development. Mainstreaming as a process focuses more on how the relevant procedures can be improved/designed to ensure that its goal can be achieved. Mainstreaming must occur as far upstream as possible in the policymaking process to avoid being an add-on, end-of-pipe solution and at the same time implemented at other levels. Mainstreaming implies that the issue being mainstreamed becomes the overriding objective.

The history of environmental mainstreaming reveals that, despite the rhetoric, it is politically difficult to implement. Very often countries and project developers trade off environmental concerns for economic ones. As long as climate change is merely integrated into policy processes, the large emission reductions needed will remain unachieved in the ICs and DCs. Mainstreaming climate change provides a vehicle for the radical redesign of development patterns. The current financial crisis may offer a window of opportunity to engage in this.

In terms of development cooperation, mainstreaming would imply redesigning the entire aid portfolio from the perspective of climate change, not just climate proofing or climate screening. However, such an approach will probably have to be donor-driven unless the donor and the partner country both agree on the mainstreaming ideas. There is thus a risk that mainstreaming climate change into development cooperation could become the new conditionality (see Chapter 2). This could be a strong argument against recommending mainstreaming climate change in development cooperation.

3.3.5 *Mainstreaming and levels of governance*

Mainstreaming of climate change occurs at various levels of governance, from local through regional to global, and involves all actors. International mainstreaming calls for restructuring systems (and related incentives) including trade, investment and aid to ensure that climate change issues are taken into account. Mainstreaming climate change in bilateral and multilateral development cooperation relations between ICs and DCs calls for ensuring that the entire development cooperation life cycle is revamped to take into account climate change (see Table 3.4). Within the domestic context, mainstreaming implies taking a climate change lens and applying it to the development process.

Incorporating climate change concerns into development cooperation is different from incorporating these concerns into national and supranational development strategies, because more than one jurisdiction is involved; there is a power asymmetry between donors and partners, and an asymmetry of issues (obligations and impacts) in relation to the environment. Further, the distribution of responsibilities at different levels of policymaking can make aid subject to a combination of high politics (where security issues are critical) and low politics (where environmental issues are relevant; Persson, 2009).

For effective mainstreaming in official development assistance, three levels need to be engaged if results are not to be counter-productive. Every level requires its own tools and measures. At the macro level, discussions focus on the total quantity of assistance and the priorities for assistance. At the meso level, analysis should ensure that resources are not diverted from national policies. At the micro level, existing

Table 3.4 *Mainstreaming climate change at various levels of governance*

Level	Within	Comment
Global	Global organizations	Integration is being discussed (see Chapter 5)
	Foreign direct investment	No obvious discussion here
	Non-ODA transfers	Some NGOs are taking this into account
	ODA	OECD DAC recommends integration (see Chapter 5)
Supranational	EU	Integration is a formal policy objective (see Chapter 6)
	EU member states	Variation in incorporation discussions (see Chapter 7)
National	ICs	Greater focus on integration than on mainstreaming, more in the EU than in the formerly communist countries or the USA
	DCs	Incorporation is being discussed, the focus is on proofing
Local	ICs	Where national governments are reluctant, local actors are active
	DCs	Progress on incorporation is slow

projects are climate proofed to make sure that greenhouse gas emissions are low and that the projects are unlikely to be affected by the impacts of climate change (see Figure 3.2).

3.3.6 Linking climate with other mainstreaming discourses

Section 3.3.1 argued that mainstreaming discourses on environment, gender and disaster are closely linked to climate change. This section develops this argument further.

The advantages of embedding climate change in the environmental discourse are that environmental issues are seen more holistically, and thus one does not prioritize climate change over others (such as local air and water pollution), and environmental discourses as a whole call on society to question existing development and the way it is measured. The disadvantages are that the focus on climate change per se gets diluted and the scarce resources are once more shared between problems.

Gender mainstreaming is relevant to climate change because first, climate change has a gender component, definitely on the impacts side, but also on the emissions side. More women than men are affected, and differently affected, as a result of weather-related disasters that may be increasing in frequency and

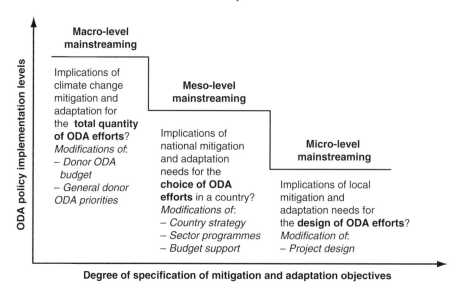

Figure 3.2 Levels of mainstreaming in ODA decision-making. Source: Gupta *et al.* (2010).

intensity because of climate change (Hannon, 2002). For climate change, this would imply understanding the different roles of men and women both in mitigation and in adaptation, and the patterns of interaction and balance of power between men and women. Gender approaches imply discussing social and political issues in a more central manner, thereby enlarging the current focus on technologies and markets. Second, gender mainstreaming is also seen as a critical goal in many development and development cooperation policies (see Chapter 4), and in environmental and disaster management policies. However, the scientific literature (e.g. IPCC-3, 2007; Röhr, 2006: 1) and climate change policymakers have ignored gender issues thus far (Hemmati, 2008:1). Gender has been taken into account only marginally in the Clean Development Mechanism, where women's groups often have a role to play in energy (e.g. Skutsch, 2004; Skutsch and Wamukonya, 2001; Gupta *et al.*, 2008). Recently, Women for Climate Justice and a Global Gender and Climate Alliance have been demanding a more gender-sensitive approach, and the climate secretariat has responded by calling for gender mainstreaming (Hemmati, 2008).

Disaster mainstreaming is relevant for mainstreaming climate change since it helps provide an approach to dealing with disasters, including weather variability and climate change. It is relevant to development cooperation, since DCs (including small island countries) are significantly more affected in terms of loss of life and physical property, if not in terms of economic losses, than ICs (World Congress on

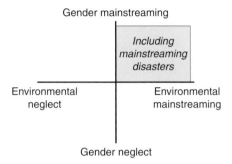

Figure 3.3 Climate mainstreaming should take into account environmental, disaster and gender mainstreaming.

Natural Disaster Mitigation, Delhi Declaration, 2004).[4] This is because poor physical infrastructure and land use management, resulting from chronic poverty, social and economic exclusion, rapid urbanization and poor planning, on the one hand '*and climate change and climate variability on the other* [emphasis Gupta]' increase vulnerability. The impacts of climate change are likely to exacerbate further the existing weather variability that affects coastal and agricultural communities, among others. Climate change impacts range from small (climatic variability in the short term that can force farmers into bankruptcy) to large disasters (such as cyclones that can wipe out or dislodge entire communities of people living in the affected areas). Thus, it makes sense for climate change mainstreaming to build on the existing work to mainstream disaster management and to use the same tools as well. Preventing disasters can be more cost-effective than coping with the impacts of disasters (UNEP–UNDP, 2007). Furthermore, disaster and gender mainstreaming are closely related since both imply taking into account issues specific to women – who are relatively invisible, often politically excluded, have different health needs, different exposure to violence and different needs for survival, livelihood and shelter (Pincha, 2008). Finally, many ICs and DCs have adopted disaster mainstreaming in their development strategies and there is a clear commitment to this concept (e.g. Chamsyah, 2007). Many ICs have also adopted disaster mainstreaming in their development cooperation strategies.

This chapter argues that, if actors mainstream climate change into development and development cooperation processes, such mainstreaming should take into account existing relevant mainstreaming processes and should not substitute for what already exists. In other words, the perspective should concentrate on the right-hand quadrant of the Figure 3.3, which can be summed up as the CEGD (climate, environment, gender and development) perspective.

---

[4] In the decade preceding 1994, 531 000 people died, 2.5 billion people were affected and property worth about USD 654 billion was damaged according to the Delhi Declaration (2004).

3.4 Operationalizing incorporation of climate change

3.4.1 Introduction

This section discusses some operational issues. It elaborates on mainstreaming adaptation and mitigation (see Section 3.4.2), the process of incorporating climate change (see Section 3.4.3), its advantages and disadvantages (see Section 3.4.4) and bottlenecks in implementation (see Section 3.4.5).

3.4.2 Mainstreaming adaptation and mitigation in development

Table 3.5 highlights the implications of incorporating adaptation and mitigation in development and development cooperation.

In terms of mitigation, mainstreaming would call for redesign and restructuring of policy processes and the prioritization of climate change impacts in all policies. In terms of development cooperation, this would imply an entire examination of the project and programme portfolio to see whether a dominant climate perspective would lead to a different approach and choice of projects and programmes.

In terms of adaptation, mainstreaming would imply that policies aim at promoting a spectrum of measures to reduce the vulnerability of society – ranging from actual hard measures (e.g. dikes) to increasing the adaptive capacity of society and ensuring that the potential impacts of climate change are taken into account. In terms of development cooperation for adaptation, mainstreaming would imply that the existing portfolio of aid agencies is reviewed to take into account climate change and other environmental issues, and gender- and disaster-related issues. Resources spent on adaptation may mean less for mitigation; resources spent on some aspects of adaptation may aggravate other aspects of adaptation – resources spent on enhancing the adaptive capacity of society might be spent at the cost of developing hard measures to deal with climate change. Resources spent on adaptation will also mean fewer resources for non-climate expenditures, such as schools and hospitals, and bread on the table for the poor. While, in general, development and adaptation may be synergetic, in the short term there may be trade-offs (Michaelowa and Michaelowa, 2008). For example, while development may imply investing in shrimp culture in coastal wetlands, this may be negatively affected by a rise in sea level (Mani *et al.*, 2008). While, in the short term, it may be more appropriate to provide drinking water to people and irrigation water to farmers, in the long term it may be appropriate to think in terms of migration or a change in occupation for the current farmers. Developing targets for adaptation and adaptive capacity is a complex task and will need specific expertise that is still scarce.

Table 3.5 *Stages of incorporating climate change into development and development cooperation and their implications for adaptation and mitigation*

Stages of incorporation	Adaptation	Mitigation
Ad hoc projects	Design individual projects that take adaptation, environmental impacts and gender issues into account	Design individual projects that take mitigation, environmental impacts and gender issues into account
Win–win	Try to identify systematically existing projects in the current portfolio that take adaptation, environmental impacts and gender issues into account	Try to identify existing projects in the current portfolio that take mitigation, environmental impacts and gender issues into account
Climate proofing: the application of check-lists to adaptation	Subject individual projects and the entire current portfolio of projects to screening for climate risk and opportunities to promote adaptation, minimize disasters and take gender issues into account	Not applicable
Climate integration	Not applicable	Subject the individual projects and the entire current portfolio of projects to screening for opportunities to reduce direct and indirect GHG emissions and promote investment in non-fossil-fuel technologies
Mainstreaming: reorganization and redesign	Redesign and reorganize development and development assistance policy and portfolio of programmes and projects from the perspective of climate change adaptation, environmental impacts, disaster management, resource use and gender	Redesign and reorganize development and development assistance policy and portfolio of programmes and projects from the perspective of climate change mitigation, environmental impacts, disaster management, resource use and gender

3.4.3 *The process of incorporating climate change*

Different processes and tools can be envisaged for the different steps of incorporating climate change into development and development cooperation. This section, building on Persson and Klein (2009), discusses the changes required at the normative, substantive and procedural levels and at agency level (actors).

At the normative level, mainstreaming climate change calls for the design of relevant principles and ideas that give rise to a change in vocabulary and language that is institutionalized through the adoption of ideas into policy documents.

At the substantive level, mainstreaming climate change calls for an effective link to the notion of sustainable development taking social, economic and environmental aspects into account, tends towards a holistic approach towards development, and emphasizes strong rather than weak sustainability.

In terms of procedural aspects, win–win and climate-proofing criteria call for check-lists, portfolio screening (Klein *et al.*, 2007), audits, reports and other technocratic approaches (Persson, 2007), and mitigation integration calls for environmental impact assessment, taking also mitigation into account, gender impact assessment, social impact assessment and disaster impact assessment. Full mainstreaming calls for political discourses at national through bilateral to international levels, moving upstream from environmental impact assessment to development design itself, using the CEGD (climate, environment, gender and development) lens, redesign and reorganization, systemic approaches, upstreaming of policy intervention and political debate.[5] It could build on environmental mainstreaming instruments such as country programming, Dialogue Papers, private-sector programming, country environmental analysis, Poverty Reduction Strategy Papers (PRSPs) and country environmental profiles. Learning from water management, it could apply a systems approach examining the relationship between humans and the environment, a strategic approach that focuses on prioritizing some issues and a stakeholder approach to ensure greater opportunities for participatory decision-making (Hooper, 2005).[6] It could build on the gender mainstreaming tools adopted by the UNDP[7] and the World Bank[8], including baseline definitions, and a gender lens approach for the following: project design and reviews; measuring performance; developing terms of reference of surveys etc.; planning and execution of programme evaluations; meetings, workshops and conferences; and

---

[5] Inspired by the gender literature, the critical question is the following: is it enough simply to modify the environmental impact assessment to include climate change, or should we be looking at an environmental bottom line (instead of a profit bottom line) and developing instead an economic impact assessment?

[6] See the tools used in the Global Water Partnership's Toolbox available online at http://www.gwpforum.org.

[7] The UNDP has a number of tools; see http://www.undp.org/women/resources.html.

[8] Operational tools (including Promising Approaches to Engendering Development, Tools for Specific Sectors, and Themes and Briefing Notes on Gender and Development) and data (including data on gender, key gender employment indicators and revised gender statistics); see http://web.worldbank.org.

language.[9] It could further adopt disaster-management tools, including risk assessment, damage-probability matrices, geographic information systems, remote sensing, microzonation and hazard maps, vulnerability atlases, and disaster-resistant technology supported by a global disaster knowledge network (Tearfund, 2005; Benson and Twigg, 2007), and adopt Benson and Twigg's (2007: 14) seven-step approach to mainstreaming disaster management: raising awareness; creating an enabling environment; development of tools, training and technical support; changes in operational practice; measuring progress; and learning and experience-sharing at all stages.

Thus, climate change mainstreaming would require the use of the CEGD lens for examining all activities in a society from scoping to implementation. It could use Benson and Twigg's seven-step approach for implementation. In terms of development cooperation, it could use existing national reports such as the PRSPs and CSPs as an entry point (Table 3.6).

In terms of actors, while ad hoc and win–win approaches may call for action by individual agents, integration calls for the development of technocratic tools and the training of policymakers and NGOs to apply these tools and critically assess the results. Mainstreaming as a strategic process calls for stakeholders from all walks of life to engage in a political discussion about what mainstreaming actually implies for existing development processes.

3.4.4 *The advantages and disadvantages: an instrumental perspective*

The advantages and disadvantages of incorporating climate change into development and development cooperation are varied. Ad hoc projects may demonstrate the need for proofing, integration and mainstreaming and can justify full-scale efforts to do so. The pilot phase on Activities Implemented Jointly (see Chapter 5) demonstrated this. However, these efforts have marginal impacts. Win–win projects may demonstrate that short- and medium-term advantages can coexist at no substantial cost. However, these remain also relatively small in scale. The Clean Development Mechanism (see Chapter 5) is an example of this.

Climate proofing may allow for cost-effective achievement of adaptation, gender, disaster management and development goals. Climate proofing for adaptation is a good practice and a no-regrets option when it comes to mainstreaming of predefined or existing projects within climate-sensitive sectors (Persson, 2007). It can benefit the poorest in DCs, and brings the research and policy communities working on these fields together – so that they can learn from and reinforce each other. There is a probability that poverty alleviation and some adaptation projects address the same subjects, but there still remains a risk of diversion of resources to large infrastructural projects.

---

[9] See http://portal.unesco.org.

Table 3.6 *Normative, substantive, procedural aspects and the roles of actors in the stages of incorporating climate change*

Stages	Normative	Substantive	Procedural	Actors
Ad hoc	No action needed	One or two demonstration projects	Development of demonstration project; implementing existing guidelines	Individual agents of change
Win–win	Identifying win–win projects (Adaptation)	Systematic search for win–win options	Implementing existing guidelines	Individual agents of change
Climate proofing		Improving cost-effectiveness of projects; systematic climate proofing of portfolio; there may be win–lose options	Portfolio screening, disaster mapping, gender impact assessment, check-lists, NAPAs	Formally appointed actors
Climate integration (mitigation)	Preventing lock-in of technologies	Systematic climate screening of emissions of portfolio; there may be win–lose options	Environmental impact assessment; climate impact assessment, Technology Needs Assessments, National Communications, Country Strategy Papers	Formally appointed actors
Climate mainstreaming	Redesign of ideological basis and involves major institutional changes	Redesign from the start of the policy cycle	Dialogue, debate; use of CEGD lens	All social actors

Mitigation integration can help achieve climate, environment and development goals; linking climate change to development issues may prevent 'legacy' issues and technological lock-in to GHG-unfriendly development pathways. Better substantive links between environmental, climate change and development projects reduce the risk of loss to reputation, and operations and credit risks, for the investing party (Seymour *et al.*, 2005). Mitigation integration brings the climate, environmental and development communities together. However, a mitigation perspective may shift resources from poverty alleviation to polluting sectors.

Mainstreaming climate change may make sense because of the substantive links between the subjects and because policy developments can be sequenced better. Furthermore, mainstreaming climate change in ODA may catalyse mainstreaming climate change in development policy. Before mainstreaming, ODA has little pockets of areas that invest in climate change mitigation and adaptation. After mainstreaming, all ODA is redesigned to take climate change into account and this potentially leverages more changes in DCs.

However, while mainstreaming climate change in development cooperation is theoretically attractive, the history of such exercises demonstrates that, even though social actors may aim at mainstreaming, policy outcomes may fall back to integration and the making of trade-offs where environmental, gender and disaster-related issues are given less importance than short-term economic goals. Efforts to achieve mainstreaming often end up not going further than integration, and in the process the item being integrated becomes invisible. See Huq *et al.* (2006) and Walby (2005), in other issue areas; and Tearfund (2006) and Persson and Klein (2009), in relation to climate change. Mainstreaming may be simply interpreted as a technocratic process of identifying win–win solutions or subjecting existing portfolios of programmes and projects to check-lists, rather than a politically challenging process that questions existing approaches and calls for new approaches (Walby, 2005). Mainstreaming is politically challenging because it upsets the existing order of society, challenges vested interests, and appears expensive and hence unaffordable. The risk that environmental objectives are made secondary to economic objectives is always present, and this perhaps can be addressed only by giving environmental objectives priority (Lafferty and Hovden, 2003). If environmental issues and climate change are prioritized, this reduces the pressure to adapt.

Furthermore, mainstreaming may hijack development cooperation processes and divert resources. Michaelowa and Michaelowa (2008) argue that, although in theory there is a close relationship between climate change and development cooperation, if development cooperation is meant to help the poorest, then the bulk of the existing climate-related projects financed by ODA or otherwise (e.g. CDM; see Chapter 5) do not help the poorest of the poor and would therefore amount to a diversion of resources (cf. Caparrós and Pereau, 2005); see Table 3.7. This argument is developed further in Chapter 10.

3.4.5 Practical challenges in mainstreaming

Full mainstreaming is difficult to implement in ICs because it challenges existing paradigms and creates new losers and winners. It is also difficult to implement in DCs that wish to develop rapidly and see mainstreaming as a bottleneck to immediate growth. Consequently, mainstreaming is a challenge to development cooperation because it also implies revisiting the existing package of assistance, which will in itself create new winners and losers, whereupon the losers will actively lobby against losing.

In terms of incorporation, bottlenecks include the lack of sufficient and relevant knowledge of context-relevant mitigation and adaptation options (Tearfund, 2006); the fact that the past framing of climate change as a global issue still has legacy problems whereby many are unable to see the effective links with existing development challenges (Gupta, 1997; Gupta and Hisschemöller, 1997; Tearfund, 2006); the lack of relevant science for adaptation strategies (Tearfund, 2006); and the difficulty of assessing who is actually engaged in the climate adaptation process (Persson and Klein, 2009). Mainstreaming calls for sophisticated and high-level coordination and coherence of organizations and their policies; but such coordination processes may often stumble on existing turf battles. Another problem is the issue of coordination as a matter of form rather than of content.

Specific challenges in adaptation interventions include the difficulty of separating responses to climate variability and impacts for those who wish to focus exclusively on the latter (Persson and Klein, 2009). While it is attractive to talk about specific adaptation policies and projects, in the long term it is critical to assess the role of development cooperation in improving the adaptive capacity of society (Persson and Klein, 2009). Furthermore, adaptation measures range from soft (e.g. education) to hard (e.g. dikes), and it is not often clear which falls into which category (McGray *et al.*, 2007).

Specific challenges in mitigation policies include the continued investment in fossil fuels, for example, in the Asia Pacific Partnership on Clean Development and Climate Change agreement involving the USA, Australia, Japan, South Korea, China and India. Although the World Commission on Dams was quite negative about large dams, in recent years, climate change has given an impetus to the large-hydro discussion, and this is now a key area of investment (Roberts *et al.*, 2008).

Solutions for some of these dilemmas may take the shape of supporting adaptation to impacts irrespective of whether they are in response to climate variability or change, and to build adaptive capacity rather than identify specific adaptation options.

Table 3.7 *Stages of incorporating climate change: advantages and disadvantages*

Stages	Advantages	Disadvantages
Ad hoc	Demonstrates the need for proofing, integration and mainstreaming	Is ad hoc and impact may be marginal
Win–win	Demonstrates that short- and medium-term advantages for development, climate change, gender, disaster management and environment can co-exist at no substantial extra cost	Is relatively small-scale and impact may be small
Climate proofing	Cost-effective achievement of adaptation, gender, disaster-management and development goals; can benefit the poorest in countries; brings development, gender, disaster-management and adaptation communities together – they learn from and reinforce each other	There is a probability that poverty alleviation and some adaptation projects benefit the same communities, but may divert resources to large-scale infrastructural projects; does not take mitigation into account
Climate integration	Cost-effective achievement of mitigation, environmental and development goals; prevents legacy issues/technological lock-in; prevents reputation loss and operation risks; brings development, environment and mitigation communities together – they learn from and reinforce each other	There is a strong probability that mitigation efforts may divert resources from poverty reduction; mitigation efforts may make projects more expensive in the short to medium term; does not take adaptation into account
Climate mainstreaming	Cost-effective achievement of adaptation, mitigation and development goals; more effective sequencing of projects possible; possible catalytic effect of mainstreaming climate change in ODA	While useful and necessary for development, may make development very expensive for the poor; may make climate change invisible and ignored, leading to soft trades within integration and climate proofing exercises; may hijack the development cooperation agenda, leading to diversion of resources; may provide an excuse for not raising additional funds; mainstreaming has an international component and everyone needs to focus on this, otherwise it may have a conditionality component

3.5 Mainstreaming: politically challenging, hijacking the development cooperation agenda or running the risk of invisibility?

This chapter concludes that the urge to mainstream climate change arises from the scientific recognition that climate change is a serious global problem encompassing all facets of human behaviour and lifestyle and that it needs more focused attention. Different social actors have different reasons for wanting to mainstream climate change into the development and development cooperation processes, and these are converging into a mainstreaming discourse. Combining these discourses may expand the constituency in favour of financial support to developing countries. *This convergence does not, however, imply that mainstreaming climate change into development cooperation is not a controversial issue. This chapter has merely hinted at the potential for diversion of resources and hijacking of the development cooperation agenda.* The nature of the controversies will come further to light in the following chapters.

Second, mainstreaming climate change comes in a long line of mainstreaming endeavours in the policy world in general and in the development cooperation world in particular. This chapter argues that climate change mainstreaming should be seen in the light of environmental mainstreaming and should be linked to gender and disaster mainstreaming through a CEGD (climate, environment, gender and development) lens.

Third, mainstreaming is a complex concept. It is the last stage in the process of incorporating climate change into policy. Full mainstreaming goes beyond integration, check-lists and climate proofing. It calls for political debates that question the fundamental premises of society, both in ICs and in DCs. This may make it relatively unpopular among those with vested interests in the existing system of wealth generation.

Fourth, there are clear net advantages to mainstreaming climate change into the development (not necessarily the development cooperation) policy debate and implementation processes. These include better and more cost-effective chances of achieving both the sustainable development goals and the climate change goals, avoidance of 'legacy' issues such as technological and institutional lock-in, and a better ability to sequence measures to achieve such goals. The negative side effect is that, even though efforts are undertaken to mainstream issues, political challenges often imply a retreat to integration and proofing, and that integration may make an issue 'invisible'. This can be dealt with by adopting a two-track approach of both mainstreaming climate change into policy and having an additional dedicated climate policy.

Finally, this chapter suggests that, if mainstreaming is to be effective, it has to be adopted in all policy arenas rather than just being confined to the development cooperation policy area, if it is adopted at all. This idea is explored further in the following chapters.

Acknowledgements

This chapter has benefited from the extensive review comments of Åsa Persson, Joanne Bayer, Jill Jaeger and Eileen Harloff.

References

Agrawala, S., ed. (2005). *Bridge over Troubled Waters: Linking Climate Change and Development*. Paris: Organisation for Economic Co-operation and Development.

Agrawala, S. and van Aalst, M. (2008). Adapting development cooperation to adapt to climate change. *Climate Policy*, **8**(2), 183–93.

Beijing Declaration (1995). Declaration of the Fourth World Conference on Women, Beijing. Available online at http://www.un.org/womenwatch/daw/beijing/platform/declar.htm.

Benson, C. and Twigg, J. (2007). *Tools for Mainstreaming Disaster Risk Reduction: Guidance Notes for Development Organizations*. Geneva: ProVention Consortium.

Bouwer, L. M. and Aerts, J. C. J. H. (2006). Financing climate change adaptation. *Disasters*, **30**(1), 49–63.

Caparrós, A. and Pereau, J.-C. (2005). *Climate Change Abatement and Development Aid in North–South Differential Games*. Mimeo. Madrid: National Council for Scientific Research.

CEC (2000). *Communication from the Commission to the Council and the European Parliament, The European Community's Development Policy*, COM(2000)212 final, 26 April 2000. Brussels: Commission of the European Communities.

(2001). *Commission Staff Working Paper, Integrating the Environment into EC Economic and Development Cooperation*, SEC(2001)609 final, 10 April 2001. Brussels: Commission of the European Communities.

Chamsyah, B. (2007). *Mainstreaming Disaster Risk Reduction in National Policies and Programmes: An Overview of National Action Plan for Disaster Risk Reduction in Indonesia*. Paper presented at the 2nd Asian Ministerial Conference on Disaster Risk Reduction, New Delhi, 7–8 November 2007.

Council of Europe (1998). *Gender Mainstreaming: Conceptual Framework, Methodology, and Conceptualisation of Existing Practices*. Strasbourg: Council of Europe.

Danida (2008). *Danish Climate and Development Action Programme: A Tool Kit for Climate Proofing Danish Development Cooperation*. Copenhagen: Ministry of Foreign Affairs of Denmark.

Delhi Declaration (2004). Declaration of the World Congress on Natural Disaster Mitigation, New Delhi, 19–22 February 2004; available online at http://www.ieindia.org/recommend/worldcongnatdisastmiti.pdf.

Dilley, M., Chen, R. S., Deichmann, U., Lerner-Lam, A. L. and Arnold, M. (2005). *Natural Disaster Hotspots: A Global Risk Analysis*. Washington, D.C.: World Bank.

ECA (2006). *Court of Auditors Special Report No. 6/2006 Concerning the Environmental Aspects of the Commission's Development Cooperation (Pursuant to Article 248(4), Second Subparagraph, EC)*. OJ C 235/1, 29 September 2009.

ECOSOC (2006). *United Nations Economic and Social Council (ECOSOC) Resolution 2006/36: Mainstreaming a Gender Perspective into All Policies and Programmes in the United Nations System*, 27 July 2006, available online at http://www.unhcr.org/refworld/docid/46c455acf.html.

EU (2005). *Joint Statement by the Council and the Representatives of the Governments of the Member States Meeting within the Council, the European Parliament and the Commission: The European Consensus on Development*, EU Document 2006/C46/01. *Official Journal of the European Union* C46, 1–19, 24 February 2006.

EU Cardiff Conclusions (1998). *Presidency Conclusions: Cardiff European Council*, 15–16 June 1998, EU Document SN 150/1/98 REV 1.

EU Cologne Report (1999). *Commission Working Paper Addressed to the European Council: The Cologne Report on Environmental Integration – Mainstreaming of Environmental Policy*, SEC(99)777 final, 26 May 1999. Brussels: Commission of the European Communities.

G-77 and China Summit Declaration (2005). *Doha Declaration*, Second South Summit 2005, G-77/SS/2005/1.

Gigli, S. and Agrawala, S. (2007). *Stocktaking of Progress on Integrating Adaptation to Climate Change into Development Co-operation Activities*. Paris: Organisation for Economic Co-operation and Development.

Gupta, J. (1997). *The Climate Change Convention and Developing Countries: From Conflict to Consensus?* Dordrecht/Boston, MA: Kluwer Academic Publishers.
  (2007). Legal steps outside the climate convention: litigation as a tool to address climate change. *Review of European Community and International Environmental Law*, **16**(1), 76–86.

Gupta, J. and Hisschemöller, M. (1997). Issue-linkages: a global strategy towards sustainable development. *International Environmental Affairs*, **9**(4), 289–308.

Gupta, J., Persson, Å., Olsson, L. *et al.* (2010). Mainstreaming climate change in development cooperation: conditions for success, in *Adaptation and Mitigation of Climate Change*, ed. M. Hulme and H. Neufeldt. Cambridge: Cambridge University Press.

Gupta, J., van Beukering, P., van Asselt, H. *et al.* (2008). Flexibility mechanisms and sustainable development: lessons from five AIJ projects. *Climate Policy*, **8**(3), 261–76.

Hannon, C. (2002). *Mainstreaming Gender Perspectives in Environmental Management and Mitigation of Natural Disasters*. Paper presented at the Roundtable Panel and Discussion organized by the United Nations Division for the Advancement of Women and the NGO Committee on the Status of Women in preparation for the 46th session of the Commission on the Status of Women on the Disproportionate Impact of Natural Disasters on Women, 17 January 2002.

Hemmati, M. (2008). *Gender Perspectives on Climate Change*. Written statement, United Nations Commission on the Status of Women, 52nd session, New York, NY, 25 February–7 March 2008; Women for Climate Justice.

Hicks, R. L., Parks, B. C., Roberts, J. T. and Tierney, M. J. (2008). *Greening Aid? Understanding the Environmental Impact of Development Assistance*. Oxford: Oxford University Press.

Hooper, B. (2005). *Integrated River Basin Governance: Learning from International Experience*. London: International Water Association Publishing.

Huq, S., Reid, H. and Murray, L. (2006). *Climate Change and Development Links*. IIED Gatekeeper Series 123. London: International Institute for Environment and Development.

IPCC (2007). *Climate Change 2007: Synthesis Report*. Cambridge: Cambridge University Press.

IPCC-2 (2007). *Climate Change 2007: Impacts, Adaptation and Vulnerability. Contribution of Working Group II to the Fourth Assessment Report of the Intergovernmental Panel on Climate Change*. Cambridge: Cambridge University Press.

IPCC-3 (2007). *Climate Change 2007: Mitigation of Climate Change. Contribution of Working Group III to the Fourth Assessment Report of the Intergovernmental Panel on Climate Change*. Cambridge: Cambridge University Press.

Klein, R. J. T., Eriksen, S. E. H., Næss, L. O. *et al.* (2007). Portfolio screening to support the mainstreaming of adaptation to climate change into development assistance. *Climatic Change*, **84**(1), 23–44.

Lafferty, W. and Hovden, E. (2003). Environmental policy integration: towards an analytical framework. *Environmental Politics*, **12**(3), 1–22.

Lasco, R. D., Pulhin, F. P., Jaranilla-Sanchez, P. A., Garcia, K. B. and Gerpacio, R. V. (2008). *Mainstreaming Climate Change in the Philippines*. Working Paper No. 62. Los Banos: World Agroforestry Centre.

Mani, M., Markandaya, A. and Ipe, V. (2008). *Climate Change: Adaptation and Mitigation in Development Programs – A Practical Guide*. Washington, D.C.: World Bank.

Maurer, C. and Bhandari, R. (2000). *The Climate of Export Credit Agencies*, New York, NY: World Resources Institute.

McGray, H., Hammill, A. and Bradley, R. (2007). *Weathering the Storm: Options for Framing Adaptation and Development*. Washington, D.C.: World Resources Institute.

Michaelowa, A. and Michaelowa, K. (2008). Climate or development: is ODA diverted from its original purpose? *Climatic Change*, **84**(1), 5–22.

OECD (1999). *Compendium on Donors' Operational Practices in Support of Environmental Goals*. Paris: Organisation for Economic Co-operation and Development.

(2001). *Donor Support for Institutional Capacity Development in Environment: Lessons Learned*. Evaluation and Aid Effectiveness No. 3. Paris: Organisation for Economic Co-operation and Development.

Persson, Å. (2007). Different perspectives on EPI, in *Environmental Policy Integration in Practice: Shaping Institutions for Learning*, ed. M. Nilsson and K. Eckerberg. London: Earthscan, pp. 25–48.

(2009). Environmental policy integration and bilateral development assistance: challenges and opportunities with an evolving governance framework. *International Environmental Agreements*, **9**(4), 409–29.

Persson, Å. and Klein, R. J. T. (2009). Mainstreaming adaptation to climate change into official development assistance: challenges to Foreign Policy Integration, in *Environmental Change and Foreign Policy: Theory and Practice*, ed. P. Harris. Cambridge: Cambridge University Press, pp. 162–77.

Picciotto, R. (2002). The logic of mainstreaming: a development evaluation perspective. *Evaluation*, **8**(3), 322–39.

Pincha, C. (2008). *Disaster Sensitive Gender Management: A Toolkit for Practitioners*. Chennai: Earthworm Books for Oxfam America and NANBAN Trust.

Rees, T. (1998). *Mainstreaming Equality in the European Union: Education, Training and Labour Market Policies*. London/New York, NY: Routledge.

Rio Declaration (1992). *Rio Declaration and Agenda 21*. Report on the UN Conference on Environment and Development, Rio de Janeiro, 3–14 June 1992, UN Document A/CONF.151/26/Rev. 1 (Vols. 1–III). New York, NY: United Nations.

Roberts, J. T. (2007). Urgent but uncertain: the dilemmas for climate change, development, adaptation and justice for development and humanitarian work. *Monday Developments: The Latest Issues and Trends in International Development and Humanitarian Assistance*. August 2007, 10–11.

Roberts, J. T., Starr, K., Jones, T. and Abdel-Fattah, D. (2008). *The Reality of Official Climate Aid*. Oxford Energy and Environment Comment November 2008. Oxford: Oxford Institute for Energy Studies.

Röhr, U. (2006). *Gender Relations in International Climate Change Negotiations*. Berlin: LIFE e.V./Genanet/WECF.

SAARC (2008). Background paper for the SAARC Workshop on Mainstreaming Disaster Risk Reduction in Development, organized by SAARC Disaster Management Centre, New Delhi in collaboration with the Disaster Management Centre (DMC), Government of Sri Lanka.

Schipper, L. and Pelling, M. (2006). Disaster risk, climate change and international development: scope for, and challenges to, integration. *Disasters*, **30**(1), 19–38.

Seymour, F., Maurer, C. and Quiroga, R. (2005). *Environmental Mainstreaming: Applications in the Context of Modernization of the State, Social Development, Competitiveness, and Regional Integration*. New York, NY: Inter-American Development Bank, Sustainable Development Department.

Skutsch, M. (2004). *CDM and LULUCF: What's in It for Women? A Note for the Gender and Climate Change Network*. Enschede: University of Twente.

Skutsch, M. and Wamukonya, N. (2001). Is there a gender angle to climate change negotiations? *Energy & Environment*, **13**(1), 115–24.

Tearfund (2005). *Mainstreaming Disaster Risk Management: A Tool for Development Organizations*. Teddington: Tearfund.

Tearfund (2006). *Overcoming the Barriers: Mainstreaming Climate Change Adaptation in Developing Countries*, Tearfund Climate Change Briefing Paper 1. Teddington: Tearfund.

UNEP (2007). *GEO Yearbook: An Overview of Our Changing Environment*. Nairobi: United Nations Environment Programme.

UNEP–UNDP (2007). *Guidance Note on Mainstreaming Environment into National Development Planning*. Nairobi/New York, NY: United Nations Environment Programme and United Nations Development Programme.

US AID (2007). *Adapting to Climate Variability and Change: A Guidance Manual for Development Planning*. Washington, D.C.: United States Agency for International Development.

   (2008). *Integrating Climate Change into Development*. Washington, D.C.: United States Agency for International Development.

van Aalst, M. (2006). *Managing Climate Risk: Integrating Adaptation into World Bank Group Operations*. Washington, D.C.: World Bank.

Walby, S. (2005). Gender mainstreaming: productive tensions in theory and practice. *Social Politics*, **12**(3), 321–43.

WCED (1987). *Our Common Future: Brundtland Report of the World Commission on Environment and Development*. Oxford: Oxford University Press.

Werksman, J. D. (1993). Greening Bretton Woods, in *Greening International Law*, ed. P. Sands. London: Earthscan, pp. 65–84.

World Bank (2008). Climate Investment Funds website and presentations, published online at http://www.worldbank.org.

Part III

Governance

4

Global governance: development cooperation

JOYEETA GUPTA

4.1 Introduction

This chapter focuses on the evolution of North–South or rich–poor development and development cooperation issues within global developmental and environmental governance. It explores the policy development within the UN, development banks and the trade bodies (see Section 4.2), and the OECD (see Section 4.3), culminating in the current discussions on 'mainstreaming' climate change into their policy processes. It identifies three critical aspects of the North–South discussion. First, it focuses on the right to development that the developing countries (DCs) have tabled since the 1960s. It shows that, despite some progress having been made, the industrialized countries (ICs) are sceptical about this right. Second, it looks at the quantitative commitment to development assistance that has continuously been on the political agenda since the 1960s. Even though the ICs continue to reiterate the quantitative target, they are unable to implement it. Third, it examines the controversial discussion on 'new and additional' resources in the context of the environmental debate (see Section 4.4).

This chapter then reflects on global governance regarding sustainable-development issues (see Section 4.5). It argues that governance on development and environmental issues, or sustainable-development governance, is diffuse and spread throughout the UN and OECD systems. It submits that there is increasing convergence in policy rhetoric amongst the various bodies in that they have adopted the concept of sustainable development and the 'mainstreaming' jargon, but there is considerable divergence in the ways in which they interpret these terms. Finally, it notes that aid resources are a mere fraction of the total investment flows and that environmental aid flows are in turn a mere fraction of total aid flows.

Mainstreaming Climate Change in Development Cooperation: Theory, Practice and Implications for the European Union, ed. Joyeeta Gupta and Nicolien van der Grijp. Published by Cambridge University Press. © Cambridge University Press 2010.

4.2 The UN: sustainable development cooperation

4.2.1 Introduction

Since World War II, the world has been divided into the first (capitalist), second (communist) and third (consisting mostly of the former colonies) worlds. Under post-war enthusiasm, the UN Charter (1945) was adopted to promote international peace and security and cooperation in solving economic, social, cultural and/or humanitarian problems as well as to promote recognition of human rights.[1] This section elaborates on the progress made on development issues, the integration of the environment into development policy and the role of the development banks, trade and investment, before drawing inferences.

This section shows that, at global level, governance on environment, development, investment and trade is distinct. The environmental critique of development and development banks led to the first set of integrative initiatives (e.g. the Global Environment Facility, Commission for Sustainable Development, United Nations Development Group, UNEP–UNDP partnership, Equator Principles). The trade arena, however, tends to be more indifferent to the environmental critique (Bryner, 1999), although there are some fair-trade initiatives. The investment arena is less affected by the development and environmental critique, and arbitration cases that discuss disputes are often subject to confidentiality rules. The funds available are an indicator of the relative influence of the various bodies. For example, UNEP has less than half a billion and UNDP USD 5 billion,[2] while the World Bank has several billions.

It is difficult to see whether anything substantial has changed since White's (1996: 273) assessment that 'Trade, financial and economic institutions, in particular, have found great difficulty in adopting effective environmental policies, reflecting the conflict between the dominant free trade and economic development ideology and subservient green ideology, resulting in what is known as the "green washing" of these institutions.' Banks and investors have to deliver profits to their shareholders. 'Poverty is good business for the World Bank. According to the Bank's own estimates, for every dollar paid by the US government, the US companies receive about US $1.10 in business. Other shareholders don't do badly either' (Agarwal *et al.*, 1999: 353).

---

[1] Para. (3): 'To achieve international co-operation in solving international problems of an economic, social, cultural or humanitarian character, and in promoting and encouraging respect for human rights and for fundamental freedoms for all without distinction as to race, sex, language or religion' (UN Charter, 1945).

[2] See http://www.undp.org/publications/annualreport2008/resources.shtml.

4.2.2 The evolution of development issues

The development debate passed through a number of phases (Dadzie, 1993). In the 1950s and 1960s, the notion of development planning was spread by ICs to DCs. In the 1970s and early 1980s, the DCs were calling for a fair framework of international economic relations (the New International Economic Order) to improve the chances of development in the post-colonial period (Tinbergen, 1976) and claimed permanent sovereignty over national resources (Schrijver, 1995). Following the debt crises of the 1970s, the International Monetary Fund and the World Bank adopted new rules to ensure fiscal discipline that were seen as conditionalities by the DCs. In the 1980s, the New International Economic Order disappeared from the international agenda, and debt was a key problem. Since the 1990s, there has been a revival of the development agenda in the context of the new environment agenda. In the 2000s, there is a focus on the Millennium Development Goals (MDGs) and aid for trade.

Cooperation on development issues: towards quantitative commitments

The success of the Marshall Plan and the huge differences between ICs and DCs inspired a policy focus towards helping DCs – via UN agencies, national development cooperation agencies and loans through the development banks. The 1950s witnessed the establishment of the early aid and technical assistance programmes. Individual countries established aid agencies. In the 1960s, the UN General Assembly adopted the United Nations Development Decade to increase the annual growth rate in DCs to 5% by 1970 and to ensure that assistance was 'approximately 1% of the combined national incomes of the economically advanced countries' (see Section 4.4.3). The UNDP and the OECD Development Assistance Committee (DAC) were established to promote development in poor countries (see Section 4.3).

Demands for the New International Economic Order

In the 1970s, the DCs lobbied for the New International Economic Order (NIEO), leading to the adoption of three instruments[3] with the following aims: to enhance DC control over multinationals, including the freedom to nationalize or expropriate property if necessary; to guarantee permanent sovereignty over natural resources; to be able to set up cooperative associations of producers of primary commodities (e.g. like OPEC); to ensure that trade is equitable and provides fair prices for commodities; to promote technology transfer and assistance to DCs; to support reductions of tariff barriers in ICs; and to substitute a new system for the existing

---

[3] The Declaration on the New International Economic Order, a Programme of Action and a Charter of Economic Rights and Duties of States (NIEO Declaration, 1974).

Bretton Woods system (see Section 4.2.4). Although the NIEO promoted equity and sovereign equality,[4] the ICs did not eventually implement their obligations because the USA felt steamrollered into accepting these documents (Schrijver, 1995). Nevertheless, the NIEO continues to be the basis of the DC position, as can be seen from the documents and statements of the G-77. During this period, DCs and some legal scholars (Garcia-Amador, 1990; Chowdhury *et al.*, 1992) argued in favour of the right to development for people and states (see Section 4.4.2).

During the 1970s and 1980s, large loans were made, and loan periods extended, to stable DCs and this led gradually to the debt crises. By the end of the 1980s, DCs had paid significantly more in debt servicing than they had received as aid and direct investment and were 61% more in debt than in 1982 (George, 1992: xv–xvi). Debt repayment and fiscal discipline under the structural adjustment programmes characterized the period. For many this was the lost development decade.

The fall of the Berlin wall and the rise of sustainable development

The 1990s started optimistically with the end of the cold war and the expectation of a peace dividend – money saved from defence could be channelled towards new global challenges. However, a new group of potential aid recipients, or partner countries, was born – the countries with economies in transition. Instead of being able to play one side against the other and get financial resources from both East and West, the East was now competing with the global South for resources from the West. Furthermore, a perception of the ineffectiveness of aid combined with right-wing ideology led to increasing donor fatigue and a decreased willingness to fund projects in DCs (Cárdenas *et al.*, 1995). Aid resources dipped to the lowest ever level in 1997. In that year, the UN Secretary General created the United Nations Development Group (UNDG), with 25 agencies and 5 observers, administered by the United Nations Development Programme (UNDP) to coordinate development cooperation activities. In 2000, 189 countries endorsed the MDGs (see Table 4.1). These goals were to be achieved through people-centred development, local ownership, global integration and international partnership. Consensus among the OECD, the UN and the International Monetary Fund led to a joint publication – *A Better World for All* – in 2000. The MDGs provided part of the rationale for ODA spending.

In 2001, the UN Economic Commission for Europe Ministerial Declaration, the 2002 World Summit on Sustainable Development and the 2002 UN Conference on

---

[4] The NIEO text states that it is based on the 'united determination to work urgently for the establishment of a New International Economic Order based on equity, sovereign equality, interdependence, common interest and cooperation among all States, irrespective of their economic and social systems which shall correct inequalities and redress existing injustices, make it possible to eliminate the widening gap between the developed and the developing countries and ensure steadily accelerating economic and social development and peace and justice for present and future generations'.

Table 4.1 *The Millennium Development Goals*

Goals
Eradicate extreme poverty and hunger
Achieve universal primary education
Promote gender equality and empower women
Reduce child mortality
Improve maternal health
Combat HIV/AIDS, malaria and other diseases
Ensure environmental sustainability
Develop a global partnership for development

Source: www.un.org/millennium.

Financing the Environment[5] reiterated the need for poverty alleviation. Furthermore, in 2005, as a response to the critique of aid (see Chapter 2), the Paris Declaration on Aid Effectiveness was adopted (see Section 4.3.4).

Towards incorporating climate change into the development agenda

Increasingly, climate change is being incorporated into the development agenda. Several UN agencies (e.g. UNDP *et al.*, 2003) have analysed the link between climate change and the MDGs. It is argued that if the poor could adopt the lifestyle of the rich this would require nine times the Earth to absorb all the carbon; one person out of 19 will face a climate disaster in DCs compared with one out of 1500 in the OECD countries; and a child born in a drought is stunted for life (UNDP, 2007). 'Accordingly, climate issues should be mainstreamed into national economic planning and budgetary processes, both to ensure macroeconomic stability and to ensure budgetary allocations for activities that minimize climate risk' (UNDP *et al.*, 2003: 20). The UNDP sets the tone for development perspectives on climate change at UN level in its coordinating role within the United Nations Development Group.

Even on the poorest continent, climate change is seen as critical. In 2007, the Economic Commission for Africa announced that it would be 'mainstreaming' climate change because climate change might derail ongoing MDG activity in, and over-tax the low adaptive capacity of, the region (ECA, 2007a). It has prepared the 'Climate for Development in Africa Programme' (ECA, 2007b), in order to assist countries to mainstream climate change into their policies, and is collaborating with the African Development Bank. Their focus is on adaptation, mitigation and negotiation assistance (ECA, 2007a).

---

[5] 'Our goal is to eradicate poverty, achieve sustained economic growth and promote sustainable development as we advance to a fully inclusive and equitable global economic system' and many of the NIEO ideas were reiterated as well as the 0.7% target (Monterrey Consensus, 2002).

4.2.3 Integrating environmental issues into the development agenda

This section, building on, but not confined to, Soroos (1999), divides the history of environmental institutions into three periods: the post-Stockholm era, the post-Rio era and the post-Johannesburg period. Possibly the credit crisis of 2008 has launched the start of a fourth period. In the pre-Stockholm era, resource regimes had been developed for the management of transboundary water regimes. A few international environmental treaties had also been adopted.

The post-Stockholm era

Following the 1972 United Nations Conference on the Human Environment and its Stockholm Declaration (1972), the United Nations Environment Programme (UNEP) was established with a limited mandate to catalyse and coordinate policies. UNEP has established databases – Global Environmental Monitoring Systems, the International Registry of Toxic Chemicals, the Global Resource Information database and Infoterra – and launched treaty negotiations following the adoption in 1981 of the Montevideo Programme to promote the development of international environmental law. UNEP's concern for linking environmental and developmental issues made it suspect for the ICs (Gosovic, 1992: 228). It was not put in charge of implementing the work set out by Agenda 21 or hosting the climate negotiations. Its budget has been fluctuating over the decades, but, following a difficult period in the 1990s, its current budget for 2008–9 is USD 152 million (UNEP, 2008).

The post-Rio era

The World Commission on Environment and Development (WCED, 1987) linked the three fields of development, environmental protection and finance together by the term 'sustainable development'. The 1992 United Nations Conference on Environment and Development and its products (two conventions, the Rio Declaration and Agenda 21, 1992) included sustainable development as a legal principle. In 1992, the UN Commission on Sustainable Development was set up in New York under the UN Economic and Social Council to oversee the implementation of Agenda 21. With limited powers and resources, it serves mostly as a forum for dialogue. In 1993, the Department for Policy Coordination and Sustainable Development was established to coordinate UN work on development, social and environmental issues, and to support the Commission on Sustainable Development. In 1997, the UN General Assembly Special Session to Review Implementation of Agenda 21 concluded that nations needed to, inter alia, integrate economic, social and environmental objectives in national policy, and committed to strengthen the UNDP and the UN Commission on Sustainable Development. Environmental integration was a key idea in this era.

Table 4.2 *Key principles of country-led mainstreaming*

Principles
Find the right entry point
Find a 'champion'
Ensure the commitment of the planning or finance team
Provide country-specific evidence
Perform integrated policy appraisals
Engage key sector agencies
Consider the environment agency capacity
Acknowledge the need for sustained support

Source: UNEP–UNDP (2007).

The post-Johannesburg era: towards 'mainstreaming' climate change

In 2000, the First Global Ministerial Environmental Forum concluded that the environment was deteriorating rapidly and that inequity was a key challenge.[6] Since then UNEP and the UNDP have joined hands to 'mainstream' (see Section 3.3.2) environmental issues into development planning, but with a focus on poverty and the environment. Their programme focuses on understanding poverty–environment linkages and their importance for pro-poor growth, integrating the environment into national developmental processes and building implementation capacity (UNEP–UNDP, 2007). They argue that environmental protection benefits the poor by ensuring livelihoods, improving local health, reducing vulnerability to environmental risks and providing access to ecosystem services. They suggest a number of steps to effective 'mainstreaming' (see Table 4.2). They argue further that indicators of successful environmental integration include the following: ensuring the inclusion of poverty–environment linkages; poverty reduction strategies at national level; relevant capacity and finance; inclusion in sectoral planning; resource mobilization; enhanced donor contributions; and improved livelihoods.

Climate change is now a key issue for all UN agencies. The UNDP–UNEP Partnership on Climate Change aims to incorporate adaptation into national development plans and UN Cooperation Frameworks and to enable countries to access carbon finance and cleaner technologies. While the UNDP focuses on capacity building and policy design, UNEP focuses on norms and technical analysis (UNEP, 2007). These two together now set the tone with respect to 'mainstreaming' climate change.

---

[6] '*Conscious* that the root causes of global environmental degradation are embedded in social and economic problems such as pervasive poverty, unsustainable production and consumption patterns, inequity in distribution of wealth, and the debt burden' (Malmö Ministerial Declaration, 2000).

4.2.4   The evolution of the role of the development banks

Investing for development

The global economic organizations (the International Monetary Fund (IMF) and World Bank) were established in 1944 to promote international monetary cooperation and exchange-rate stability and to assist in the development of members, respectively. In 1948, the IMF became a specialized agency of the UN. In addition, several regional developmental banks were set up.[7] In 1974, a Joint Ministerial Committee was formed by the Board of Governors of the World Bank and the IMF on the Transfer of Real Resources to Developing Countries to advise on poverty alleviation, debt and the impact of the economic policies of ICs on DCs (Colas, 1994: 339).

Responding to the environmental critique

Environmental critique[8] of development bank projects in the 1970s and 1980s (Werksman, 1993; White, 1996; Hicks *et al.*, 2008), and the negative impacts of the debt crises on the environment (Miller, 1991), social resilience and forestry (George, 1992), combined with international environmental law developments put pressure on the development banks to adopt a green vocabulary and apply environmental and social impact assessments to their projects. In 1990, the World Bank, in collaboration with UNEP and the UNDP, established the Global Environment Facility (GEF) to finance several of the issues identified at the Rio Conference – climate change, depletion of the ozone layer, and protection of biodiversity and international waters. The GEF has been subjected to considerable criticism since then on its internal policies (GEF, 1994), its relationship with the environmental treaties (Gupta, 2006; Mace, 2005) and its ability to green the activities of the World Bank (White, 1996). The World Bank also launched mechanisms such as debt-for-nature swaps and the Carbon Fund.

Incorporating environment and climate change into bank activities

In 2003, the development banks and financial institutions adopted the Equator Principles.[9] Project financing in excess of USD 50 million must meet specific

---

[7] The best-known regional banks are the Inter-American Development Bank (established in 1959), the African Development Bank (1966), the Asian Development Bank (1966), the Caribbean Development Bank (1970), and the European Bank for Reconstruction and Development (1991). Furthermore, several regional groupings of countries have established international financial institutions to finance various projects or activities in areas of mutual interest. The largest and most important of these is the European Investment Bank, which was established in 1958 by the Treaty of Rome.

[8] 'The Bank and the affiliated international institutions known as the Bretton Woods group, have been accused of bankrolling ecological and economic disaster in the developing world, by promoting developmental projects that have denuded forests, depleted soils, and increased dependence on unsustainable energy sources' (Werksman, 1993: 65).

[9] See http://www.equator-principles.com/index.shtml.

procedures. These principles imply that these banks must proactively examine the environmental risk of their investments through, inter alia, stakeholder involvement in their policy processes.

Since the 1990s, the World Bank has set up funds to deal with climate change. The GEF was clearly an early fund. The Climate Investment Fund, established in 2008, comes as the latest in a series of funds including the Clean Technology Fund, the Strategic Climate Fund, the Pilot Programme for Climate Resistance and the Forest Investment Programme.

Since 2005, the banks have been adopting the rhetoric of 'mainstreaming' climate into development work. The World Bank (2006) adopted the Clean Energy for Development Investment Framework (CEIF) in 2006 and sees climate change as a 'critical pillar of the development agenda' (Zoellick, 2007). In 2008, the World Bank's Strategic Framework for Climate Change and Development emphasized (a) support for economic growth, poverty reduction and achieving the MDGs; (b) energy access in meeting the above goals; and (c) adaptation and adaptive capacity for sustaining the other two goals.

In a World Bank report, Mani *et al.* (2008: Preface) state that 'The need to "mainstream" climate policy into development goals is well-recognized within the World Bank, as well as at the national level and among other donor agencies, individually and collectively.' Such incorporating efforts are in relation to development and poverty reduction policy. Adaptation support should not be seen as an end itself, but rather as support towards the development goals (Mani *et al.*, 2008: ix). They argue that 'in low-income countries, a primary focus on growth and poverty reduction can increase climate resilience by helping these countries diversify their economies'. Mani *et al.* (2008) offer guidelines on (a) general measures to incorporate climate change into development policy; (b) how general macro-economic policy impacts on mitigation and adaptation; (c) how to address pathways by which climate change impacts on development assistance; and (d) how specific policies can mainstream climate change into development policy. They have developed an entire scheme that shows how adaptation and mitigation actions may be affected by countrywide and sectoral policies. They include a list of policies and indicators of progress for each specific policy that World Bank officials need to take into account. Measures to mainstream are shown in Table 4.3, which shows that 'mainstreaming' is to be achieved through enhanced liberalization, trade and the application of impact assessments.

The European Investment Bank (EIB) is the EU's long-term lending bank and is committed in its lending policies to contribute to the EU policy objectives. EIB lending outside the EU is governed by a series of mandates which build upon the EU's external cooperation and development policies. The EIB provides loans to the countries in the Mediterranean Neighbourhood and the Africa, Pacific and

Table 4.3 *Transmission channels for mainstreaming*

	Policies	Measures
Economy-wide reforms	Macro-economic stability	Improve fiscal performance; reduce debt; enhance trade
	Improve investment climate	Improve investment procedures; enhance private participation; liquidate state enterprises
	Improve public financial management	Strengthen management; improve procurement and civil service reform
	Governance reforms	Improve public administration and fiduciary standards; reduce corruption; improve judiciary
	Social protection	Policies for the poor; pension reform; support for SMEs; credit supply
	Decentralization	Fiscal decentralization; strengthen local capacity
	Competition and property rights	Liberalize services; reform state-owned utilities
	Modernize rural economy	Diversify production
Sectoral reforms	Agriculture	Ensure plan planting, practices and infrastructure take climate impacts into account
	Forestry	Sustainable management for forests and production forests; protect biodiversity
	Mining	Energy efficiency; Environment Impact Assessments
	Fisheries	Sustainable fisheries
	Environmental management	Financing; environmental impact mitigation plans; regulation; monitoring; forest impact assessments
	Education	Financing; transparency; educating parents; monitoring school performance
	Health	Financing and health care, especially for climate-related diseases.
	Infrastructure	Reforms in transport, urban, ports, water and telecommunications sectors
	Energy	Tariff reform; privatization; competition; investment in renewables; energy reforms
	Finance	Increase resilience; implement reforms in rural finance and insurance
	Tourism	Improve tourist destinations with good environmental practice

Source: Based on Mani *et al.* (2008).

Caribbean regions, and engages in economic cooperation with Asia and Latin America, which amounted to 2.6 billion euros in 2008. The EIB's approach to climate change issues is described in 'The EIB Statement of Environmental and Social Principles and Standards' of 2009. This statement mentions, inter alia, that the EIB promotes the renewable energy sector, optimizes the scope for energy efficiency in all the projects it is financing and aligns its operations with other EU climate-policy investment priorities. It furthermore declares that, for carbon-intensive projects, it incorporates the costs of such emissions into the financial and economic analyses that inform its financing decision. Furthermore, the EIB is committed to developing its own knowledge and expertise on climate change risk management, and, where risks are identified, it requires the project developer to identify and apply adaptation measures to ensure the sustainability of the project. With regard to countries outside the EU, the statement explicitly formulates that 'The EIB is committed to supporting environmentally sustainable, clean energy growth paths in countries outside the EU, including the promotion of the transfer and development of clean technologies, as well as the establishment and development of financial mechanisms for cost-effective climate change mitigation, such as the carbon market.' In concrete terms, this means that the EIB participates in a number of global carbon funds and is developing a carbon-footprint methodology to assess its project portfolio.

Other development banks are also incorporating climate change. The Asian Development Bank has adopted policies to 'mainstream' climate change mitigation and adaptation into its policies,[10] the African Development Bank has adopted the Energy Framework, and the Inter-American Development Bank has chosen to focus on climate change. These banks are also linking the MDGs to climate change (UNDP *et al.*, 2003; GEF, 2005), which justifies such efforts at incorporation.

The evolution of the trade and investment agenda

Countries have traded with each other for a long time. Governance promoting free trade and removing trade barriers at national level as a means to create global prosperity was promoted through the General Agreement on Tariffs and Trade (GATT, 1947), which was ratified initially by 23 countries. Discussions on free trade were nevertheless accompanied by continuing agricultural subsidies in ICs, reducing the market access of DCs. This has been a sore point in IC–DC relationships. Since aid flows are a fraction of trade flows, in the past many have argued that aid would not be necessary if trade could be made more fair, and if the ICs would phase out the subsidies on their own production processes.

---

[10] Tearfund (2006) also reports that funds from the Asian Development Bank have been used to develop a Climate Change Adaptation Programme for the Pacific (CLIMP), which, in combination with a World Bank report on *Cities, Seas and Storms*, has led to a series of activities to incorporate adaptation into national policy.

Trade did not initially take environmental aspects into account. In 1995, the World Trade Organization was established to promote free trade (WTO, 1994). While it supports sustainable development and member states may adopt environmental regulations under specific circumstances, and there have been disputes involving the environment (e.g. the shrimp–turtle case),[11] in general its contribution to addressing environmental concerns has been quite limited. The North American Free Trade Agreement also tried to integrate environment and trade goals (Esty, 1999: 192). At the same time, many DCs have tried to resist pressure to include environmental aspects within trade.

In 2002, the Monterrey Declaration on Financing the Environment recognized the concerns of DCs, stating

We acknowledge the issues of particular concern to developing countries and countries with economies in transition in international trade to enhance their capacity to finance their development, including trade barriers, trade-distorting subsidies and other trade-distorting measures, particularly in sectors of special export interest to developing countries, including agriculture; the abuse of anti-dumping measures; technical barriers and sanitary and phytosanitary measures; trade liberalization in labour intensive manufactures; trade liberalization in agricultural products; trade in services; tariff peaks, high tariffs and tariff escalation, as well as non-tariff barriers; the movement of natural persons; the lack of recognition of intellectual property rights for the protection of traditional knowledge and folklore; the transfer of knowledge and technology; the implementation and interpretation of the Agreement on Trade-Related Aspects of Intellectual Property Rights in a manner supportive of public health; and the need for special and differential treatment provisions for developing countries in trade agreements to be made more precise, effective and operational.

(Monterrey Consensus, 2002: para. 28) [footnote omitted]

The focus on competitive advantage leads countries to minimize domestic costs and this often leads to the externalization of costs. This reduces the incentive for DCs to make their products and services more expensive and thus less competitive and reduces the incentive for IC agencies not to sell existing and older technologies to DCs (technology dumping).

In the area of investment, in the post-colonial period, DCs expropriated foreign investments often because they favoured permanent sovereignty over natural resources. Capital-exporting countries then sought to protect their own investments abroad through private international law and regional and bilateral investment agreements. The multilateral investment agreement negotiations launched by the OECD in 1995 failed for lack of support, and since then more than 3000 bilateral and regional investment agreements have been negotiated. These protect the rights of investors and very few of these give environmental issues serious consideration.

---

[11] *US-Import Prohibition on Certain Shrimp and Turtle Products*, WTO Appellate Body Report WT/DS57/AB/R, 12 October 1998, available online at http://www.wto.org.

State–investor agreements often have negative impacts on the environment in DCs leading, inter alia, to policy freezing (Tienhaara, 2008).

4.3 The evolution of OECD policy

4.3.1 Introduction

Since 1960, the OECD's Development Cooperation Directorate and its Development Assistance Committee (DAC) have made policy on development cooperation. They support the donor countries in making effective aid policy. The DAC looks at the role of development cooperation in assisting DCs to participate in the global economy and overcoming poverty.

This section presents the following arguments. The OECD DAC has developed and promoted ideas about how development cooperation can be improved. It is now moving towards incorporating climate change adaptation into development cooperation activities. It is a donor forum that does not actively integrate either the UN or the partner countries into its policymaking processes. The Paris Declaration is an exception. There are few publications analysing the effectiveness of the OECD DAC, perhaps because of an implicit perception that the OECD DAC is not influential. Nonetheless, the OECD DAC is influential insofar as member states try to align their policies with OECD DAC recommendations, inter alia, to avoid unpleasant monitoring reports. Finally, the OECD DAC includes only those OECD members that contribute aid, and furthermore does not provide a forum for non-OECD countries that provide or should provide aid.

4.3.2 The evolution of policy within the OECD DAC

In 1960, the OECD established the Development Assistance Group with 11 members[12] and this evolved into the Development Assistance Committee. It provided a forum for consultation on how resources were to be made available and how the flow of long-term funds could be improved. It could make recommendations on aid and invite other countries and bodies to participate in its processes. Table 4.4 provides a history of the OECD work on aid, while the section below highlights only some issues.

The 1961 Resolution of the Common Aid Effort (OECD, 1961) states that the OECD members are conscious of the aspirations of the DCs to improve 'standards of life for their peoples' and 'convinced of the need to help the less

---

[12] The initial members were Belgium, Canada, France, Germany, Italy, Portugal, the UK, the USA and the Commission of the European Economic Community. Subsequently Japan and the Netherlands joined.

Table 4.4 *The history of DAC work on aid*

Year	Event
1961	Development Assistance Committee and Development Department established
1961	Resolution of the Common Aid Effort provides rationale for aid
1962	Aid reviews begin
1964	Development Centre and DAC UN Conference on Trade and Development Working group established
1965	DAC supports contention that aid target should be 1% of income
1966	Guidelines for the Coordination of Technical Assistance approved
1967	Data on assistance as percentage of national income published for first time
1969	Official Development Assistance (ODA) concept adopted
1969	Recommendation from Pearson Report that ODA should be 0.7% of GNI
1970	Effort to untie aid fails
1972	ODA defined for first time
1975	OECD Declaration on Relations with Developing Countries adopted
1977	Decision to promote Economic Growth and Meeting Basic Human Needs
1978	Recommendation to increase the average grant element target of ODA from 84% to 86%
1979	Guidelines for Improving Aid Implementation
1980	Aid effectiveness becomes an issue
1981	New look at North–South relations; gender reporting
1983	Adopts principle on gender
1985	*Twenty-Five Years of Development Co-operation* published
1986	Policies on improving and coordinating aid and environmental assessment
1988	Principles for Project Appraisal adopted
1989	Development cooperation redesigned
1991	Emphasis on human rights and democracy
1993	Only aid to DCs (Part I countries) is seen as ODA; aid to countries in transition is not
1995	'Development Partnerships in the New Global Context' (good governance and gender)
1996	Strategy on Shaping the 21st Century: The Contribution of Development Co-operation
2001	Recommendation on Untying ODA to the Least Developed Countries
2002	The development of a Shared Development Agenda; The DAC Guidelines
2003	Rome Declaration on Harmonization
2004	Development and security important post 9/11
	Combating corruption, DAC joins New Partnership for Africa's Development
2005	Paris Declaration on Aid Effectiveness
2006	Focusing aid for trade
2006	Framework for Common Action around Shared Goals (environmental mainstreaming)
2008	Accra Agenda for Action

Source: Based on OECD (2006a).

developed countries help themselves by increasing economic, financial and technical assistance and by adapting this assistance to the requirements of the recipient countries'. Consequently, it recommended that the volume of resources should be increased and 'assistance provided on an assured and continuing basis would make the greatest contribution to sound economic growth in the less-developed countries'. The group agreed that such assistance should be in grant or soft-loan form and recommended 'that a study should be made of the principles on which governments might most equitably determine their respective contributions to the common aid effort having regard to the circumstances of each country, including its economic capacity and all other relevant factors' (OECD, 1961).

In 1972, the OECD defined ODA as follows:

ODA consists of flows to developing countries and multilateral institutions provided by official agencies, including state and local governments, or by their executive agencies, each transaction of which meets the following test: a) it is administered with the promotion of the economic development and welfare of developing countries as its main objective, and b) it is concessional in character and contains a grant element of at least 25% (calculated at a rate of discount of 10%).

In 1980, the publication of *The Important but Elusive Issue of Aid Effectiveness*, made effectiveness a critical issue. In 1985, the DAC published *Twenty-Five Years of Development Co-operation* and concluded that, although in Sub-Saharan Africa aid has not been very successful, in other countries it has contributed to enhancements of growth and exports to the OECD countries. The document recognized that, since aid is given to countries with problems, it could not be seen as a sort of investment banking with high economic returns. A year later it adopted some ideas that were later integrated into the MDGs.

In 1989, the DAC's *Development Co-operation in the 1990s* recognized the vicious circle of poverty and environmental degradation, and recommended promoting sustainable economic growth, broader participation in productive processes and a more equitable sharing of their benefits and environmental sustainability. In 1996, the OECD endorsed the *Shaping the 21st Century* report that ultimately led to the adoption of the MDGs. On the basis of this report, the OECD decided to focus on economic growth and basic human needs, and noted that the latter was an important element of economic growth and would lead to greater equity. The focus on the poor was not motivated purely by charitable concerns but arose from a productivity orientation. In 2006, the OECD decided to use aid to promote trade, which had until then been only about 25% of all aid and was expected to rise considerably.

Table 4.5 *OECD members and donors: inconsistencies*

OECD members (30)	Non OECD DAC (7)
Australia, Austria, Belgium, Canada, Czech Republic, Denmark, Finland, France, Germany, Greece, Hungary, Iceland, Ireland, Italy, Japan, South Korea, Luxembourg, Mexico, Netherlands, New Zealand, Norway, Poland, Portugal, Slovakia, Spain, Sweden, Switzerland, Turkey, UK, USA	Czech Republic, Hungary, Iceland, Mexico, Poland, Slovakia, Turkey

4.3.3 OECD membership and the DAC

Although the number of OECD members is increasing, new members do not automatically become DAC members (see Table 4.5).

4.3.4 Current OECD policy on aid delivery: the Paris Declaration

The debate on aid effectiveness began taking shape in the late 1990s, in response to critiques from various sides (see Chapter 2). After the International Conference on Financing for Development in Monterrey in 2002, leaders of the major multilateral development banks and international and bilateral organizations, and representatives of donor and partner countries, decided to convene a High Level Forum (HLF) on harmonization of aid practices. This Forum took place in Rome in 2003, and led to the Rome Declaration on Harmonization. Shortly after this meeting, the OECD DAC created a Working Party on Aid Effectiveness and Donor Practices to promote, support and monitor progress on harmonization and alignment, and to prepare further activities. Two years later, in 2005, the second HLF was hosted by the French government, resulting in the adoption of the Paris Declaration on Aid Effectiveness by over 100 representatives of ICs and DCs, and other organizations.[13] The ministers stated that, while the volumes of aid must increase in order to achieve the Millennium Development Goals, also the effectiveness of aid must

[13] The DCs that endorsed the Paris Declaration include Albania, Bangladesh, Benin, Bolivia, Botswana, Burkina Faso, Burundi, Cambodia, Cameroon, China, the Comoros, Congo (DR), the Dominican Republic, Egypt, Ethiopia, Fiji, The Gambia, Ghana, Guatemala, Guinea, Honduras, India, Indonesia, Jamaica, Jordan, Kenya, Kuwait, the Kyrgyz Republic, Laos, Madagascar, Malawi, Malaysia, Mali, Mauritania, Mongolia, Morocco, Mozambique, Nepal, Nicaragua, Niger, Pakistan, Papua New Guinea, the Philippines, Rwanda, Saudi Arabia, Senegal, South Africa, the Solomon Islands, Sri Lanka, Tajikistan, Timor-Leste, Tanzania, Thailand, Turkey, Uganda, Tunisia, Vanuatu, Vietnam, Yemen and Zimbabwe. It also included OECD members that are listed as DCs under the Convention – Mexico and South Korea.

increase significantly, in order to support efforts by partner countries to strengthen governance and improve development performance (Paris Declaration, 2005: para. 1).

The Paris Declaration included a statement of resolve, partnership commitments and indicators of progress. The statement of resolve focused on the need to scale up effective aid that is predictable; provide assistance aligned to DC policies; enhance accountability; ensure donor harmonization; adopt indicators, timetables and targets; and monitor and evaluate implementation. The partnership commitments focused on DC ownership of projects, alignment to national policies, donor harmonization, result orientation and mutual accountability. The Paris Declaration adopted 12 indicators of progress to be monitored and implemented; see for progress reports OECD (2007; 2008; 2009).

In 2008, the third HLF took place in Accra, with the participation of about 1700 participants from DCs, ICs, UN and multilateral institutions, global funds, foundations and civil society organizations. The Forum adopted the Accra Agenda for Action, which aims to accelerate and deepen the implementation of the Paris Declaration. However, thus far, practice shows that the implementation process is accompanied by serious difficulties (e.g. Hyden, 2008; OECD, 2009).

4.3.5 *Current discussions on incorporating climate change in the OECD*

The first initiatives for integrating climate change concerns into development cooperation were provided by the introduction of the Rio markers for measuring aid to the environment in 1998 and the Policy Statement on Integrating the Rio Conventions in Development Co-operation issued by the OECD DAC High Level Meeting on 16 May 2002 (Table 4.6). This statement was accompanied by extensive policy guidelines to stimulate such a process. Subsequently, the Paris Declaration on Aid Effectiveness called on donors to harmonize environmental impact assessments such that they go beyond health and social issues to include global environmental issues, including climate change (Paris Declaration, 2005: para. 40). It also called for common application procedures and approaches and for developing capacity to implement them, as well as for other cross-cutting issues such as gender.

In 2006, the OECD adopted a Declaration on Integrating Climate Change Adaptation into Development Co-operation (OECD, 2006b). The document recognized that the poor will be particularly vulnerable to the impacts of climate change 'as environmental "costs" at the global, national and local levels, bear heaviest upon the poor'. It recognized that helping vulnerable countries and peoples contributes to achieving the MDGs and that adaptation is not a 'stand-alone' agenda. The climate agenda also needed to be integrated into development policymaking, including Poverty Reduction Strategy Papers; and it observed that such adaptation could be

Table 4.6 *OECD DAC policy in relation to incorporating climate change*

2002 Integrating the Rio Conventions into Development Co-operation: Policy Statement and DAC Guidelines
2005 Paris Declaration on Aid Effectiveness
2006 Declaration on Integrating Climate Change Adaptation
2006 Framework for Common Action around Shared Goals (environmental mainstreaming)
2008 Statement of Progress on Integrating Climate Change Adaptation into Development Co-operation
2008 Accra Agenda for Action
2009 Joint High Level Meeting on Environment and Development: policy guidance on integrating climate change adaptation into development cooperation adopted

synergistic with efforts on other global environmental fronts. The Declaration argued further that

The development co-operation agencies of OECD countries have a long experience working with developing country partners on poverty alleviation and on reducing the human and economic losses from natural climate-related disasters. Environment agencies have particular expertise on climate change impacts and adaptation that can be brought to bear in developing country contexts. This combined experience can be harnessed to tackle additional challenges posed by climate change.

The countries then declared that they would work to integrate climate change adaptation into development planning and assistance, both within their own governments and in activities undertaken with partner countries, especially via country assistance strategies, sectoral policy frameworks, Poverty Reduction Strategies, long-term investment plans, technical consultations and sector reviews, as well as strategic and project-level environmental impact assessments and National Adaptation Programmes of Action (NAPAs).

The OECD also announced that it aimed to develop and apply appropriate tools (e.g. screening tools) and other methodologies to assess exposure of relevant development activities, long-term development plans and investments to climate risks, as well as tools to increase the resilience of relevant sectoral activities, decision systems and tools relevant to local planning needs, to develop good practices, and to implement periodic monitoring. These practical tools were eventually provided by the policy guidance on integrating climate change adaptation into development cooperation which was endorsed, together with a Policy Statement, during an OECD High Level Meeting on Environment and Development in May 2009.

Resources for incorporating climate change

There have been resources devoted to climate-related activities since the 1950s. However, they have clearly not been noted as such. A study of the period 1980–99 concerning 427 000 projects showed that donor countries spent about USD 1 billion annually during those 20 years on climate-related activities (Hicks *et al.*, 2008). A study of 20 000 projects per year from 2000 to 2006 showed that on average 16 433 projects per year were devoted to climate change, including energy efficiency projects, renewable energy projects, carbon sequestration, land management, adaptation studies, and disaster prevention and adaptation policy such as building dikes (Roberts *et al.*, 2008). Until 2005, less than 2% of foreign aid was spent on climate-change-related issues according to Roberts *et al.* (2008), though others argue that this is a conservative figure.

4.4 Sustainable development cooperation: contested rights and commitments

4.4.1 Introduction

This section highlights some relevant discussions at UN level. It examines the history of the right to development and its current status, the quantitative target of 0.7% of GNI for development assistance and the discussions on environmental assistance.

This section focuses on the controversy around the right to development within certain UN fora, especially whether it implies (a) a right to development cooperation and (b) a right to redesign global institutions. Furthermore, a parallel discussion on the commitment to provide assistance to DCs demonstrates rhetorical commitment and goodwill regarding the 0.7% target. During the last 40 years, however, the resources for development assistance have been below 0.7% levels and currently the resources appear to fall considerably short of what may be needed to fulfil the MDGs. Environmental assistance appears to come over and above the minimum needed for meeting the MDGs and, although there is scope for double counting, the assistance is seen to be in the nature of USD 125 billion annually. Lastly, discussions on the right to development, development, development aid and investment appear to be spread throughout the UN system and, although the UN Development Group is expected to coordinate these discussions, there do not appear to be significant links between the different activities (cf. Piron, 2002).

4.4.2 The right to development

Post-war optimism was reflected in the non-legally binding UN General Assembly Universal Declaration on Human Rights (UNGA, 1948) and two legally binding

Covenants in 1966, one on political rights (promoted by the Western countries – first-generation human rights) and one on economic and social rights (promoted by communist states and DCs – second-generation human rights; Kirchmeier, 2006). In 1969, the General Assembly adopted the Declaration on Social Progress and Development (UNGA, 1969).[14] Although the USA accepted this at the time, in 1987 it withdrew its support (Alston and Simma, 1988).

Since achieving independence, DCs have insisted on the right to development – the third generation of human rights (Garcia-Amador, 1990). The UN Commission on Human Rights (1977) requested the Secretary General to conduct a study 'of the international dimensions of the right to development as a human right in relation with other human rights based on international cooperation, including the right to peace, taking into account the requirement of the New International Economic Order and the fundamental human needs'. A 1978 workshop of The Hague Academy of International Law and the United Nations University articulated the right to development in international law. In 1981, the Banjul African Charter on Human Rights and Peoples' Rights adopted the right to development and defined it more as a right of peoples, not individuals. The right to development is linked to the concept of absolute poverty, an idea developed by the UNDP; see Table 4.7.

In 1986, the Office of the UN High Commission on Human Rights adopted the Declaration on the Right to Development (1986),[15] which built upon the 1969 Declaration. Its Article 1 reads as follows:

The right to development is an inalienable human right by virtue of which every human person and all peoples are entitled to participate in, contribute to, and enjoy economic, social, cultural and political development, in which all human rights and fundamental freedoms can be fully realized.

The human right to development also implies the full realization of the right of peoples to self-determination, which includes, subject to the relevant provisions of both International Covenants on Human Rights, the exercise of their inalienable right to full sovereignty over all their natural wealth and resources.

The Declaration mandates states to create national and international conditions favourable to the realization of the right to development. The right to development in international law is an attempt at perfecting democracy (de Waart, 1992; Chowdhury *et al.*, 1992; VerLoren van Themaat and Schrijver, 1992). In 1993, at the World Conference on Human Rights, the Vienna Declaration and Programme of Action (UNGA, 1993), which was adopted by 171 countries, stated that

The World Conference on Human Rights reaffirms the right to development, as established in the Declaration on the Right to Development, as a universal and inalienable right and an

---

[14] The Declaration was accepted with 119 votes for, none against and 2 abstentions (Cuba and Malta).
[15] A/RES/41/128, 4 December 1986.

Table 4.7 *The evolution of the right to development*

Year	Event	Issue
1948	Universal Declaration on Human Rights	Sets the stage for human rights issues (Western perspective)
1960s	DCs seeking NIEO	Sets the stage for demanding a change in the global order (Southern, non-aligned movement perspective)
1966	Covenant on Political Rights	Legally binding, first-generation rights (Western demand)
1960s	Covenant on Social–Economic Rights	Legally binding, second-generation rights (communist and DCs)
1969	Declaration on Social Progress and Development	Adopted by all countries except Cuba and Malta
1970s	Articulation of the concept by developing and industrialized country experts	Articulation of the right to development – third-generation rights
1981	Banjul Charter	Adoption of the right as the right of peoples by African countries
1986	UN Declaration on the Right to Development	Adoption by UN Human Rights Commission, opposed by USA, eight states abstained from voting; mentions NIEO
1993	Vienna Declaration and Programme of Action (Art. 10)	Adopted by 172 countries at World Conference on Human Rights
1998	Working group on the Right to Development	Monitors progress made at UN level on this right
2000	Millennium Declaration (Art. 11)	Adopted by 147 countries
2001	Durban Declaration and Programme of Action (Arts. 19 and 28)	Discussed the right to development in the context of racism, racial discrimination, xenophobia and related intolerance
2008	UN Human Rights Council	Establishment of a three-year process to study human rights with respect to climate change, water and sanitation

integral part of fundamental human rights. As stated in the Declaration on the Right to Development, the human person is the central subject of development. While development facilitates the enjoyment of all human rights, the lack of development may not be invoked to justify the abridgement of internationally recognized human rights. States should cooperate with each other in ensuring development and eliminating obstacles to development. The international community should promote an effective international cooperation for the realization of the right to development and the elimination of obstacles to development. Lasting progress towards the implementation of the right to development requires effective development policies at the national level, as well as equitable economic relations and a favourable economic environment at the international.

(Vienna Declaration, 1993: para. 10)

The right to development should be fulfilled so as to meet equitably the developmental and environmental needs of present and future generations.

(Vienna Declaration, 1993: para. 11)

In 2000, the Millennium Declaration stated that 'We will spare no effort to free our fellow men, women and children from the abject and dehumanizing conditions of extreme poverty, to which more than a billion of them are currently subjected. We are committed to making the right to development a reality for everyone and to freeing the entire human race from want.' (para. 11). In 2002, the International Law Association adopted the New Delhi Declaration on Sustainable Development that attempts to combine a number of development and environmental objectives (ILA, 2002); see Chapter 2.

In 2000, the UN's Independent Expert on the Right to Development proposed the establishment of 'development compacts' between the international community (donors, international organizations, etc.) and states, creating reciprocal obligations to ensure a right to a process of development as against outcomes of development. A Working Group on the Right to Development monitors progress on this right. This right is discussed at the UN General Assembly (Third Committee), the Commission (and now Council) on Human Rights, Charter-based mechanisms, the Committee on Economic, Social and Cultural Rights, the UN Conference on Trade and Development, and LDC meetings.

In 2008, the UN Council on Human Rights launched a process to define a human right with respect to climate change and so, even in the world of rights, climate change is moving towards centre stage.

This right to development is not seen as legally binding, even though 171 countries have assented to it. Piron (2002) argues that for ICs this is a right of individuals and groups within countries; that is comprehensive, including all other rights; for which nations have primary responsibility; that calls for good governance principles, but should not deal with economic, trade and investment issues (for which there already exist appropriate forums); that cannot be used to justify the violation of other rights; that is consistent with the use of human-rights indicators; and that does not include the proposed development compact.

In contrast, the DCs see the right to development as a right of states as well as individuals; that is new and separate; with implications for technology transfer and assistance as well as for reforming trade and investment rules at the global level; that prioritizes the needs of humans but should not be reduced to development indicators; and that is complemented by the global development compact and follow-up mechanisms. The DCs see the adoption of the Millennium Development Goals, particularly the eighth goal on partnership for development, as articulating the right to development. The working group on the right to development monitors the progress made on implementing the right also in terms of the implementation of the MDGs.

According to Adam (2006: 3), this dualistic nature of the 'Right to Development' – a right within the state and between states – is on the other hand the biggest obstacle to its acceptance by the international community, which fears that it may be interpreted as meaning 'a right to development assistance'. Kirchmeier (2006) argues that ICs are afraid that a right to development might amount to a 'right to everything'. For ICs, this right could mean redesign of the trade and investment regimes, re-evaluation of the rights and responsibilities of multinationals, redesign of 'ecospace' rights of the DCs and a legally binding obligation on the part of the ICs in terms of developmental and environmental assistance: 'In past years, the Right to Development was interpreted by some as creating an international legal obligation on the part of ICs to provide development assistance to DCs. Such a legally binding obligation is rejected by ICs, and is not supported by an analysis of the status of the Right to Development under international law' (Piron, 2002: 5). Hence, Germany and the USA accept the right, but more in terms of its domestic implications, while the UK and EU recognize certain international responsibilities, but do not see these as flowing out of the right to development (Kirchmeier, 2006). The EU participates constructively in talks on the right to development, whereas the IMF does not engage in these, the World Bank sends representatives and the UNDP tries to provide content (Piron, 2002).

From the DC perspective, the right to development is part of the New International Economic Order for a level playing field and global equity. Such a right empowers them and makes them partners in development, not just recipients of aid. Such a right strengthens their negotiating position vis-à-vis trade, banking and investment negotiations.

4.4.3 The 0.7% commitment

Central to the right to development is the political commitment to provide development assistance (Cárdenas *et al.*, 1995). The genesis of the 0.7% target can be traced back to the World Council of Churches proposal of 1958 to provide 1.0% of national income to DCs in the form of grants and loans. This was reiterated by the UN General Assembly in 1960 (UNGA, 1960), by the UN Conference on Trade and Development in 1964 (UNCTAD, 1964), by the Second UN Development Decade (1970; cited in Cárdenas *et al.*, 1995: 190) and by the UN General Assembly in 1970 (UNGA, 1970). By 1975, new definitions called on ICs to ensure that official assistance was 0.7% of national income. Sweden (in 1974) was the first country to reach this target. The Netherlands (1975), Norway (1976) and Denmark (1978) achieved the target soon thereafter. In 2002, the target was adopted by the Monterrey Declaration on Financing for Development (Monterrey Consensus, 2002: para. 42) and the World Summit on Sustainable Development (WSSD, 2002: Art. 82). At Gleneagles, the G8 summit stated that the 0.7% target 'would greatly assist the funding of national and regional initiatives to combat poverty and hunger' (G8, 2005: para. 7). See Table 4.8.

At Monterrey in 2002, countries highlighted the rationale[16] behind ODA and stated that

> In that context, we urge developed countries that have not done so to make concrete efforts towards the target of 0.7 per cent of gross national product (GNP) as ODA to developing countries and 0.15 to 0.20 per cent of GNP of developed countries to least developed countries, as reconfirmed at the Third United Nations Conference on Least Developed Countries, and we encourage developing countries to build on progress achieved in ensuring that ODA is used effectively to help achieve development goals and targets. We acknowledge the efforts of all donors, commend those donors whose ODA contributions exceed, reach or are increasing towards the targets, and underline the importance of undertaking to examine the means and time frames for achieving the targets and goals.
>
> *(Monterrey Consensus, 2002: para. 42)*

Six years after Monterrey, the Heads of State meeting at the Monterrey follow-up conference in Doha (Doha Declaration, 2008: para. 42) concluded that, although ODA had increased by 40% since 2001, most of these flows were for debt relief and humanitarian assistance; and that some countries had met their 0.7% target and others had set timetables for fulfilling their longstanding commitments, such as the European Union, which has agreed to provide, collectively, 0.56% of GNP for ODA by 2010 and 0.7% by 2015 and to channel at least 50% of collective aid increases to Africa, while fully respecting the individual priorities of member states in development assistance. Furthermore, the Doha Declaration welcomed the more than doubling of ODA by the USA.

---

[16] This rationale included that such assistance can help create an environment for private-sector activity and lead to growth; and that it can generate resources for education, health, public infrastructure development, agriculture and rural development, and to enhance food security. If the MDGs are to be achieved, additional ODA resources will be needed.

Table 4.8 *The history of the 0.7% target*

Year	Venue/proposer	Percentage of national income
1958	World Council of Churches	1% (included private income)
1960	UN General Assembly Resolution 1522	1%
1964	UNCTAD meeting	1%
1965	OECD DAC reaffirms support for UNCTAD target	1%
1967	G-77 asks for separate minimum target for official flows	
1968	Tinbergen, chair UN Committee on Development Planning proposal	0.75% (including only official concessional and non-concessional flows) by 1972
1969	World Bank – Pearson Commission Report: Partners in Development, based on new OECD definitions, methods and data	0.7% (including only official concessional flows (ODA)) by 1975–80
1970	International Development Strategy for the Second United Nations Development Decade, UN General Assembly Resolution 2626	0.7% (although DCs kept arguing for 1%)
1975	UN General Assembly Resolution 3517	0.7%
2002	International Conference on Financing for Development	0.7%
2002	World Summit on Sustainable Development	0.7%
2005	EU-15	0.7% by 2015
2005	G8 Gleneagles Summit	0.7%
2008	UN Summit	0.7%
2009	G20	Respective ODA commitments

Source: Based on Cárdenas *et al.* (1995).

In 2009, the G20 reiterated the need to achieve the ODA pledges (G20 London Summit, 2009). This is to be seen in the light of the past performance of donors (see Figures 4.1 and 4.2).

In 2007, the 15 EU countries that are DAC members together provided 59.5% of total net ODA, at 0.39% of their combined GNI. The share provided by these EU

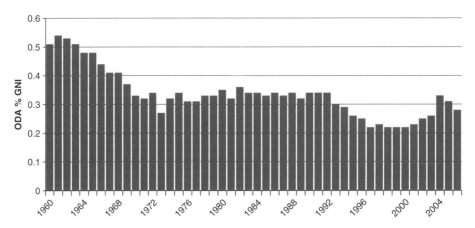

Figure 4.1 Net ODA as a percentage of GNI of all DAC countries, 1960–2007, based on OECD DAC statistical databases.

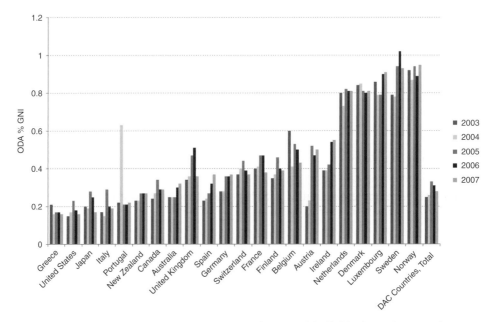

Figure 4.2 Net ODA as a percentage of GNI of individual DAC countries, 2003–2007, based on OECD DAC statistical databases.

Table 4.9 *Additional resources needed for meeting MDGs*

MDGs	USD billions	Reference
Global	50	Zedillo *et al.* (2001)
For 30 African countries	20–25	African Development Bank (2002)
For poverty goal	54–63	Devarajan *et al.* (2003)
For poverty goal	15–46 plus 100% debt cancellation	Greenhill (2002)
For other goals	16.5	Greenhill (2002)
Global	100	Oxfam (2002)
Global	50–80	Vandemoortele (2002)
Global	73 in 2006, 135 in 2015	UN Millennium Project (2005)
Range	50–135	

Source: Based on Clemens *et al.* (2007).

donor countries is projected to increase to some 65% by 2010, in contrast with the projected further decline in US ODA disbursements (OECD DAC, 2008).

The history of the 0.7% target reveals that, although there is a political and moral perception both by donors and by partner countries of the need to hold on to this target, donors do not see this as legally binding, but as to some extent a combination of altruistic (solidarity) and self-serving goals (creation of markets, via e.g. aid for trade, and security). The political commitment to help DCs achieve the Millennium Development Goals has led to many calculations regarding how much additional money this will cost, and estimates range from USD 50 billion to USD 135 billion (see Table 4.9).

4.4.4 *Environmental assistance: new and additional assistance*

In addition, there have been discussions on environmental assistance. In 1972, the Environment Fund was an early mechanism to fund UN activities on the environment (Stockholm Declaration, 1972); its resources were to be additional to, not at the cost of, development activities.[17] Subsequent funds included those for the 1983 International Tropical Timber Agreement, the 1987 Basel Convention on the Control of the Transboundary Movement of Hazardous Wastes and their Disposal and the 1990 Montreal Protocol on Substances that Deplete the Ozone Layer.

---

[17] '*Decides* that, in order to ensure that the development priorities of developing countries shall not be adversely affected, adequate measures shall be taken to provide additional financial resources on terms compatible with the economic situation of the recipient developing country, and that, to this end, the Executive Director, in co-operation with competent organizations, shall keep this problem under continuing review' (Stockholm Declaration, 1972: xxvii).

The critical question that can be raised concerns what the rationale is for assisting DCs in the area of the environment. Franck (1995: 381) argues, with respect to the Montreal Protocol, that it was 'both fair and, ultimately, cheaper, for the rich nations to help the poor to adapt to the changes that global ozone layer protection will require of them'. The alternative was for ICs to reduce their own emissions and prevent the DCs from causing any. 'Aside from the inherent unfairness of such a "trumping" policy of unilateral global standard-setting, it would have provoked widespread rebellion against controls, probably fuelled more by outrage at the illegitimacy of the means by which the controls are set and enforced than by any essential self-interest of violators' (Franck, 1995: 386).

The proliferation of funding mechanisms for the environment suggests that such assistance is seen as essential to global environmental governance. However, much of the funding is voluntary and dependent on the goodwill of donors. Agenda 21 (1992: para. 33.20) estimated that the development and environmental issues of the 1990s would cost about USD 600 billion annually for the period 1993–2000. It suggested that a quarter of this amount should be funded by ICs – amounting to USD 125 billion. The ensuing funding mechanism, the Global Environment Facility, focused on only 4 (and later 6) of the 40 issues identified and had a very limited budget.

4.5 Key challenges in sustainable development governance

The governance processes on environment and development reveal a number of trends.

Diffuse governance

Global governance on sustainable development is dispersed across organizations (the UN, the Bretton Woods institutions, the WTO, the OECD etc.) and has different principles (Biermann and Bauer, 2005; Thai *et al.*, 2007). While the initial focus was on helping DCs develop, the focus in the mid 1980s was on structural adjustment and the Washington Consensus, and since 2000 the focus has been on the MDGs. After 50 years, these communities have united partly under the banner of the UN Development Group established in 1997 and under the 2005 Paris Declaration on Aid Effectiveness, and each body is currently seeking to incorporate climate change into its development policy. The most dominant bodies in global development policy are the UN Commission on Sustainable Development, ECOSOC and the UN Development Group. The first has limited power, the second has limited interest and the third is not much more than a common forum. This means that there is a major vacuum at global level in this area. The most influential body in the area of development cooperation policy is the OECD. However, it neither has the mandate to represent non-OECD countries that are or should be donors nor includes partner countries in its policymaking process.

Convergence in rhetoric, but divergence in interpretation

There is increasing convergence in rhetoric about sustainable development govern-ance. While developed countries are willing to accept certain norms, they are less willing to apply these norms, seeking refuge in diverging interpretations. Although there is convergence in the rhetoric on sustainable development and 'mainstream-ing' climate change, and regarding coordination processes, the trade and invest-ment regimes have remained outside the purview of the all-encompassing discussions on the subject (Zelli, 2007; Tienhaara, 2009). Yet most international financial transfers occur in trade and investment regimes (see Table 1.2). Furthermore, sustainable development and its integrative nature are controversial (Pallemaerts, 1993) and the underlying agendas have not been reconciled. 'The backdrop of the general unwillingness of the North to make the sacrifices necessary to finance the development urged by the South has not changed greatly throughout the UN period, but the UN has contributed to increased understanding of the issues' (Roberts and Kingsbury, 1993: 56). The question thus is whether the 'mainstream-ing' discussion is merely 'greenwash' or is perhaps even more pernicious in that it overrides other key development challenges. Is not the research undertaken with respect to climate change far greater than all that is undertaken in relation to other MDGs? While the development banks have adopted the vocabulary of 'main-streaming' climate change, a key issue is whether this will make a difference.

Consistent financial commitment for development and *failure to provide it*

Since 1960, there has been a political commitment to providing 0.7% of national income as assistance (see Table 4.6). However, in terms of actual implementation, the OECD DAC countries have never really gone beyond 0.4%. There have been continuing recognitions of this failure (e.g. Monterrey Consensus, 2002; Paris Declaration, 2005; Roberts *et al.*, 2008). However, there is also continuing resistance to recognizing this target as legally binding and as following from the contested right to development. While the DCs try to move the discussions out of the charity sphere, the ICs avoid this discussion and instead frame it more in terms of partnership.

Small environmental resources

There is a certain commitment to provide environmental resources and this is probably more in line with the interests of the ICs. The United Nations Conference on Environment and Development calculated that USD 125 billion was necessary on an annual basis to deal with environmental and developmental issues. At the time, this could scarcely have taken the impacts of climate change into account, since information was minimal regarding costs. However, the Global Environment Facility has had limited resources in comparison. There is also a mismatch between the (4–6) goals of the financial mechanism established and the 40 issues raised by the United Nations Conference on Environment and Development. Finally, there is competition between the financial institutions (e.g. the GEF) and the treaties they serve (Mace, 2005), and between the financial institutions and the UN system (e.g. in terms of the voting system), and relatively austere use of funds within the UN (Childer and Urquhart, 1994: 78; White, 1996: 261).

From the fear of evaluation to accountability fixations

A key issue that arises is the lack of comprehensive evaluation of aid (see Chapter 2). Khosla (2001) argued that the implementation of Agenda 21 was never really evaluated properly. OECD activities and the performance of the OECD have also not been the subject of scientific research: '[I]ndustrial countries have been reluctant to institute a procedure that might generate embarrassing criticisms of their performance, while developing countries have been leery of a reporting system that could become the basis for imposing environmental conditions for economic assistance' (Soroos, 1999: 40–1). However, in recent years, the demand for accountability has led to a new shift towards a large number of assessment programmes (van den Berg and Feinstein, 2009).

Aid versus other flows: a drop in the ocean?

Aid is a small part of all trade and financial flows (see Table 1.1). Increasingly, the low aid flows are leading to a call to generate new resources through the use of market mechanisms (see Chapter 5). However, a critical issue is whether market mechanisms can help identify the alternative development model that countries need to develop in order to achieve the MDGs and sustainable development and to mainstream climate change in development cooperation. There are two groups in this debate: one argues that incremental changes to the system through appropriate uses of market forces will lead to major results; while the other school argues that systemic change is necessary (Miller, 1991: 41). Sustainable development and 'mainstreaming' climate change are ideas that no one can challenge, but can lead to merely symbolic policies (Gupta and Hisschemöller, 1997).

4.6 Conclusions

This chapter has provided an overview of the evolution of policy at UN and OECD level on development cooperation and mainstreaming climate change in development and development cooperation. It has argued that global governance is diffuse and that the most influential think tank in this area is probably the OECD. However, this body represents neither DCs nor non-OECD donors. Furthermore, there is a convergence in rhetoric towards coordinating policies on development and development cooperation and in mainstreaming climate change in both processes. However, the convergence in rhetoric does not necessarily demonstrate a convergence in terms of a change in behaviour. The DCs have favoured a right to development over the last four decades and, although this has now been adopted, it remains contested by ICs. Although ICs have voluntarily committed to providing resources to DCs over the last six decades and have continuously reiterated the need to do so, they have not met their own targets.

Most importantly, mainstreaming climate change in aid is unlikely to have significant impacts if the bulk of the investment flows do not also take such mainstreaming into account. This reflects the difference in ideological perspective about whether free trade and some amount of environmental internalization (though not in the trade and investment world) will be adequate to achieve the goal of sustainable development.

Acknowledgements

This chapter builds upon the critical and constructive review comments of P. J. I. M. de Waart, Jill Jaeger and Philipp Pattberg.

References

Adam, E. (2006). Preface to Kirchmeier, F. (2006). *The Right to Development: Where Do We Stand?* Dialogue on Globalization Occasional Papers No. 23. Geneva: Friedrich-Ebert-Stiftung.

African Development Bank (2002). *Achieving the Millennium Development Goals in Africa: Progress, Prospects and Policy Implications. Global Poverty Report.* Abidjan: African Development Bank.

Agarwal, A., Narain, S. and Sharma, A. (1999). *Green Politics: Global Environmental Negotiations.* New Delhi: Centre for Science and Environment.

Agenda 21 (1992). Report on the UN Conference on Environment and Development, Rio de Janeiro, 3–14 June 1992, UN Document A/CONF.151/26/Rev. 1 (Vols. I–III).

Alexandria Conference (2008). International Workshop Evaluating Climate Change and Development, GEF and World Bank, Alexandria, June 2008. (Major papers have been published in Feinstein, O. and van den Berg, R., eds. (2009). *Evaluating Climate Change and Development.* World Bank Series on Development. New Brunswick/London: Transaction Publishers.)

Alston, Ph. and Simma, B. (1988). Session of the UN Committee on Economic, Social and Cultural Rights. *American Journal of International Law*, **82**, 603–15.

Biermann, F. and Bauer, S., eds. (2005). *A World Environment Organization: Solution or Threat for Effective International Environmental Governance.* Aldershot: Ashgate Publishers.

Bryner, G. C. (1999). Agenda 21: myth or reality, in *The Global Environment: Institutions, Law and Environment*, ed. N. J. Vig and R. S. Axelrod. London: Earthscan, pp. 157–89.

Cárdenas, E. J., Di Cerisano, C. S. and Avalle, O. (1995). The changing aid environment: perspectives on the official development assistance debate. *ILSA Journal of International & Comparative Law*, **2**(1), 189–201.

Childer, E. and Urquhart, B. (1994). *Renewing the United Nations System*. Uppsala: Dag Hammarskjold Foundation.

Chowdhury, S. R., Denters, E. M. G. and de Waart, P. J. I. M., eds. (1992). *The Right to Development in International Law*. Dordrecht: Martinus Nijhoff Publishers.

Clemens, M. A., Kemp, C. J. and Moss, T. J. (2007). The trouble with the MDGs: confronting expectations of aid and development success. *World Development*, **35**(5), 734–51.

Colas, B. (1994). *Global Economic Co-operation: A Guide to Agreements and Organisations*. Deventer: Kluwer Law and Taxation Publishers.

Dadzie, K. (1993). The UN and the problem of economic development, in *United Nations, Divided World: The UN in International Relations*, ed. A. Roberts and B. Kingsbury. Oxford: Clarendon Press, pp. 397–426.

Devarajan, S., Easterly, W. R. and Pack, H. (2003). Low investment is not the constraint on African development. *Economic Development and Cultural Change*, **51**(3), 547–71.

de Waart, P. J. I. M. (1992). Implementing the right to development: the perfection of democracy, in *The Right to Development in International Law*, ed. S. R. Chowdhury, E. M. G. Denters and P. J. I. M. de Waart. Dordrecht: Martinus Nijhoff Publishers, pp. 191–212.

Doha Declaration (2008). Doha Declaration on Financing for Development: Outcome Document of the Follow-up International Conference on Financing for Development to Review the Implementation of the Monterrey Consensus. Doha, Qatar, 29 November–2 December 2008. Available online at http://www.un.org/esa/ffd//doha/documents/Doha_Declaration_FFD.pdf.

ECA (2007a). *Climate Change and the Broad Mandate of the Economic Commission for Africa*. United Nations Joint Press Kit for the UN Climate Change Conference in Bali, December 2007, UN Document DPI/2483.

   (2007b). *Climate for Development in Africa Programme*. Addis Ababa: Economic Commission for Africa.

Esty, D. (1999). Economic integration and the environment, in *The Global Environment: Institutions, Law and Environment*, ed. N. J. Vig and R. S. Axelrod. London: Earthscan, pp. 190–209.

Franck, T. M. (1995). *Fairness in International Law and Institutions*. Oxford: Oxford University Press.

G8 (2005). *The Gleneagles Communiqué, Statement of the G8 at the Gleneagles Summit in 2005, Gleneagles*.

G20 London Summit (2009). *Global Plan for Recovery and Reform: The Communiqué from the London G20 Summit*.

Garcia-Amador, F. V. (1990). *The Emerging International Law of Development: A New Dimension of International Economic Law*. New York, NY: Oceana Publications.

GATT (1947). General Agreement on Tariffs and Trade. Available online at http://www.wto.org/english/docs_e/legal_e/gatt47_01_e.htm.

GEF (1994). *Report of the Independent Evaluation of the Global Environment Facility, Pilot Phase*. Washington, D.C.: World Bank, UNDP and UNEP, Global Environment Facility Secretariat.

   (2005) *Achieving the Millennium Development Goals: A GEF Progress Report*. Washington, D.C.: Global Environment Facility.

George, S. (1992). *The Debt Boomerang: How Third World Debt Harms Us All*. London: Pluto Books.

Gosovic, B. (1992). *The Quest for World Environmental Cooperation; The Case of the UN Global Environment Monitoring System*. London: Routledge.

Greenhill, R. (2002). *The Unbreakable Link: Debt Relief and the Millennium Development Goals*. London: New Economics Foundation

Gupta, J. (2006). The Global Environment Facility in its North–South context, in *Contemporary Environmental Politics: From Margins to Mainstream*, ed. P. Stephens, J. Barry and A. Dobson. London: Routledge, pp. 231–53.

Gupta, J. and Hisschemöller, M. (1997). Issue-linkages: a global strategy towards sustainable development. *International Environmental Affairs*, **9**(4), 289–308.

Hicks, R. L., Parks, B. C. Roberts, J. T. and Tierney, M. J. (2008). *Greening Aid? Understanding the Environmental Impact of Development Assistance*. Oxford: Oxford University Press.

Hyden, G. (2008). After the Paris Declaration: taking on the issue of power. *Development Policy Review*, **26**(3), 259–74.

ILA (2002). *Resolution 3/2002 of the International Law Association: The New Delhi Declaration of Principles of International Law Relating to Sustainable Development*.

Khosla, A. (2001). *The Road from Rio to Johannesburg*. UNED Forum Millenium Papers, Issue 5. London: UNED Forum.

Kirchmeier, F. (2006). *The Right to Development: Where Do We stand?* Dialogue on Globalization Occassional Papers No. 23. Geneva: Friedrich-Ebert-Stiftung.

Mace, M. J. (2005). Funding for adaptation to climate change: UNFCCC and GEF developments since COP-7. *Review of European Community and International Environmental Law*, **14**(3), 225–46.

Malmö Ministerial Declaration (2000). *Declaration of the first Global Ministerial Environment Forum, Malmö*.

Mani, M., Markandaya, A. and Ipe, V. (2008). *Policy and Institutional Reforms to Support Climate Change Adaptation and Mitigation in Development Programs: A Practical Guide*. Washington, D. C.: World Bank.

Miller, M. (1991). *Debt and the Environment: Converging Crises*. New York, NY: United Nations Publications.

Monterrey Consensus (2002). *Monterrey Consensus of the International Conference on Financing for Development*, UN Document A/Conf/198. Available online at http://www.un.org/esa/ffd/monterrey.

NIEO Declaration (1974). *United Nations General Assembly Declaration on the New International Economic Order*, UN Document A/RES/S-6/3201. New York, NY: United Nations.

OECD (1961). *Resolution of the Common Aid Effort Mandate*. Mandate of the Development Assistance Committee, as decided by the Ministerial Resolution of 23 July 1960. Paris: Organisation for Economic Co-operation and Development.

(2006a). *DAC in Dates: The History of the OECD's Development Assistance Committee*. Paris: Organisation for Economic Co-operation and Development.

(2006b). *Declaration on Integrating Climate Change Adaptation into Development Co-operation*. Paris: Organisation for Economic Co-operation and Development.

(2007). *Aid Effectiveness. 2006 Survey on Monitoring the Paris Declaration. Overview of the Results*. Paris: Organisation for Economic Co-operation and Development.

(2008). *Aid Effectiveness. 2008 Survey on Monitoring the Paris Declaration. Making Aid more Effective by 2010*. Paris: Organisation for Economic Co-operation and Development.

(2009). *Aid Effectiveness. A Progress Report on Implementing the Paris Declaration*. Paris: Organisation for Economic Co-operation and Development.

OECD DAC (2008). *Development Co-operation Report 2008*. Paris: Organisation for Economic Co-operation and Development.

Oxfam (2002). *Last Chance at Monterrey: Meeting the Challenge of Poverty Reduction.* Briefing paper No. 17. Oxford: Oxfam.

Pallemaerts, M. (1993). International environmental law from Stockholm to Rio: back to the future, in *Greening International Law*, ed. P. Sands. London: Earthscan, pp. 1–19.

Paris Declaration (2005). *Paris Declaration on Aid Effectiveness: Ownership, Harmonisation, Alignment, Results and Mutual Accountability.* High Level Forum on Aid Effectiveness, Paris, 28 February–2 March 2005. Available at http://www.oecd.org.

Piron, L.-H. (2002). *The Right to Development: A Review of the Current State of the Debate.* London: Department of International Development.

Roberts, A. and Kingsbury, B. (1993). Introduction: the UN's roles in international society since 1945, in *United Nations, Divided World: The UN in International Relations*, ed. A. Roberts and B. Kingsbury. Oxford: Clarendon Press, pp. 1–63.

Roberts, J. T., Starr, K., Jones, T. and Abdel-Fattah, D. (2008). *The Reality of Official Climate Aid*. Oxford Energy and Environment Comment. Oxford: Oxford Institute for Energy Studies.

Schrijver, N. (1995). *Sovereignty over Natural Resources: Balancing Rights and Duties in an Interdependent World*. Ph.D. thesis, University of Groningen.

Second UN Development Decade (1970). *Second United Nations Development Decade.* Report of the Second Committee, GA 8124, para. 43. New York, NY: United Nations.

Soroos, M. S. (1999). Global institutions and the environment: an evolutionary perspective, in *The Global Environment: Institutions, Law and Environment*, ed. N. J. Vig and R. S. Axelrod. London: Earthscan, pp. 27–51.

Stockholm Declaration (1972). *Report of the UN Conference on the Human Environment*, Stockholm, 5–16 June, 1972; UN Document A/Conf.48/14/Rev. 1. New York, NY: United Nations.

Tienhaara, K. (2008). *The Expropriation of Environmental Governance: Protecting Foreign Investors at the Expense of Public Policy*. Ph.D. thesis, VU University Amsterdam.

Tinbergen, J. (1976). *Reshaping the International Order: A Report to the Club of Rome.* New York, NY: E. P. Dutton.

UN Charter (1945). Charter of the United Nations (San Francisco), 26 June 1945, and amended on 17 December 1963, 20 December 1965 and 20 December 1971, ICJ Acts and Documents No. 4.

UN Commission on Human Rights (1977). *Resolution 4 (XXXIII) of 21 February 1977.*

UNCTAD (1964). *United Nations Conference on Trade and Development, Geneva, 23 March–16 June 1964: Proceedings*. New York, NY: United Nations.

UN Declaration on the Right to Development (1986). *UN High Commission on Human Rights, Declaration on the Right to Development, United Nations General Assembly Resolution 41/128 of 4 December 1986*. Available online at http://www.un.org/en/documents.

UNDP (2007). *Human Development Report 2007/2008: Fighting Climate Change – Human Solidarity in a Divided World*. New York, NY: United Nations Development Programme.

UNDP, UNEP, World Bank *et al.* on behalf of the Poverty–Environment Partnership (2003). *Poverty and Climate Change: Reducing the Vulnerability of the Poor through Adaptation, Poverty–Environment Partnership 2003*. New York, NY: United Nations Development Programme.

UNEP (2007). Note by Executive Director on the Cooperation between the United Nations Environment Programme and the United Nations Development Programme. UN Document UNEP/GC/24/INF/19.

(2008). *Annual Report, 2008*. Nairobi: United Nations Environment Programme.

UNEP–UNDP (2007). *Guidance Note on Mainstreaming Environment into National Development Planning*. Nairobi/New York, NY: United Nations Environment Programme and United Nations Development Programme.

UNGA (1948). *Universal Declaration of Human Rights*. Adopted by the United Nations on 10 December 1948. Available online at http://www.un.org/en/documents/udhr.

  (1960). *Accelerated Flow of Capital and Technical Assistance to the Developing Countries*. United Nations General Assembly Resolution 1522, 15 December 1960. Available online at http://www.un.org/documents.

  (1969). *Declaration on Social Progress and Development*. United Nations General Assembly Resolution A/RES/2542, 11 December 1969. Available online at http://www.un.org/documents/ga/res/24/ares24.htm.

  (1970). *International Development Strategy for the Second United Nations Development Decade*, United Nations General Assembly Resolution 2626, UN Document A/8028/1970. Available online at http://www.un.org/documents.

  (1993). *Vienna Declaration and Programme of Action, World Conference on Human Rights*, UN Document A/Conf.157/23. New York, NY: United Nations.

UN Millennium Project (2005). *Investing in Development: A Practical Plan to Achieve the Millennium Development Goals*. London: Earthscan.

Vandemoortele, J. (2002). *Are the MDGs Feasible?* New York, NY: UNDP Bureau for Development Policy.

van den Berg, R. and Feinstein, O., eds. (2009). *Evaluating Climate Change and Development*. World Bank Series on Development, Volume 8. New Brunswick/London: Transaction Publishers.

VerLoren van Themaat, P. and Schrijver, N. (1992). Principles and instruments for implementing the right to development within the European Community and in Lomé IV states, in *The Right to Development in International Law*, ed. S. R. Chowdhury, E. M. G. Denters and P. J. I. M. de Waart. Dordrecht: Martinus Nijhoff Publishers, pp. 89–112.

WCED (1987). *Our Common Future: Brundtland Report of the World Commission on Environment and Development*. Oxford: Oxford University Press.

Werksman, J. D. (1993). Greening Bretton Woods, in *Greening International Law*, ed. P. Sands. London: Earthscan, pp. 65–84.

White, N. D. (1996). *The Law of International Organisations*. Manchester: Manchester University Press.

World Bank (2006). *Clean Energy and Development: Towards An Investment Framework*. Washington, D.C.: World Bank.

WSSD (2002). *Johannesburg Plan of Action, World Summit on Sustainable Development*, Johannesburg, 26 August–4 September 2002. Available online at http://www.un.org/esa/sustdev/documents/WSSD_POI_PD/English/POIChapter4.htm.

WTO (1994). *Marrakesh Agreement Establishing the World Trade Organization*. Available online at www.wto.org.

Zedillo, E. and the Panel Members (2001). *Recommendations of the High-Level Panel on Financing for Development*. UN Document A/55/1000. New York, NY: United Nations.

Zelli, F. (2007). The World Trade Organization: free trade and its environmental impacts, in *Handbook of Globalisation and the Environment*, ed. Thai, K. V., Rahm, D. and Coggburn, J. D. London: Taylor and Francis, pp. 177–215.

Zoellick, R. (2007). Speech at the Conference of the Parties to the UN Climate Change Conference in Bali, December 2007. See page v of http://siteresources.worldbank.org/INTCC/Resources/SFCCD_Concept_and_Issues_Paper_Consultation_Draft_27March2008sm.pdf.

5

Global governance: climate cooperation

JOYEETA GUPTA, HARRO VAN ASSELT AND
MICHIEL VAN DRUNEN

5.1 Introduction

Following the discussion of the global context for development and development cooperation, this chapter focuses on cooperation within the climate change context. It addresses the key issues of international climate change cooperation between rich and poor countries. To this end, it first explains the major elements of consensus in the Climate Convention in terms of principles and commitments (see Section 5.2). It then examines the evolving nature of the climate change deal between industrialized countries (ICs) and developing countries (DCs) and explains why many DCs have difficulties with this process (see Section 5.3). It then examines the resources needed in the regime, with a particular focus on the needs of DCs, compares those with what is available, and discusses the principles on sharing resources and the need for ideas to generate additional funding. It also looks at the relationship between the mechanisms (see Section 5.4) and at the key market mechanisms both for technology transfer and for generating adaptation resources (see Section 5.5), before drawing conclusions (see Section 5.6).

5.2 Principles and mechanisms: the consensus of 1992

5.2.1 Introduction

Climate change is intrinsically a North–South issue (see Section 1.2.3). This section explains the consensus on the division of responsibilities between ICs and DCs that underlies the climate regime. It argues that the climate regime was developed according to a leadership paradigm, which substituted for the liability and 'polluter pays' principles, and was initially accepted by the DCs in the constructive spirit in which it was offered. The principle of common but differentiated responsibilities and respective capabilities was accepted, together with a number of other supporting ideas about helping the most vulnerable countries. Furthermore, although the need

Mainstreaming Climate Change in Development Cooperation: Theory, Practice and Implications for the European Union, ed. Joyeeta Gupta and Nicolien van der Grijp. Published by Cambridge University Press. © Cambridge University Press 2010.

for DCs to develop was recognized, this fell short of recognizing the right to develop. Although there was an expectation that resources would be made available to DCs, there was a greater focus on market mechanisms.

5.2.2 *The preambular text and principles*

The Climate Convention provides the consensus on how responsibilities should be divided between countries, including four elements of critical value. First, the climate regime is designed according to the leadership paradigm (Gupta, 1998), in line with ideas promoted in earlier declarations.[1] The Climate Convention calls on ICs to 'take the lead in combating climate change and the adverse effects thereof' (FCCC, 1992: Art. 3(1); see also Art. 4(1)). The idea of leadership has been invoked often since 1992.

Second, the Climate Convention recognizes the common but differentiated responsibilities and respective capabilities of countries. This principle indicates that the responsibilities are divided among countries in relation to how much a country contributed to causing the problem and its respective ability to pay. This is in line with the initial declarations on climate change, which focused on the need for countries to pay according to their capability (UNGA, 1988), for ICs to compensate vulnerable countries for the ' abnormal burden' that they would have to bear (Hague Declaration, 1989) and for a euphemistic 'special responsibilities' approach (Bergen Declaration, 1990). This idea is emphasized in the principle on the circumstances of particularly vulnerable countries (FCCC, 1992: Art. 3(2)).

Third, an element of the leadership paradigm is the notion of the need of DCs to grow. The Climate Convention mentions that DCs need to grow 'to meet their social and developmental needs' (FCCC, 1992: Preamble para. 3), that these are legitimate priority needs (FCCC, 1992: Preamble para. 23), that all countries have the right to and should promote sustainable development (FCCC, 1992: Art. 3(4)), and that an open economic system will enable DCs to achieve sustainable economic growth (FCCC, 1992: Art. 3(5)). Although it recognizes the need to grow and the right to sustainable development, it falls short of recognizing the right to development (see Section 4.4.2; see also Box 5.1).

Fourth, the leadership paradigm effectively displaces the 'polluter pays' and the 'no harm' principles. The Climate Convention does not explicitly include the 'polluter pays' principle, the concept of no harm or the concept of liability. It recalls that states have the sovereign right to exploit their own resources and the responsibility to ensure that 'activities within their jurisdiction or control do not cause

---

[1] Notably the Noordwijk Declaration on Climate Change (1990: para. 7), the Bergen Declaration on Sustainable Development (1990: para. 11) and the Ministerial Declaration of the Second World Climate Conference (SWCC Declaration, 1990: para. 5).

Box 5.1 Climate change, the right to development and human rights

In 2008, the UN Human Rights Council stated that it was '[c]oncerned that climate change poses an immediate and far-reaching threat to people and communities around the world and has implications for the full enjoyment of human rights'. It noted the past human rights declarations and the findings of the IPCC reports, recalling both the 1986 Declaration on the Right to Development and the particular vulnerability of the world's poor, and called on the Office of the High Commissioner for Human Rights to prepare a study on human rights and climate change, which it did in 2009.[2]

There are at least two dimensions to the relationship between climate change and human rights. The first is the right to development for which DCs have fought long and hard (see Section 4.4.2). The 1986 Declaration on the Right to Development sees this right as an 'inalienable human right by virtue of which every human person and all peoples are entitled to participate in, contribute to, and enjoy economic, social, cultural and political development' and includes 'the exercise of their inalienable right to full sovereignty over all their natural wealth and resources'. The Climate Convention hints at this by including references to the legitimate needs of DCs (FCCC, 1992: Preambular paras. 3 and 23, and Art. 4(5)). Some literature focuses on the right to develop and its implications for the right to emit greenhouse gas emissions (Baer *et al.*, 2008; Gupta and van Asselt, 2007). The UNDP (2007: 50), however, dismisses this approach, arguing that 'the presumed "right to emit" is clearly something different than the right to vote, the right to receive an education or the right to enjoy basic civil liberties'. However, while a right to pollute is difficult to justify, a right to basic emissions as a corollary to the right to development is not so easy to dismiss.

The second dimension is the obligation to ensure that no violations of human rights will occur. Arts (2009) elaborates on how children's rights recognized in international law will most probably be violated by the impacts of climate change. She argues that from this perspective there is a clear and compulsory agenda for action to reduce emissions and protect children. However, climate change is not just about future generations, but very much also about current generations. Although it is difficult to distinguish between climate variability and change, the UNDP (2007: 60) argues that inaction on climate change would represent a very immediate violation of universal human rights.

The report prepared for the Human Rights Council[3] argues that human rights and the environment are interdependent and interrelated and that there is no explicit recognition of a right to environment in human rights declarations, that there are implicit links and that climate change will probably affect the rights of women, children and indigenous peoples,[4] leading to displacement and security risks.[5] Nevertheless, proving this in legal terms may be complex, but not necessarily impossible. The report concludes that 'The physical impacts of global warming cannot easily be classified as human rights violations, not least because climate change-related harm often cannot clearly be attributed to acts or omissions of specific States. Yet, addressing that harm remains a critical human rights concern and obligation under international law. Hence, legal protection remains relevant as a safeguard against climate change-related risks and infringements of human rights resulting from policies and measures taken at the national level to address climate change.'[6]

---

[2] A/HRC/10/61, 15 January 2009.
[3] UN Human Rights Council, Report of the Office of the United Nations High Commissioner for Human Rights on the relationship between climate change and human rights, A/HRC/10/61; 15 January 2009.
[4] Report of the Office of the United Nations High Commissioner for Human Rights, above n 1, Section IIc.
[5] Report of the Office of the United Nations High Commissioner for Human Rights, above n 1, Section IId and e.
[6] Report of the Office of the United Nations High Commissioner for Human Rights, above n 1, Para 96.

damage to the environment of other States or areas beyond the limits of national jurisdiction' (FCCC, 1992: Preamble para. 8). It does not, however, accord this idea, which was included in the Stockholm Declaration on the Human Environment (1972), the Rio Declaration and Agenda 21 (1992), the status of a principle. Furthermore, the inclusion of the cost-effectiveness concept (FCCC, 1992: Art. 3 (3)) within the precautionary principle limits the extent to which action should be taken. Thus, the initial framing of the issue in political terms was careful to avoid any notion of transboundary pollution and liability, and defined issues in terms of 'neutral factual statements' (Bodansky, 1993: 498). The Climate Convention recognized the need to help DCs on the basis of the idea of common but differentiated responsibilities and respective capabilities, committed ICs to provide assistance to DCs (FCCC, 1992: Arts. 3(2), 4(3), 4(4), 4(5), 4(8), 11 and 21) and made the implementation of DC obligations under the Convention subject to financial and technological assistance from the ICs (FCCC, 1992: Art. 4(7)).

5.2.3 The cooperative mechanisms

The Climate Convention set up a financial mechanism, whose operating entity was the Global Environment Facility (FCCC, 1992: Arts. 11 and 21). Since then, other funds have been created to provide finance to DCs to reduce their growth of emissions and to help in adaptation. In 1997, the Kyoto Protocol called for the establishment of the Adaptation Fund, which became operational in 2008. In 2000, the Special Climate Change Fund was established to finance adaptation and the transfer of technologies in the fields of energy, transport, industry, agriculture, forestry and waste management, and to help DCs diversify their economies. Another fund established was the Least Developed Country Fund, to help the poorest countries to prepare and implement National Adaptation Programmes of Action. In 2004, the Global Environment Facility set up the Strategic Priority on Adaptation. While most funds are financed by voluntary contributions from the richer ICs listed in Annex II to the Climate Convention, the adaptation fund is financed by a compulsory levy on the Clean Development Mechanism (CDM) (see Section 5.3.4; see also Table 5.5 for an elaboration of the funds). However, these funds have not been able to provide the 'predictable and adequate levels of funding' called for in the decisions of the Conference of the Parties (Marrakesh Accords, 2001: Arts. 1.b and 1.d of Dec. 7/CP7).

The Climate Convention introduced the concept of project-based emissions trading – joint implementation – without defining it (Gupta, 1997). In 1995, the concept was adopted as Activities Implemented Jointly at the first Conference of the Parties. In 1997, this concept transformed into Joint Implementation (between ICs) and the Clean Development Mechanism (CDM) (between ICs and DCs) in the Kyoto Protocol. These market mechanisms enable investors to purchase emission-reduction credits, from countries with economies in transition and DCs, respectively, in return for financial investment in emission reduction. The Convention called for technology transfer to the DCs (FCCC, 1992: Art. 4(5)) and the Kyoto Protocol and the Marrakesh Accords (2001) further elaborated on this idea.

The official climate and environmental funds are lower than those that can potentially be raised by market mechanisms. For example, the GEF has a total budget of USD 3 billion and leveraged USD 14 billion for *all* its activities (UNDP, 2007: 169), while, in 2006 alone, the CDM raised USD 5.2 billion for climate mitigation activities (UNDP, 2007: 169). Compare also the ability of the CDM to generate funds for adaptation – between USD 160 and 950 million in the period until 2012, while only about USD 170 million has been raised or pledged in the other two climate funds (Müller, 2007: 3).

5.3 The evolving North–South deal: the controversies

5.3.1 Introduction

Since climate change is essentially a North–South problem (see Section 1.2.3), it is important to understand the nature of the North–South deal made in the Climate Convention. Chapter 1 argued that the climate change problem was initially framed as a global, future, abstract and technological problem that subsequently evolved into a multi-level, current, real development problem (see Section 1.3). This section argues that the nature of the political deal between ICs and DCs has also evolved since the 1990s. Initially, in the constructive Rio days the leadership paradigm implied *both* emission reductions *and* 'new and additional' assistance to DCs. This evolved into emission reduction *via* assistance to DCs through project-based emissions trading, as well as adaptation assistance via the CDM. As resources commensurate with the nature of the problem proved more difficult to raise, there was a trend to 'mainstream' climate change in development cooperation and to transfer responsibility to other social actors (see also Section 1.6 and Chapter 4). Throughout these discussions, a key issue has been the interpretation of the 'new and additional' resources that would be made available.

5.3.2 Phase 1: emission reductions and assistance to the South

The 1992 leadership paradigm implied that the North should lead in addressing the problem, and the South should follow (see Section 5.2). The North was expected to lead by reducing its own emissions (FCCC, 1992: Arts. 3(1) and 4(1)) and by providing 'new and additional financial resources' to DCs (FCCC, 1992: Arts. 4(3), 4(4), 4(5), 4(7), 4(8), 11 and 21).

The term 'new and additional' can be found in the Climate Convention and the Kyoto Protocol. The Convention states that IC Parties 'shall provide new and additional financial resources to meet the agreed full incremental costs incurred by developing countries in complying with their obligation under Article 12, paragraph 1' (FCCC, 1992: Art. 4(3)).[7] The Kyoto Protocol also states that ICs should '[p]rovide new and additional financial resources to meet the agreed full costs incurred by developing country Parties' in implementing certain activities such as emission inventories and should 'provide financial resources, including for the transfer of technology, needed by the developing country Parties to meet the agreed full incremental costs of advancing the implementation of existing commitments in Article 4' (KP, 1997: Art. 11).

From the DC perspective, new and additional implied financing over and above the ODA commitment of 0.7% of GNP was promised repeatedly to the DCs (see Table 4.6). Hence, for example, 'Argentina sustains that any funding for adaptation must not be counted towards meeting the UN agreed target of 0.7% for aid. Developed countries have delivered just USD 48 million to international funds for least-developed country adaptation, and have counted it as aid. This practice undermines international development and poverty alleviation efforts, upon which current official development assistance is currently founded' (Government of Argentina, 2008; see also Government of China, 2008; Government of India, 2008). The Government of India calls for an additional 0.5% of GNI, while the G-77 and China together ask for an additional 0.5%–1% of GNI. The UNDP (2007: 193) clarifies that '[t]he critical starting point is that adaptation financing has to take the form of new and additional resources. That means that the international effort should be supplementary to the aid targets agreed at Gleneagles and supplementary to the wider aspiration of achieving an aid-to-GNI level of 0.7% by 2015'. The Commission on Climate Change and Development (2009: 18) states that, 'By additional, we mean additional to the commitment of paying 0.7 percent of gross national income as ODA.' However, most ICs interpret this as simply implying over and above existing ODA, since pledges to increase their aid state that the additional aid will go to climate change.

Thus, the Climate Convention emphasizes four points. First, that the ICs would take the lead to reduce their own emissions. Second, that they would provide 'new and additional resources' to the DCs, although such new and additional resources would be used to finance only the agreed full incremental costs of achieving global benefits. Third, that '[t]he extent to which developing country Parties will effectively implement their commitments under the Convention will depend on the effective implementation by developed country Parties of their commitments

---

[7] The articles referred to focus on measures such as the making of national inventories of GHG emissions, national programmes, the promotion of technology transfer and sustainable management of natural resources, cooperation on adaptation, integration of climate change considerations into national policies, and cooperation in research, information, education and in the preparation of national communications.

under the Convention related to financial resources and transfer of technology and will take fully into account that economic and social development and poverty eradication are the first and overriding priorities of the developing country Parties' (FCCC, 1992: Art. 4(7)). Fourth, that the interests of the most vulnerable countries have to be taken into account.

5.3.3 *Phase 2: emission reduction via assistance to the South*

By 1995, the Conference of the Parties revealed that various ICs were seeking to reduce their own emissions partly via the use of market mechanisms financed by the 'new and additional' resources. This argument has four elements to it.

First, the targets in the Climate Convention appeared not to be legally binding because of a linguistic trick (Sands, 1992; Bodansky, 1993; Gupta, 1997). Most ICs were thus not on track to reducing their greenhouse gas emissions to 1990 levels by 2000. The Kyoto Protocol called for a joint 5.2% reduction of emissions by 2012 with respect to emission levels in 1990, but, because the largest emitter at the time – the USA – did not participate, emission reductions were considerably lower than they should have been when compared with what was seen as necessary in the Toronto Declaration (1988), or even what IPCC-1 (1990: xviii) saw as necessary if stabilization at 1990 concentration levels were to be achieved in 2000. Complicating matters was the aftermath of the US Byrd–Hagel Resolution (1997), which put pressure on the large DCs to adopt meaningful action and was seen by DCs to be in conflict with Article 4(7), in particular, and the spirit of the original 1992 deal.

Second, the targets can be realized through cost-effective measures undertaken in DCs through the flexibility mechanisms of the Kyoto Protocol. Since a part of the targeted reductions can be achieved through investments in DCs, the actual reduction of emissions within the ICs in 2008–12 falls considerably short of the formal existing target. The offsetting character of the flexibility mechanisms is problematic (Boyd *et al.*, 2007). For instance, the Netherlands seeks to achieve 50% of its emission-reduction goal under the Kyoto Protocol through the flexibility mechanisms. Of the EU's target of a reduction of 8% for 15 EU Member States, 5% will probably be met by the CDM (Henningsen, 2008).

Third, the CDM was initially seen as the critical vehicle for technology transfer and assistance to DCs. However, there have been doubts about the additionality of the projects and the real technology-transfer content in these projects driven primarily by the cost-effectiveness of projects (Haites *et al.*, 2006; Schneider, 2007; Wara and Victor, 2008).

Fourth, the CDM was seen as generating both new and additional financial resources for mitigation since it effectively engaged social actors. This is important, because the voluntary resources for mitigation activities in the funding mechanisms are relatively low.

5.3.4 Phase 3: assistance via assistance to the South

After 1997, when the CDM was conceived, and there were limited resources for adaptation, attention focused on creating an Adaptation Fund through voluntary contributions as well as a percentage of the proceeds of the Certified Emission Reductions. (An exception was subsequently made for projects involving LDCs.) This part is the most reliable source of funding. This is highly ironic because it raises the costs of doing projects in the South vis-à-vis the North. Although there have been discussions at the negotiations about taxing the other 'flexibility mechanisms' that operate on a North–North basis, there is considerable resistance to this idea.

5.3.5 Phase 4: assistance via development cooperation

As the impacts of climate variability (whether attributable to climate change or not) become more obvious, and since raising resources to deal with climate change is not easy (see Section 5.4), attention is shifting to what can be done with the existing and available resources. By 2007, there was a strong movement in the ICs to merge development aid and climate aid as a way to leverage more funding for climate mitigation and adaptation, but also as an additional argument to boost the flagging interest in providing development aid (Table 3.1; see Chapter 4 on global govern-ance shifts towards mainstreaming, Chapter 6 on EU policy and Chapter 7 on the policies of EU Member States). This merger has been presented in terms of the discourse on mainstreaming and integration; however, the term 'mainstreaming' has often been used loosely in relation to the definition provided in Chapter 3. This merger is viewed quite critically by the DCs in view of the debate on 'new and additional resources' (Government of Argentina, 2008; Government of China, 2008: Government of India, 2008; G-77 and China Proposal, 2008). At the same time, ICs are trying to shift responsibility to other social actors through the search for market mechanisms to generate additional resources for climate change assis-tance (see Table 5.7). While the small island states see any effort to generate additional resources for adaptation as laudable, the larger DCs are holding out for a fixed official climate assistance commitment.

The change in the interpretation of the leadership paradigm has not gone unnoticed by the DCs and the issues mentioned above are still being debated today. Table 5.1 provides an overview of the different phases that can be dis-cerned, and discusses the transformation of the leadership paradigm in the context of implications for the South and the effects on the idea of 'new and additional'

Table 5.1 *The evolving leadership paradigm*

	Explanation	Implications for the South	Implications for aid
Leadership paradigm (1990–6)	The North leads through emission reduction at home, and technological and financial cooperation with the South.	The South can learn from the mistakes of the North and will also be helped to do so. The North–South approach paid off, if briefly.	North reduced emissions and provides new and additional assistance to developing countries.
Conditional leadership paradigm; leadership lost (1996–2001)	The EU-led coalition waits for the USA-led coalition to ratify the Kyoto Protocol, the USA-led coalition waits for the key developing countries, and the developing countries wait for the North to take action.	Hardening of the developed country position means that developing countries will have to negotiate skilfully and demonstrate their commitment in order to find a way out. Otherwise, it is the developing countries that stand to lose the most.	North reduces emissions via new and additional assistance.
Partial restoration of leadership paradigm (2001–3)	The USA withdraws. The EU is forced to choose between two alternatives – withdrawal and further action. The first is morally	The North–South approach is no longer, in itself, a fruitful approach. This can only serve to	North finds costs of reducing emission heavy because of non-participation of the USA and money is perceived as being diverted

	repugnant, thus the EU goes for the second option and inspires many other developed countries, including Japan, to do so. The large developing countries follow quickly.	aggravate the fragile goodwill of the EU. A new constructive and cooperative mode has to be found.	from development to environment aid.
Leadership competition (2003–7)	The USA initiates a large number of programmes and actively sets itself up as competitor to the UN FCCC regime.	This competition can have the potential of causing confusion among developing countries as to the seriousness of the problem and the appropriate way of dealing with it.	Discussion on mainstreaming climate change in aid discussions.
Rise of new powers (2007–)	Russia, China, India, South Africa and Brazil become serious actors.	Prior to the Copenhagen meeting of 2009, large developing countries offered to take on voluntary commitments; poor countries felt excluded.	As yet unknown.

Source: Updated from Gupta (2003).

resources. It shows that the emergence of the idea of mainstreaming climate change into development assistance can be seen in relation to the waning leadership of the ICs.

5.4 The resource gap

5.4.1 Introduction

This section reviews the amount of resources needed for dealing with the climate change problem and compares that with the resources available. It discusses potential options for raising new resources. Finally, it examines the relationship among the various funding mechanisms. What is clear is that there is a major gap between the resources available and those needed; there is a proliferation of ideas about how the gap can be closed to some extent; and the proliferation of funds and mechanisms does not add up to a comprehensive system for financing measures in the developing world.

5.4.2 The resources needed

Chapter 1 explained the intended role of technology transfer in assisting DCs to leapfrog their way to sustainable development. Although there are challenges, the identification of appropriate technologies and the purposeful transfer and adaptation of these technologies to meet local circumstances is a critical part of the climate change agreement (FCCC, 1992: Art. 4(5)). At the same time, no real measures were taken to stop or discourage on-going technology transfers of older and more affordable technologies. It was hoped initially that climate-relevant technology transfer would simply occur either via the funding mechanism or through project-based emissions trading. Although bureaucratic steps have been taken,[8] and scientific research undertaken (Metz *et al.*, 2000), climate-relevant technology flows have been slow. This is because technologies are privately owned and expensive and their transfers need to be funded; capacity-building activities have to be promoted in the DCs to ensure that only appropriate technologies are imported; the dumping of older and cheaper technologies continues; and ICs compete in marketing their own technologies. The bottom line is that technology transfer has failed to achieve its goals (UNDP, 2007: 162).

On the related issue of capacity building, discussions finally took off at Marrakesh in 2001. Substantively, it was agreed that there was no single best

---

[8] A consultative process on technology transfer was started in 1998 with the Buenos Aires Plan of Action, consisting of a Technology Needs and Needs Assessment Framework, an Expert Group on Technology Transfer, a Technology Transfer Clearing House and country-specific assessments.

Table 5.2 *Additional investment and financial flows needed for adaptation in 2030, by sector, for DCs*

Sector	Areas/adaptation measures considered	Global cost (2005 USD bn)	Proportion needed in DCs (%)	Cost for DCs (2005 USD bn)
Agriculture, forestry and fisheries	Production and processing, R&D, extension services	14	50	7
Water supply	Water-supply infrastructure	11	80	9
Human health	Treating increased incidence of cases of diarrhoea, malnutrition and malaria	5	100	5
Coastal zones	Beach nourishment and dikes	11	45	5
Infrastructure	New infrastructure	8–130	25	2–32
Total		49–171		28–58

Source: Based on FCCC Secretariat (2008a). The last column and the bottom row have been derived from the first two columns.

practice for capacity building – there was no 'one size fits all' formula. Capacity building was seen as a continuous, progressive and iterative process. For maximum effect, the process needed to be integrated into existing processes and programmes as well as focused simultaneously on all related conventions. In 2004, the capacity-building efforts were reviewed, and a compilation was completed in 2007. However, most of these efforts turned out to be stand-alone projects or add-ons to existing projects, financed by the GEF's enabling activities or technical support resources. The DCs have argued that these capacity-building exercises should be based on an assessment of DC needs and not necessarily on what the ICs visualize as necessary (see Chapters 8 and 9).

The bottom line is that, without mobilizing financial resources, neither technology transfer nor adaptation will be addressed. Both ICs and DCs will need resources to mitigate emissions and adapt to climate change. The ICs are expected to generate their own resources. The DCs are likely to need help in generating the resources they need. This section first examines the resources needed, before elaborating on how these can be shared and divided between countries.

Recent documents provide estimates with respect to the costs of adaptation and mitigation. A paper on investment and financial flows presented to the Conference of the Parties in 2008 (FCCC, 2008a) assesses global adaptation costs for five sectors as shown in Table 5.2. The derived costs for the DCs amounts to USD 28–58 billion annually for adaptation purposes in these five sectors.

Table 5.3 *Additional investment and financial flows needed for mitigation in 2030, by sector, especially for DCs*

Sector	Areas/mitigation measures considered	Global cost (2005 USD bn)	Proportion needed in DCs (%)	Cost for DCs (2005 USD bn)
Fossil-fuel supply	Lower production due to reduced demand and greater use of other fuels	−59	54	−32
Power supply	Lower fossil-fired generation capacity; more renewables; carbon dioxide capture and storage; nuclear energy; hydropower	−7	49	−3.4
Industry	Greater energy efficiency; carbon dioxide capture and storage; reduced emissions of non-CO_2 greenhouse gases	36	54	19
Buildings	Greater energy efficiency	51	28	14
Transportation	More fuel-efficient vehicles; greater use of biofuels	88	40	35
Waste	Capture and use of methane from landfills and wastewater plants	1	64	0.6
Agriculture	Reduced methane emissions from crops and livestock	35	37	13
Forestry	Reduced deforestation and forest degradation; sustainable forest management	21	99	21
Technology, R&D	Double the amount that is currently being spent in this area	35–45	–	
Total net additional		200–210		68

Source: Based on FCCC Secretariat (2008a). The last column has been derived from the first two columns.

The costs of mitigation for the DCs amount to about USD 68 billion per year (see Table 5.3). However, this is based on greater use of biofuels, while the recent biofuel controversy reveals that this approach might not be so easy to exploit without negative impacts on the developing world. Furthermore, by excluding the research costs in the South, the proposal assumes that Northern technologies can be transferred without adaptation to local circumstances and that no local solutions are needed; this is counter-productive and will foster dependency in the long term.

While the World Bank (2006) estimates for adaptation range from USD 10 billion to USD 40 billion, Oxfam (2007) argues that, even if countries are able to accelerate their efforts to reduce emission of greenhouse gases, annual adaptation costs may amount to USD 50 billion. Stern (2007) calculates that the cost of climate change is in the range 5%–20% of annual GDP while the cost of action would be 1% of GDP to achieve stabilization at 550 ppm CO_2-equivalent. In a scenario in which no measures are taken, the costs of impacts would amount to a loss of 5% of annual global GDP from now to eternity. This 5% loss is a conservative estimate and does not include a broad range of social and environmental impacts.

The UNDP (2007: 167–8) also comes up with some estimates of costs. The cost of stabilization at 450 ppm CO_2-equivalent (thus going further than Stern (2007)) comes to about 1.6% of annual world GDP. The cost of adopting a low-carbon future in, for example, India is estimated at USD 5 billion annually for the period 2012–17. The UNDP states that the cost of electricity modernization itself amounts to about USD 165 billion until 2010, rising by 3% annually until 2030. The DCs currently have access to only half this amount. The UNDP concludes that USD 25–50 billion will be needed annually to assist DCs to adopt low-carbon technology. Such transfers will have to be supported by the development of national knowledge and science. In terms of adaptation, the UNDP calculates that climate proofing existing development aid will cost about USD 44 billion per year by 2015; adapting poverty programmes to take adaptation into account will cost USD 40 billion per year; and disaster relief will need an additional investment of USD 2 billion annually, adding up to about USD 86 billion per year by 2015 (about 0.2% of OECD GDP in 2015).

Behrens (2009) concludes that mitigation costs consistent with keeping emissions within a 2 °C trajectory amount to USD 58–1338 billion, averaging about USD 415 billion annually, of which half would need to be spent in the DCs. Opschoor (2009) sums up various efforts and concludes that about USD 200 billion will be needed for mitigation and USD 70 billion for adaptation. See Table 5.4.

5.4.3 The resources available and the institutional framework

In comparison with the resources needed, the resources available are quite limited and the institutional framework is disputed. The GEF, which was established in 1990, became the (interim) operating entity for all the climate funds. Over the years, three issues – focus, approach and modalities, and resources – have developed as critical issues.

In terms of focus, although the Climate Convention did not state so, until 1996 the GEF provided resources for mitigation but not for adaptation activities. Some research activities were then funded, but it was only with the establishment of the

Table 5.4 *The costs of adaptation (A) and mitigation (M)*

Costs (USD bn/yr)	A/M	Time/level	Reference
9–41	A	Present	World Bank (2006)
4–37	A	Present (550 CO_2e)[a]	Stern (2007)
50	A	Present	Oxfam (2007)
86	A	2015 (450 CO_2e)[a]	UNDP (2007)
28–67	A	2030	FCCC Secretariat (2007a)
49–171 for all countries; 28–58 for DCs	A	2030	FCCC Secretariat (2008a: para. 61)
70	A	2 °C, 2030 and beyond	Opschoor (2009)
75–100 for DCs	A	2 °C by 2050	EACC (2009)
25–50	M	2015 (450 CO_2e)[a]	UNDP (2007)
58–1338; average 415; 50% for DCs	M	2 °C	Behrens (2009)
200–210 for all countries; 68 for DCs	M	25% below 2000 levels in 2030	FCCC Secretariat (2008a: paras. 16 and 60)
200	M	2 °C, 2030 and beyond	Opschoor (2009)
100–140	A/M	2030	Müller (2008: 19)

[a] CO_2e, ppm CO_2-equivalent.
Source: Based on Mani *et al.* (2008), FCCC Secretariat (2008a), UNDP (2007), Opschoor (2009) and Behrens (2009).

LDC fund that dedicated resources for the preparation of National Adaptation Programmes of Action were provided. In 2004, the GEF established the Strategic Priority for Adaptation. The Adaptation Fund that became operational in 2008 is also intended to fund adaptation activities.[9]

In terms of approach and modalities, the choice of the GEF as the interim funding mechanism for the Climate Convention has been controversial from the start for a number of reasons (Gupta, 1995; Mace, 2005). Most importantly, the nature of the supervision of the GEF is under debate: is the GEF answerable to the Conference of the Parties or to its own Board? This has been a critical issue over the last 18 years. In 2007, discussions about the role of the GEF led to the consensus that meetings of the Board of the Adaptation Fund should take place in Bonn, not Washington, D.C.; applications for funding should be directed to the Board of the Adaptation Fund, not to the implementing agencies of the GEF; and that the GEF should use national

[9] This fund has a 16-member Board, a secretariat (the GEF) and a trustee (The World Bank). The mandate of the Adaptation Board is to identify strategic priorities that the Conference of the Parties must accept, design operational policies, set criteria for project selection, establish rules of procedure, monitor and review activities, establish committees, panels and working groups as required, and to bear the responsibility for the monetization of certified emission reductions (CMP.3 Decision on Adaptation Fund, see http://unfccc.int/files/meetings/cop_13/application/pdf/cmp_af.pdf).

Table 5.5 *Existing adaptation funds and amounts delivered by 2008 (USD million)*

Fund	Period	Amount
Funds under the Climate Convention		
LDC Fund	2001–	172
Special Climate Change Fund	2004–	91
Adaptation Fund	2008–12	50
Strategic Priority on Adaptation	2004–	50
Funds of the Banks		
GEF: Small Grants Programme	1992–	39
Pilot Programme for Climate Resilience	2009–12	0
Funds of NGOs		
Rockefeller Foundation: Climate Change Initiative	2007–	70 for 2007–11
European Funds		
European Commission – Global Climate Change Alliance	2008–10	Est. 28 annually
German Climate Initiative (Ministry of Environment)	2008–12	Est. 50 annually
Total		429.5

Source: Derived from van Drunen *et al.* (2009).

experts, simplify the incremental cost principle, adopt the lessons from Piloting an Operational Approach to Adaptation, improve access to funds, and report to the Conference in time for it to be able to examine the report carefully before the meetings start, to ensure that the agreed full costs of DCs are covered in relation to Article 12(1) and to report on these (FCCC Secretariat 2007b: 33).

Furthermore, resources within the GEF for climate change activities have been limited. These have been supplemented by other funding schemes. Adaptation efforts are financed through various funds that provide grants and/or loans. Table 5.5 shows that, between 1992 and 2008, USD 429.5 million was spent on adaptation funding.

In addition to this, several funds have been set up by NGOs, such as the Rockefeller Foundation Climate Change Initiative, and governments, such as the EU's Global Climate Change Alliance (see Chapter 6) and Germany's Climate Initiative. Furthermore, there are disaster-preparedness funds, and funds of the International Red Cross that are provided as relief to countries during (weather) disasters. Some of this comes from ODA money (Bouwer and Aerts, 2006).

However, first, the resources available are a fraction of what is needed and resources committed on paper are quite different from what is actually spent. For example, the UNDP (2007: 204) argues that by mid 2007 only USD 26 million had actually been spent on adaptation – which is roughly what the UK spends on flood defence every week. Second, most funds are dependent on voluntary contributions, except the German Climate Initiative, which is based on auctioning emission allowances in the EU's Emissions Trading System. Third, the only other scheme

Box 5.2 Reducing farmers' vulnerability: weather insurance for Malawi

Africa is particularly vulnerable to climate variability. In 2008, some 25 million people in sub-Saharan Africa faced weather-related food crises. Notably, in Southern Africa drought and adverse weather caused much of the recent hunger experienced there (Hess and Syroka, 2005). But climate change is set to make many of Africa's problems worse. Scientists argue that many of Africa's dry areas will become drier and its wet areas wetter. It is forecast that by 2020 between 75 and 250 million Africans will suffer from too much or too little water. In some African countries, yields from rain-fed agriculture could fall by half by 2020.

The recent introduction of weather-related insurance programmes in Africa, including an index-based insurance pilot programme in Malawi, is offering smallholder farmers some security against the high risks of flood and drought. Such programmes could prove key in terms of assisting DCs in adapting to climate change and thus reducing disaster risk. For example, their implementation could encourage farmers to develop agricultural practices that are capable of withstanding changing climates.

The Malawi scheme offers index-based weather insurance to smallholder groundnut farmers. While conventional crop insurance is written against actual losses, this scheme is written against a physical trigger, such as insufficient rainfall at key points in the growing season. In other words, the insurer will pay out if rainfall falls below a specified level regardless of crop damage.

Under this scheme, participating farmers are provided with improved agricultural inputs (i.e. seed) before the rainy season through a contract that specifies (1) an index-based weather insurance component, in which the premium is calculated on the probability of a payout, and (2) a loan component provided by a bank. At the end of the season the farmer owes the lending institution an amount equal to the cost of agricultural inputs, plus insurance premium, plus interest and taxes. If rains are good (as measured at a nearby weather station operated by the meteorological service), then the insurance company keeps the premium and farmers pay back the loan with proceeds from the (presumably good) harvest. If measured rains are below certain trigger values (based on critical stages of the groundnut growing season), then the insurance company pays part, if not all, of the loan to the bank.

By improving farmers' creditworthiness, this scheme also improves their ability to access credit for investing in higher-yield/higher-return crops. Banks typically have considered lending to farmers with no collateral as being excessively risky because of the high risk of loan default in the aftermath of droughts. By coupling bank loans with index-based weather insurance, farmers can receive the requisite credit for seeds and other agricultural inputs, and they can expect a net gain after repayment of the coupled loan–insurance contract.

Almost 900 farmers took part in the first Malawi pilot project and more than 2500 signed up for the second season. In a survey of participating farmers carried out by the IIASA, 86% said they would participate again, more than two-thirds indicated that they had encouraged other farmers to join the scheme the following season, and 62% answered that they were better able to cope with drought and food shortages if they were in the insurance scheme. But the survey also highlighted less successful aspects of the scheme. Many participating farmers, for example, did not fully understand the index-based system and did not wholly trust the organizations in charge. More than a quarter of survey respondents did not trust the weather-station data, and other concerns ranged from the quality of the seed to insufficient payment from insurance (Bayer *et al.*, 2007).

Scaling up the Malawi scheme thus faces significant challenges, but the organizing body, the National Smallholder Farmers' Association of Malawi (NASFAM) is currently working to address these, particularly in terms of increasing trust and improving communication.

The ability to estimate the effects of climate change on the near- and long-term future of micro-insurance schemes serving the poor should prove invaluable to insurers and international development communities as they seek to scale up nascent weather insurance systems in Africa and beyond. Indeed, by providing technical assistance and infrastructure, such as weather forecasting stations, international donors can further help to expand successfully the micro-insurance schemes. Furthermore, donors can subsidize the insurance payments of the very poor and provide insurance back-up capital for such insurance pools. The particular challenge is to ensure the affordability of these schemes for the poor, which provides an opportunity for the donor community to re-orient from post-drought to pre-drought assistance.

By Joanne Linnerooth-Bayer

with a regular source of income is the Adaptation Fund, but it is primarily funded by a levy on North–South cooperation under the climate change agreement. The bottom line is that there is a major shortfall in resources (Müller and Hepburn, 2006; Oxfam, 2007; UNDP, 2007; Roberts *et al.*, 2008; Van Drunen *et al.*, 2009). Mitigation efforts are primarily funded by the limited resources of the GEF and through the CDM (see Section 5.5).

5.4.4 Raising new resources

A critical question is – who should bear the costs? In the absence of assistance, the South, by default, will have to bear the full burden of adaptation and mitigation. It would be inappropriate to expect the DCs to bear the full brunt of adaptation costs, since the problem has been caused primarily by the ICs. Asking the DCs to bear the costs of mitigation may also imply major risks to the ICs because emissions may spiral out of control. This is why the Climate Convention adopted the principle of common but differentiated responsibilities and respective capabilities. Different scholars interpret the principle differently.

Opschoor (2009) compares recent publications that attempt to allocate country shares on the basis of different interpretations of the principle of common but differentiated responsibilities. For example, Oxfam (2007) proposed an Adaptation Finance Index based on the greenhouse gas emissions of countries, their human development index and their population. Table 5.6 shows that shares for the ICs range from 51% to 100% depending on whether current or cumulative emissions are taken into account and other factors. Also note that the shares of China and India are quite substantial.

Such principles could be used to derive a value for official climate assistance for these countries. Clearly, most ICs may experience political challenges in raising these resources. The literature and negotiation documents helpfully suggest a number of innovative approaches to raise additional resources (see Table 5.7). These continue along the lines started by the Brundtland report (WCED, 1987), which argued that other sources of financing may be more predictable and automatic, and suggested revenues from the use of the international commons (e.g. international transport), from trade (e.g. on specific products) and from international financial measures (e.g. a link with special drawing rights).

Three types of approaches are possible: focusing on official contributions; focusing on raising resources from the market; and a hybrid approach. The first is more consistent with intergovernmental negotiations, and is easier to guarantee and evaluate. Most DCs favour this approach both as showing commitment to addressing the problem and as an essential ingredient for building mutual trust. The third approach would allow states to raise resources through market mechanisms and channel these via official climate funding.

This chapter argues that official climate assistance has a role to play and is justified by the principle of common but differentiated responsibilities and respective capabilities, since this principle applies to states. While Table 5.6 suggests that scientifically appropriate amounts and percentages can be determined for each individual IC, a simpler percentage approach, as the DCs suggest, may be easier to implement in the long term and avoids wrangling between ICs regarding how the principles should be interpreted. This does not exclude the potential for also examining other sources of revenue.

Furthermore, a key element of assistance is the idea of self-help. Assisting countries and peoples to help themselves builds on local ownership and commitment to specific concepts. For example, weather-variability risk insurance can provide considerable relief to small farmers worldwide (see Box 5.2), and, while it is essentially funded by local farmers, the reinsurance could perhaps be internationally funded (cf. Bouwer and Vellinga, 2002).

Table 5.6 *Keys for country shares in global warming and associated responsibilities and capabilities*

	World GHG emissions 2000 (%)	Emissions per capita CO$_2$ in 2000 (tonnes)	Cumulative emissions 1850–2002 (%)	GDP per capita in 2010 (1000 USD, estimated)	Oxfam AFI (CO$_2$) (% of total Annex 1 c. 2000)	RCI[a] (CO$_2$) (% of global 2010)	RCI[a] (CO$_2$) (% of global 2030)	IVM/PBL scenarios (% of global)
USA	20.6	20.4	29.3	45.6	43.7	33.1	25.5	18.5–21
EU-25	14.0	8.5	26.5	30.7	31.6 (EU-17)	25.7	19.6	18–29
Japan	3.9	9.5	4.1	33.4	12.9	7.8	5.5	4.8–10.3
Russia	5.7	10.6	8.1	15.0	0	3.8	4.6	3.3–4.7
China	14.7	2.7	7.6	5.9	0	5.5	15.2	6.7–15.1
India	5.6	1.0	2.2	2.8	0	0.5	2.3	0.5–5.5
Brazil	2.5	2.0	0.8	9.4	0	1.7	1.7	2.2–3.8
South Africa	1.2	7.9	1.2	10.1	0	1.0	1.2	0.7–1.0
Annex 1	52	11.4	76	30.9	100	77	61	51–70
DCs	48	2.1	24	5.1	0	23	39	30–49
World	100	3.9	100	9.9	100	100	100	100

[a] Responsibility Capacity Index (Baer *et al.*, 2008).
Source: Opschoor (2009), based on Bradley and Bamert (2005), Baer *et al.* (2008) and Dellink *et al.* (2008).

Table 5.7 *Possible options for raising new and additional resources*

Scheme	Description
Formal government	
Official climate assistance	Many DC governments feel that assistance should be in the form of percentage of national income. Proposals include 0.5% of GNI of ICs (Government of India, 2008), 0.5%–1.0% (G-77 and China proposal, 2008), or in accordance with common but differentiated responsibilities principle; see also Mexican Multilateral Climate Change Fund Proposal (Müller, 2008; Bouwer and Aerts, 2006: 60).
Non-Compliance Fund	Countries in non-compliance with their emissions reduction target must pay into a fund to assist DCs for clean development and possibly adaptation (Brazilian proposal, see FCCC Secretariat, 2007a).
Taxes	
Carbon tax	On energy use in OECD: at USD 3/t CO_2 on OECD energy – USD 40 billion (UNDP, 2007: 195). On energy use globally: USD 2/t CO_2, with a tax exemption of 1.5 t CO_2 per capita (Government of Switzerland, 2008).
Carbon Auction Levy Fund	Levy on the auctioned amount in EU ETS: a levy of USD 3/t CO_2 – USD 570 million (UNDP, 2007: 195).
Adaptation levy on JI and ET	This will level the playing field between these mechanisms and the CDM and generate resources.
Levy on flights	Globally: at USD 7 per flight – USD 14 billion per year (UNDP, 2007: 195). At USD 1–5, depending on criteria (Müller and Hepburn, 2006). Globally: but non-Annex I countries pay less, and LDCs not at all (Tuvalu Adaptation Blueprint).
Levy on bunker fuels	This levy will apply to the fuels used by the shipping and airline industry.
Global Climate Financing Mechanism	Bonds are issued to frontload the flow of aid (Michel, 2008).
Weather-variability risk insurance	Locally, regionally and globally. This does not provide funds, but assists self-help; outsiders could fund a reinsurance scheme. Examples include the Caribbean Catastrophe Risk Insurance Facility and the Munich Climate Insurance Initiative (MCII, 2008). See also Box 5.2.
Tobin tax	A Tobin currency transaction tax of 0.1% could raise resources (FCCC Secretariat, 2007a).
Special drawing rights	International liquidity could be provided by the IMF to member countries, and developed countries could make special drawing rights available for, inter alia, adaptation projects in the South (FCCC Secretariat, 2007a).

5.4.5 *The relationship between the funds*

In the area of mitigation, there are limited funds for promoting greenhouse-gas-friendly projects through the financial mechanism of the Climate Convention established in 1992, the Special Climate Change Fund established in Marrakesh and the CDM. The World Bank has also set up a number of funds to deal with climate change.

However, most investments in DCs take place via foreign direct investment and most probably do not take climate change into account. In 2005 and 2006, USD 500 billion of net private capital flows to DCs took place annually (Miller, 2008). Sometimes these are supported by export credits from the ICs that again may but do not necessarily take climate change into account; see, for court cases on export credits, Gupta (2007). In addition, traditional ODA money is invested in DCs. However, only a small percentage may be relevant to climate change mitigation efforts. About 6%–10% of all ODA focused on energy efforts between 1997–2005 and, although there is a shift towards projects with lower greenhouse gas emissions, this shift is fragile (Tirpak and Adams, 2008). The recent increases are attributable to increased spending on large hydro projects (Roberts *et al.*, 2008). Figure 5.1 presents the relationship between these funds.

In relation to adaptation funds, there are also some small funds available (see Figure 5.2). FDI efforts may aggravate adaptation efforts, since they do not normally take climate change impacts into account and are not climate proofed. Under the Climate Convention, there are four funds for climate change – the Adaptation Fund, the LDC Fund, the Special Climate Change Fund and the Strategic Priority on Adaptation of the GEF. Most of these funds receive resources from ODA funding

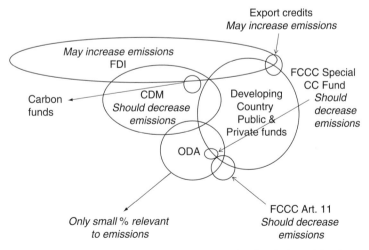

Figure 5.1 The relationship between the funds for mitigation.

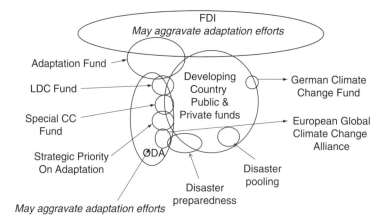

Figure 5.2　The relationship between adaptation funds. Source: inspired by Bouwer and Aerts (2006).

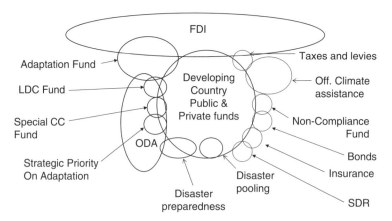

Figure 5.3　Existing and potential funds.

but can also include resources from non-ODA countries. The Adaptation Fund includes resources generated by the CDM. However, out of the ODA funds that are not specifically dedicated to climate change, only a marginal amount is spent on adaptation, and this has been decreasing since 2002 (Roberts *et al.*, 2008). In addition, disaster-preparedness funds devote some resources for adaptation. Other funds include the EU Global Climate Change Alliance and the German Climate Change Initiative. The bulk of the resources for adaptation will probably have to come from the DCs themselves.

However, if all the new proposals for funds materialize, there will be a mixture of funds in the international arena (see Figure 5.3). These multiple funds, all with their own management systems and criteria for both raising and disbursing resources, may be complicated and overlapping in remit, and may possibly ignore some crucial needs in some parts of the world.

5.5 Project-based emissions trading: the challenges

5.5.1 Introduction

A key mechanism for technology transfer to DCs is the Clean Development Mechanism. This section explains why the CDM has both an attractive and a problematic side. There are four challenges with respect to the CDM. First, the mechanism aims at both reducing emissions and generating additional sustainable development benefits. However, while there are some who question the ability of the instrument to do the former, many question the ability of the instrument to generate these additional benefits. Second, the CDM should leverage major changes in society by introducing 'additional' projects and ideas that can then be replicated in society. While CDM projects are mostly being implemented in the larger DCs, they have a small impact on these countries because they are a small part of the total investments. Third, capacity building for CDM projects is being partly funded by ODA resources and this is controversial. Finally, while a market solution has considerable potential, it will leave a large gap (Roberts *et al.*, 2008).

5.5.2 The CDM and sustainable development

The CDM is a market mechanism that aims to secure participation by the private sector in implementing the climate agreements. In order to meet DC concerns, it was also specifically suggested that such projects should explicitly meet sustainable development criteria determined by host countries (KP, 1997: Art. 12).

Although there were reasons why DCs were highly sceptical about project-based emissions trading in the 1990s, they ultimately accepted the concept of the CDM in 1997 (van Asselt and Gupta, 2009). Enthusiasm to participate in the CDM has exploded and more than 128 Designated National Authorities existed in 2008, and at least 948 CDM projects have been registered and 85 049 697 million certifications of emission reductions have been issued.

While the CDM is expected to contribute to a reduction of emissions, and emission reductions are not credited until they are verified, there are some who doubt how additional such reductions are (Michaelowa and Purohit, 2007; Henningsen, 2008). But the main issue here is the question of sustainable development. This chapter argues that the sustainable development returns of CDM projects can be disputed because (a) host countries determine what sustainable development is, and criteria range from merely applying existing environment impact assessments to applying more concrete criteria (Gupta *et al.*, 2008); and (b) the evidence shows that directly linked sustainable development impacts are more likely to be achieved than the indirect sustainable development impacts, which focus on embedding the project in the local community (Sutter and Parreño, 2007; Gupta *et al.*, 2008). Some investor countries and NGOs are trying to avoid this race to the bottom by developing transparent criteria on the basis of which projects can be evaluated as sustainable or not. The sustainable development benefits are not part of the contractual agreements and, hence, the contracts do not succeed or fail depending on whether the sustainable development component is achieved. Although prior to approval of a CDM project it is critical for the investor to clarify what sort of sustainable development benefit is aimed for, there is little incentive to ensure that these benefits are ultimately achieved.

Investor countries that wish to promote sustainable development could pay a premium for such benefits, or less for CDM credits that are not accompanied by such benefits. Sustainability benefits can be achieved if the sustainable development component is translated into measurable goals that take local aspects into account (Humphrey, 2004); if the ability of local partners to demand and ensure that a sustainable development component is included is enhanced (e.g. Ellis and Kamel, 2007); and/or if these sustainable development components are an integral part of the contract. At present there are ongoing discussions of the ways to improve and enhance this instrument through the addition of ideas such as discounting, multiplication, positive and negative lists, demand and supply quotas, ineligibility and preferential treatment (FCCC Secretariat, 2008b).

5.5.3 The CDM and ODA

The CDM was visualized as a market mechanism that would engage other social actors in the process of generating resources and reducing emissions. However, a critical problem has been the use of ODA resources for issues related to this instrument. There are two arguments in favour of using ODA resources for the capacity-building or other components of the CDM. First, since CDM resources are typically following the foreign direct investment flows, they tend to be concentrated in the richer DCs, thus the use of ODA resources for the CDM may help to ensure the necessary capacity building that will allow other DCs access to these resources. Furthermore, the use of ODA resources could leverage synergy with other existing ODA projects and thereby provide more successful projects (Michaelowa and Michaelowa, 2008).

However, there are three arguments against the use of ODA resources for CDM projects. First, at an ideological and practical level, the idea of subsidizing a market mechanism does not make sense. If some countries do not offer the market an attractive option for investments, capacity building is not going to make the country more attractive by itself. Even though 35 Designated National Authorities have been established in Africa, 20 of these countries do not have even one project. The Conference of the Parties in Nairobi in 2006 emphasized the need for capacity building and the UNDP and UNEP have started investing in such capacity building with the help of resources from some ICs, and others recommend developing a regional centre for excellence to promote such projects. However, whether such capacity building will counter the fundamental structural issues that make these DCs unattractive to investors remains to be seen.

Second, average ODA levels are below 0.7% of GNI and have not yet met the development needs of the DCs. Using such resources for the CDM is thus inappropriate in principle. Third, the use of such resources for ODA is focused more on indirectly supporting the purchase of credits for ICs and less on the priorities of the DCs and, hence, is likely to lead to a diversion of resources from ODA goals to climate change goals (Yamin, 2005; Michaelowa and Michaelowa, 2008; Gupta, 2009). For example, it is argued that 'Unless development assistance for energy increases in the coming years, the influence of multilateral banks will diminish and their ability to encourage sustainable energy projects will decline' (Tirpak and Adams, 2008: 135). Statements of this kind may lead to policy diverting development resources to climate mitigation goals.

Hence, the Conference of the Parties to the Climate Convention decided in Marrakesh in 2001 that 'Public funding for clean development mechanism projects from Parties in Annex I is not to result in the diversion of official development assistance and is to be separate from and not counted towards the financial obligations of Parties included in Annex I.' (Marrakesh Accords, 2001: Dec. 17/CP7). The question is – who will screen for this? The Designated Operation Entity is not expected to do so (Yamin, 2005). China screens all CDM projects to check the source of resources (Ming *et al.*, 2008). The OECD DAC (2004) has announced that 'The value of any CERs [certified emission reductions] received in connection with an ODA financed CDM project should lead to a deduction of the equivalent value from ODA. The DAC should also rule out the possibility of counting as ODA funds used to purchase CERs.'

Meanwhile, capacity building is exempted from this rule, and NGOs also feel strongly that such capacity building is needed to enable Africa to participate in the CDM. Curiously, however, a lot of capacity building has taken place in rapidly developing China, making the argument that capacity building is needed flawed. The Canadians set up programmes in 2003, the ADB in 2004, the World Bank in 2005, France and Germany in 2006 and Japan and the UNDP in 2007. These programmes were located all over China to engage the local population in developing CDM projects (Schroeder, 2008).

CDM and leveraging change in the large DCs

Market mechanisms tend to follow the pattern set by foreign direct investments. Most foreign direct investment is going to China and India. Predictably, China has 35.47% and India 25.77% of the CDM projects,[10] and China will probably generate the largest number of credits (59.11%).[11] Given that both countries have a high credit-generation potential, the market tends to focus on these two countries. While both countries were initially hesitant about engaging in this mechanism (Tangen *et al.*, 2001; Gupta, 2004; Zhang, 2006), they now participate and have set up their own institutional frameworks for doing so. However, foreign aid is 'minuscule' in relation to the GDP of these countries (Howell, 2007: 22) and CDM investments likewise (Srivastava *et al.*, 2007). These investments may thus have limited catalytic impacts. On the other hand, since the CDM is very visible in China and India it may have a certain radiating power. Catalysing change in other countries depends on how effectively CDM projects lead to a change in the development paths in partner countries (Bode and Michaelowa, 2003; Winkler and Thorne, 2002; Boyd *et al.*, 2007).

Other CDM-related issues

The CDM is attractive since it potentially offers DCs technologies and other measures that can help reduce their rate of growth of emissions at a price that is affordable to them. The CDM is also attractive to investors since it offers cost-effective emission-reduction options to the ICs and a viable way of exporting technologies to new markets.

However, the CDM has a number of problematic features. At an abstract level, it is problematic because of its character as an offset mechanism allowing ICs to increase their emissions, thereby reducing the pressure on them to find ways to reduce their own emissions. It is problematic because it allows ICs to export their production and consumption patterns to DCs and because it, in effect, commodifies pollution rights (cf. Juma, 1995; van Asselt and Gupta, 2009). This is problematic because at the negotiation level there is different bargaining power, processes are time-consuming and expensive (Krey, 2005; Chadwick, 2006) and many of these agreements use the bilateral investment treaties to protect the interests of the investors. It is problematic at project level, because the baselines, though methodologically sound, may be disputed; because additionality is increasingly difficult to prove (Pearson and Shao Loong, 2003; Michaelowa and Purohit, 2007; Haya, 2007; Schneider, 2007); and because only about one-third of the projects include a claim to transfer technology (Haites *et al.*, 2006). Relying on CDM and other market

---

[10] See http://cdm.unfccc.int/Statistics/Registration/NumOfRegisteredProjByHostPartiesPieChart.html.
[11] See http://cdm.unfccc.int/Statistics/Registration/AmountOfReductRegisteredProjPieChart.html.

mechanisms to address the climate change problem may be very tricky. As Roberts *et al.* (2008: 2) submit,

While they hold promise to raise substantial funds for reallocation in the future, in the short term, such market-based approaches remain very uncertain tools to address a problem of such magnitude as climate change. It is unrealistic to expect that a massive global 'public good' like avoiding climate change – where everyone benefits from its being addressed, but each have incentives to shirk responsibility – could be solved with market mechanisms alone.

5.6 Conclusions

This chapter, in discussing the nature of global climate cooperation between ICs and DCs, argues five points. First, initially the DCs were willing to go along with the idea of the leadership paradigm – the notion that the North will reduce its own emissions and provide resources to the South. However, two decades later, the North has scarcely reduced its own emissions since the target level is low, the USA has not participated in the Kyoto Protocol and Northern emissions are being offset through CDM in the South. The continuing reluctance of the ICs (e.g. Russia and the USA) to adopt targets commensurate with what the IPCC reports see as necessary if the global community is to stay within a 2 °C temperature limit was demonstrated once more during the negotiations in Copenhagen in December 2009.

Second, the North has been gradually backtracking on its definition of its own financial and technological commitments. This is neither in line with what has been agreed to, in the principle of common but differentiated responsibilities and respective capabilities, nor in line with the article that DC commitments depend on resources from the ICs (FCCC, 1992: Art. 4(7)). The new and additional resources promised are not seen by the average IC to be over and above their 0.7% commitment for development cooperation, and instead the focus is on mainstreaming climate change into development cooperation while using a levy on North–South cooperation within the CDM for generating funds for adaptation. The efforts of the DCs to link the climate change issue with human rights – both in order to guarantee some space for themselves to grow and in order to access a reliable source of funding for mitigation and adaptation efforts – are now being supported by the UN Human Rights Council.

Third, consequently there is a major gap between the resources needed to address the climate change problem and the resources that are available. The funds needed are of the order of billions; the funds available are of the order of millions, thus falling considerably short of what is necessary. Furthermore, the primary source of adaptation funding is the CDM, and this is a significant example of how limited the IC commitment is to raise additional resources to address adaptation to climate change caused by the North. The proposal to level the playing field by implementing a tax on emissions trading and joint implementation is still being negotiated. Ironically, as long as the CDM grows, the 2% tax on its proceeds fund will ensure resources for the adaptation fund. At Copenhagen, the Accord that was noted offered to generate USD 30 billion by 2012 and 100 billion annually by 2010. Whether this will be implemented remains to be seen.

Fourth, there is no shortage of ideas about raising the resources needed to deal with the problem. Such ideas can actively engage all social actors in addressing the climate change problem. The proliferation of dedicated funds and management systems suggests that large quantities of resources are being generated; however, together they do not add up to a comprehensive system for promoting technology transfer to stimulate mitigation and/or resources for adaptation. The big fear is that, if these ideas are not integrated into a common system, the help will remain fragmented and possibly self-serving. This is probably why the G-77 and China prefer official climate aid, a fixed and reliable sum of money spent in accordance with specific principles to ensure that the global community can together reduce global greenhouse gas emissions and adapt to the impacts.

Fifth, the ICs promoted the innovative CDM to engage private-sector participation in addressing global problems in a sustainable manner as an alternative way to raise resources and reduce emissions. This market mechanism is growing rapidly, especially in countries where markets and the economy are reasonably well developed. The CDM aims both to help the ICs engage the market to reduce their greenhouse gas emissions cost-effectively and to help DCs to achieve sustainable development. However, it is problematic insofar as (a) emission reductions might not be additional; (b) the indirect sustainable development component is elusive; (c) ODA resources are being diverted for CDM capacity building; (d) the CDM might not be able to leverage major changes in the large DCs where it is at present concentrated since it is a very small percentage of investments in these countries; (e) it bypasses poorer countries and weaker markets (such as in Africa) as well as the poorer segments in society; and (f) market mechanisms will probably not be able to deliver the huge reductions needed.

Had the climate change problem initially been framed as a transboundary pollution problem, solutions might have been designed in terms of the 'polluter pays' principle (which would call on countries to internalize greenhouse gas emissions into the cost equation domestically) and the liability and compensation principles (which would have called on the biggest polluters to compensate those who will suffer even though their contribution to causing the problem is limited). Such a framing would have led to drastic emission reductions in the ICs and to a funding system to compensate the most vulnerable DCs, and would have put emerging economies on notice that they too would be held liable in the future. Instead, the discussion of climate change in terms of the leadership discourse, in a time of reluctant leadership rather than statesmanship, has weakened the motivation of all to address the problem seriously.

Acknowledgements

This chapter has benefited from the comments of Ton Bresser and Karin Arts.

References

Arts, K. (2009). A child rights perspective on climate change, in *Climate Change and Sustainable Development: New Challenges for Poverty Reduction*, ed. M. A. Salih. Cheltenham: Edward Elgar, pp. 79–93.

Baer, P., Athanasiou, T., Kartha, S. and Kemp-Benedict, E. (2008). *The Greenhouse Development Rights Framework: The Right to Development in a Constrained World*. Berlin: Heinrich Boll Foundation.

Bayer, J., Suarez, P., Linnerooth-Bayer, J. and Mechler, R. (2007). *Feasibility of Risk Financing Schemes for Climate Adaptation: The Case of Malawi*. Report to the World Bank's Development Economics Research Group. Washington, D.C.: World Bank.

Behrens, A. (2009). Financial impacts of climate change mitigation. *Climate Change Law Review*, **3**(2), 179–87.

Bergen Declaration (1990). *UN/ECE Bergen Ministerial Declaration on Sustainable Development in the ECE Region*, Bergen. UN Document A/Conf.151/PC/10, Annex I.

Bodansky, D. (1993). The United Nations Framework Convention on Climate Change: a commentary. *Yale Journal of International Law*, **18**, 451–588.

Bode, S. and Michaelowa, A. (2003). Avoiding perverse effects of baseline and investment additionality determination in the case of renewable energy projects. *Energy Policy*, **31**(6), 505–17.

Bouwer, L. and Vellinga, P. (2002). Changing climate and increasing costs: implications for liability and insurance, in *Climatic Change: Implications for the Hydrological Cycle and for Water Management*, ed. M. Beniston. Dordrecht: Kluwer Academic Publishers, pp. 429–44.

Bouwer, L. M. and Aerts, J. C. J. H. (2006). Financing climate change adaptation. *Disasters*, **30**(1), 49–63.

Boyd, E., Hultman, N. E., Roberts, T. *et al.* (2007). *The Clean Development Mechanism: An Assessment of Current Practice and Future Approaches for Policy*. Tyndall Centre Working Paper 114. Oxford: Environmental Change Institute Oxford and Tyndall Centre for Climate Change Research.

Bradley R. and Baumert, K., eds. (2005). *Navigating the Numbers: Greenhouse Gas Data and International Climate Policy*. Washington, D.C.: World Resources Institute.

Byrd–Hagel Resolution (1997). *US Senate Resolution 98 on Global Warming*. Congressional Record, October 3, 1997; Senate, page S10308-S10311. Available from Congressional Records online at http://www.gpo.gov.

Chadwick, B. P. (2006). Transaction costs and the Clean Development Mechanism. *Natural Resources Forum*, **30**(4), 256–71.

Commission on Climate Change and Development (2009). *Closing the Gaps: Report of the Commission on Climate Change and Development*. Stockholm: Swedish Ministry of Foreign Affairs.

Copenhagen Accord (2010). *Decision Noted by the Parties to the Fifteenth Conference of the Parties to the Climate Change Convention.* FCCC/CP/2009/L.7, 18 December 2009.

Dellink, R., Dekker, T., den Elzen, M. *et al.* (2008). *Sharing the Burden of Adaptation Financing: Translating Ethical Principles into Practical Policy.* IVM Report R08/05. Amsterdam: Institute for Environmental Studies (IVM).

EACC (2009). *The Costs to Developing Countries of Adapting to Climate Change: New Methods and Estimates, The Global Report of the Economics of Adaptation to Climate Change Study.* Washington, D.C.: World Bank, available at http://siteresources. worldbank.org/INTCC/Resources/EACCReport0928Final.pdf.

Ellis, J. and Kamel, S. (2007). *Overcoming Barriers to Clean Development Mechanism Projects.* Paris: Organisation for Economic Co-operation and Development & International Energy Agency (IEA).

FCCC (1992). *United Nations Framework Convention on Climate Change.* Signed 9 May 1992, in New York, NY; entered into force 21 March 1994. Reprinted in (1992). *International Legal Materials* **31**(4), 849.

FCCC Secretariat (2007a). *Investment and Financial Flows to Address Climate Change.* Bonn: Secretariat to the United Nations Framework Convention on Climate Change. Available online at http://unfccc.int/files/cooperation_and_support/ financial_mechanism/application/pdf/background_paper.pdf.

   (2007b). *Decision 7/CP.13 on Additional Guidance to the Global Environment Facility, Report of the 13ᵗʰ Conference of the Parties, Bali, 3–15 December 2007*, UN Document FCCC/CP/2007/6/Add. 1 of 14 March 2008.

   (2008a). *Investment and Financial Flows to Address Climate Change: An Update. Technical Paper.* UN Document FCCC/TP/2008/7, 26 November 2008. Available online at http://unfccc.int/resource/docs/2008/tp/07.pdf.

   (2008b). *Analysis of Possible Means to Reach Emission Reduction Targets and of Relevant Methodological Issues: Technical Paper.* UN Document FCCC/TP/2008/2, 6 August 2008. Bonn: Secretariat to the United Nations Framework Convention on Climate Change. Available online at http://unfccc.int/resource/docs/2008/tp/02.pdf.

G-77 and China Proposal (2008). *Financial Mechanisms for Meeting Financial Commitments Under the Convention*, 25 August 2008. Bonn: Secretariat to the United Nations Framework Convention on Climate Change.

Government of Argentina (2008). *Argentina's Views on Enabling the Full, Effective, and Sustained Implementation of the Convention through Long-Term Cooperative Action Now, up to, and beyond 2012.* Available online at http://unfccc.int/files/ kyoto_protocol/application/pdf/argentinabap300908.pdf.

Government of China (2008). *China's Views on Enabling the Full, Effective and Sustained Implementation of the Convention through Long-Term Cooperative Action Now, up to and beyond 2012.* Available online at http://unfccc.int/files/kyoto_protocol/ application/pdf/china_bap_280908.pdf.

Government of India (2008). *Government of India Submission on Financing Architecture for Meeting Financial Commitments under the UNFCCC.* Available online at http:// unfccc.int/files/kyoto_protocol/application/pdf/indiafinancialarchitecture171008. pdf.

Government of Switzerland (2008). *Funding Schemes for Bali Action Plan: A Swiss Proposal for Global Solidarity in Financing Adaptation*, Ad Hoc Working Group on Long-Term Cooperative Action Under the Convention (AWG-LCA), 21 August 2008. Available online at http://unfccc.int/files/kyoto_protocol/application/pdf/ Switzerland_funding. pdf.

Gupta, J. (1995). The Global Environment Facility in its North–South context. *Environmental Politics*, **4**(1), 19–43.

(1997). *The Climate Change Convention and Developing Countries: From Conflict to Consensus?* Dordrecht: Kluwer Academic Publishers.

(1998). Leadership in the Climate Regime: inspiring the commitment of developing countries in the post-Kyoto phase. *Review of European Community and International Environmental Law*, **7**(2), 178–88.

(2003). The role of non-state actors in international environmental affairs. *Heidelberg Journal of International Law*, **63**(2), 459–86.

(2004). India and climate change: the challenge of sustainable development, in *Climate Change: Five Years After Kyoto*, ed. V. Grover. Oxford/New Delhi: IBH Publishing, pp. 309–32.

(2006). The Global Environment Facility in its North–South context, in *Contemporary Environmental Politics: From Margins to Mainstream*, ed. P. Stephens, J. Barry and A. Dobson. London: Routledge, pp. 231–53.

(2007). Legal steps outside the Climate Convention: litigation as a tool to address climate change. *Review of European Community and International Environmental Law*, **16**(1) 76–86.

(2009). Climate change, development and evaluation: can flexibility instruments promote sustainable development? in *Evaluating Climate Change and Development*, ed. O. Feinstein and R. van den Berg. New Brunswick, NJ: Transaction Publishers.

Gupta, J. and van Asselt, H. (2007). *Towards a Fair and Equitable Framework for Post-2012 Climate Change Policy: An ICCO Policy Brief.* Utrecht: Interkerkelijke Organisatie voor Ontwikkelingssamenwerking (ICCO).

Gupta, J., van Beukering, P., van Asselt, H. *et al.* (2008). Flexibility mechanisms and sustainable development: lessons from five AIJ projects. *Climate Policy*, **8**(3), 261–76.

Hague Declaration (1989). *Declaration of the Hague*, Meeting of Heads of State, 11 March 1989.

Haites, E., Duan, M. S. and Seres, S. (2006). Technology transfer by CDM projects. *Climate Policy*, **6**(3), 327–44.

Haya, B. (2007). *Failed Mechanism: How the CDM is Subsidizing Hydro Developers and Harming the Kyoto Protocol.* Berkeley, CA: International Rivers.

Henningsen, J. (2008). *EU Energy and Climate Policy: Two Years On*, EPC Policy Paper 55. Brussels: European Policy Centre.

Hess, U. and Syroka, J. (2005). *Risk, vulnerability and development.* Presentation at BASIX Quarterly Review & Insurance Meeting, Hyderabad, 21 October 2005.

Howell, J. (2007). Civil society in China: chipping away at the edges. *Development*, **50**(3), 17–23.

Humphrey, J. (2004). The Clean Development Mechanism: how to increase benefits for developing countries. *IDS Bulletin*, **35**, 84–9.

IPCC-1 (1990). *Climate Change: The IPCC Scientific Assessment.* Cambridge: Cambridge University Press.

Juma, C. (1995). Review, in *Research Project Reviews and Commentaries of the Report: International Policies to Address the Greenhouse Effect.* Working Paper 4. Amsterdam: Department of International Relations and Public International Law (University of Amsterdam) and Institute for Environmental Studies (VU University Amsterdam), pp. D1–9.

KP (1997). Kyoto Protocol to the United Nations Framework Convention on Climate Change. Signed 10 December 1997, in Kyoto; entered into force 16 February 2005. Reprinted in (1998) *International Legal Materials* **37**(1), 22.

Krey, M. (2005). Transaction costs of unilateral CDM projects in India: results from an empirical survey. *Energy Policy*, **33**, 2385–97.

Mace, M. J. (2005). Funding for adaptation to climate change: UNFCCC and GEF developments since COP-7. *Review of European Community and International Environmental Law*, **14**(3), 225–46.

Mani, M., Markandaya, A. and Ipe, V. (2008). *Policy and Institutional Reforms to Support Climate Change Adaptation and Mitigation in Development Programs: A Practical Guide*. Washington, D.C.: World Bank.

Marrakesh Accords (2001). Marrakesh Accords and Marrakesh Declaration, 7th Conference of the Parties to the UNFCCC, 29 October–10 November 2001. UN Document FCCC/CP/2001/13. Available online at http://unfccc.int/cop7/documents/accords_draft.pdf.

MCII (2008). Munich Climate Insurance Initiative. Available online at http://www.climate-insurance.org.

Metz, B., Davidson, O. R., Martens, J.-W., van Rooijen, S. N. M. and McGrory, L. v. W., eds. (2000). *Methodological and Technological Issues in Technology Transfer*. Cambridge: Cambridge University Press.

Michaelowa, A. and Michaelowa, K. (2008). Climate or development: is ODA diverted from its original purpose? *Climatic Change*, **84**(1), 5–22.

Michaelowa, A. and Purohit, P. (2007). *Additionality Determination of Indian CDM Projects: Can Indian CDM Project Developers Outwit the CDM Executive Board?* Zurich: Climate Strategies.

Michel, L. (2008). *Written Statement by Mr Louis Michel, Commissioner for Development and Humanitarian Aid, European Commission. Development Committee, 13 April 2008*, Document DC/S/2008–0006. Washington, D.C.: World Bank.

Miller, A. (2008). Financing the integration of climate change mitigation into development. *Climate Policy*, **8**(2), 152–69.

Ming, L., Gupta, J. and Kuik, O. (2008). Will CDM in China make a difference? Paper presented at the Berlin Conference on the Human Dimensions of Global Environmental Change, IHDP Social–Ecological Research Programme on Long-Term Policies: Governing Social–Ecological Change, Berlin, 22–23 February 2008.

Müller, B. (2007). *Nairobi 2006: Trust and the Future of Adaptation Funding*. Oxford: Oxford Institute for Energy Studies.

 (2008). *International Adaptation Finance: The Need for an Innovative and Strategic Approach*. Oxford: Oxford Institute for Energy Studies.

Müller, B. and Hepburn, C. (2006). *IATAL: An Outline Proposal for an International Air Travel Adaptation Levy*. Oxford: Oxford Institute for Energy Studies.

Noordwijk Declaration (1990). Noordwijk Declaration on Climate Change, in *Noordwijk Conference Report: Volume I*, ed. P. Vellinga, P. Kendall and J. Gupta. The Hague: Netherlands Ministry of Housing, Physical Planning and Environment.

OECD DAC (2004). *ODA Eligibility Rules for Expenditure under the Clean Development Mechanism*. DAC Document DCMDAC/Chair(2004)4/final. Paris: Organisation for Economic Co-operation and Development.

Opschoor, H. (2009). *Sustainable Development and a Dwindling Carbon Space*. Public Lecture Series 2009, No. 1. The Hague: Institute of Social Studies.

Oxfam (2007). *What's Needed in Poor Countries, and Who Should Pay?* Oxford: Oxfam International.

Pearson, B. and Shao Loong, Y. (2003). *The CDM: Reducing Greenhouse Gas Emissions, or Relabelling as Usual*. Penang: Third World Network and CDM Watch.

Rio Declaration and Agenda 21 (1992). *Rio Declaration and Agenda 21*. Report on the UN Conference on Environment and Development, Rio de Janeiro, 3–14 June 1992, UN Document A/Conf.151/26/Rev. 1 (Vols. 1–III). New York, NY: United Nations.

Roberts, J. T., Starr, K., Jones, T. and Abdel-Fattah, D. (2008). *The Reality of Official Climate Aid*. Oxford Energy and Environment Comment November 2008. Oxford: Oxford Institute for Energy Studies.

Sands, P. (1992). The United Nations Framework Convention on Climate Change. *Review of European Community and International Environmental Law*, **1**(3), 270–7.

Schneider, L. (2007). *Is the CDM Fulfilling Its Environmental and Sustainable Development Objectives? An Evaluation of the CDM and Options for Improvement*. Berlin: Öko-Institut.

Schroeder, M. (2008). Long-term capacity development for local climate governance: what next after PDD development? Paper presented at the Berlin Conference on the Human Dimensions of Global Environmental Change, IHDP Social–Ecological Research Programme on Long-Term Policies: Governing Social–Ecological Change, Berlin, 22–23 February 2008.

Srivastava, L., Ramanathan, K., Hasan, S. *et al.* (2007). *Modernising the Indian Electricity Sector*. New Delhi: The Energy Research Institute Publications.

Stern, N. (2007). *The Economics of Climate Change: The Stern Review*. Cambridge: Cambridge University Press.

Stockholm Declaration (1972). *Report of the UN Conference on the Human Environment*, Stockholm, 5–16 June, 1972; UN Document A/CONF.48/14/Rev. 1. New York, NY: United Nations.

Sutter, C. and Parreño, J. C. (2007). Does the current Clean Development Mechanism (CDM) deliver its sustainable development claim? An analysis of officially registered CDM projects. *Climatic Change*, **84**(1), 75–90.

SWCC Scientific Declaration (1990). *Scientific Declaration of the Second World Climate Conference*. Geneva: World Meteorological Organization.

Tangen, K., Heggelund, G. and Buen, J. (2001). China's climate change positions: at a turning point? *Energy and Environment*, **12**(2/3), 237–52.

Tirpak, D. and Adams, H. (2008). Bilateral and multilateral financial assistance for the energy sector of developing countries. *Climate Policy*, **8**(2), 135–51.

Toronto Declaration (1988). Conference statement of the conference The Changing Atmosphere: Implications for Global Security, organized by the Government of Canada, Toronto, 27–30 June 1988.

UNDP (2007). *Fighting Climate Change: Human Solidarity in a Divided World. UNDP Human Development Report 2007–2008*. New York, NY: Palgrave Macmillan.

UNGA (1988). *United Nations General Assembly Resolution 43/196 of 20 December 1988 on a United Nations Conference on Environment and Development*, UN Document A/RES/43/196.

van Asselt, H. and Gupta, J. (2009). Stretching too far? Developing countries and the role of flexibility mechanisms beyond Kyoto. *Stanford Environment Law Journal*, **29**(2), 311–78.

van Drunen, M., Bouwer, L., Dellink, R. *et al.* (2009). *Financing Adaptation in Developing Countries: Assessing New Mechanisms*. Netherlands Research Programme on Scientific Assessment and Policy Analysis for Climate Change (NRP-CC-WAB) Report 500102025. Available online at http://www.rivm.nl/bibliotheek/rapporten/500102025.pdf.

Wara, M. W. and Victor, D. G. (2008). *A Realistic Policy on International Carbon Offsets*. Rep. PESD Working Paper 74. Stanford, CA: Programme on Energy and Sustainable Development (PESD).

WCED (1987). *Our Common Future: Brundtland Report of the World Commission on Environment and Development*. Oxford: Oxford University Press.

Winkler, H. and Thorne, S. (2002). Baselines for suppressed demand: CDM projects contribution to poverty alleviation. Paper presented at the Annual Conference of the Forum for Economics and Environment, February 2002, Cape Town, South Africa.

World Bank (2006). *Clean Energy and Development: Towards an Investment Framework*. Washington, D.C.: World Bank.

Yamin, F. (2005). The European Union and future climate policy: is mainstreaming adaptation a distraction or part of the solution? *Climate Policy*, **5**, 349–61.

Zhang, Z. (2006). Toward an effective implementation of Clean Development Mechanism projects in China. *Energy Policy*, **34**, 3691–701.

6

Incorporating climate change into EU development cooperation policy

NICOLIEN VAN DER GRIJP AND THIJS ETTY

6.1 Introduction

This chapter investigates the extent to which the European Union (EU)[1] has incorporated climate change into its development cooperation policy. It describes the EU's development cooperation policy, its evolution and its main principles (see Section 6.2), and focuses on the process of delivering aid (see Section 6.2.5). Subsequently, this chapter explains how the policy areas of climate change and development cooperation have become linked over time (see Section 6.3), elaborates on how current EU policy incorporates climate change into development cooperation (see Section 6.3.2), and outlines the policy tools that are employed to incorporate environmental issues in general, and climate change in particular, into development cooperation programming (see Section 6.3.3). Finally, it assesses the current status of climate incorporation in EU development cooperation (see Section 6.4), and draws conclusions (see Section 6.5).

6.2 The EU and its development cooperation policy

6.2.1 The evolution of EU development cooperation policy

The EU, represented by the European Commission, is a unique actor in development cooperation. It is both a bilateral donor (providing direct support to developing countries (DCs)), and a multilateral organization with a coordinating role for the development aid policies of its 27 Member States (see also Chapters 2, 4 and 7).[2]

---

[1] Other than in quotations, this chapter will for simplicity refer to the European Union (EU) rather than the European Community (EC), even when discussing what are strictly (still) elements of 'EC law'. This terminological simplification also anticipates the replacement and succession by the EU of the EC's legal capacity, together with its legal personality in international law, following the entry into force of the 2007 Lisbon Treaty (Art. 1(3) Lisbon Treaty).
[2] The OECD DAC refers to the latter role as a 'federating' function. See OECD DAC (2007).

Mainstreaming Climate Change in Development Cooperation: Theory, Practice and Implications for the European Union, ed. Joyeeta Gupta and Nicolien van der Grijp. Published by Cambridge University Press. © Cambridge University Press 2010.

Whereas, since the very inception of European unity, in 1957,[3] the Member States have supported the economic and social development of their (former) colonies and dependent territories (see Section 6.2.3), and development cooperation was politically launched as a policy area by the Paris Summit in 1972 (together with EU environmental policy),[4] it was not until 1992 that a constitutive legal framework for EU global development cooperation was adopted. The Maastricht Treaty on European Union incorporated a new, dedicated section into the Treaty, establishing the European Community (EC): currently Title XX, comprising Articles 177–81 EC.

Article 177(1) EC sets out the objectives of EU development cooperation policy:

Community policy in the sphere of development cooperation, which shall be complementary to the policies pursued by the Member States, shall foster:

• the sustainable economic and social development of the developing countries, and more particularly the most disadvantaged among them,
• the smooth and gradual integration of the developing countries into the world economy,
• the campaign against poverty in the developing countries.

Moreover, the Maastricht Treaty instituted the principles of coordination and complementarity: Article 180 EC stipulates that the Community and its Member States are to coordinate their development cooperation policies.

The provisions of the Development Cooperation Title XX have stayed basically the same in the amendment Treaties of Amsterdam (1997) and Nice (2001), save for minor changes.[5] This may soon change. The Lisbon Treaty (signed in 2007) entails a substantial rewriting of these provisions, building on the draft amendments in the failed treaty establishing a constitution for Europe (2004). It would set development cooperation squarely within the framework of the EU's external action, and would upgrade the status of poverty eradication to the only primary objective of EU development cooperation policy.[6] In the General Provisions on the Union's External Action (Title V) it is furthermore set out that the EU is to foster the sustainable economic, social and environmental development of developing

---

[3] The Treaty of Rome establishing the European Economic Community (EEC), signed 25 March 1957.

[4] In October 1972, the Heads of State and Government of the then six EEC Member States and the accession candidates meeting in Paris endorsed the Commission's 1971 *Memorandum on a Community Development Cooperation Policy*, which proposed to extend the existing development policy of 'association with dependent overseas countries and territories' to development cooperation for developing countries more broadly. The summit also adopted the Commission's first environmental action programme, although at that time (similarly to the status of development cooperation) there was not yet a legal basis in the Treaty for environmental policy; this was introduced by the Single European Act (SEA) amendment treaty, in 1986.

[5] Most importantly, the change of the decision-making procedure (to co-decision) in Article 179 EC.

[6] According to Williams (2005: 361–70), deleting the reference to the least-developed states constitutes a sweeping change (if not *de facto* at least *de jure*), and harbours the risk of the poorest countries/LDCs 'losing out', or at least losing their priority claim to 'sustainable economic and social development' under Article 177(1) EC.

countries. The environmental aspect of sustainable development was previously lacking in Article 177(1) EC, although the integration principle in Article 6 EC did 'compensate' for this (see Section 6.3.1).

While the treaty's legal framework has not changed to date, the actual development cooperation *policy* has been drastically reformed since the late 1990s. This process was initiated by the Development Council in response to the highly critical outcomes of an evaluation of EU development instruments and programmes, which it had requested in June 1995. Three reports had highlighted the lack of an overall strategy for EU development policy, as well as the multiplicity and ambiguity of its objectives, and the lack of any hierarchical order therein.[7] Other major points of critique included the long delays in policy implementation and the rigidity and slowness of Commission procedures.

With a view to improving policy performance, the Council invited the Commission to elaborate a proposal for an overall policy statement and action plan that would enhance the relevance, efficiency, effectiveness and visibility of EU development assistance.[8] In response, the Commission published the European Community Development Policy in April 2000, aiming to reduce and eventually eradicate poverty, within the broader context of sustainable development.[9] The environment was identified as a supportive, cross-cutting issue, which needs to be integrated into all thematic priority themes in order to make development sustainable.[10] In concrete terms, this reform aimed to process pledged funds more effectively, both in the EU's own internal procedures and in the implementation of the development programmes in practice. To this end, the reform encompassed improvements in international administrative procedures and closer coordination with other donors on the ground and in multilateral fora. Furthermore, the management of development programmes has been decentralized from the EU's headquarters to the dozens of Commission delegations abroad.

In order to elaborate upon sustainable development, the Commission articulated a proposal for an EU strategy comprised of both an internal dimension (A Sustainable Europe for a Better World: A European Union Strategy for Sustainable Development)[11] and, following the request of the 2001 Göteborg European Council, an

---

[7] Global Evaluations Reports: Evaluation of European Union Aid Managed by the Commission to ACP Countries, Commission ref. ACP 951338; Evaluation of EU Development Aid to Asia and Latin American Countries, Commission ref. ASA 951401; and Evaluation of Aspects of EU Development Aid to the Mediterranean Region, Commission ref. MED 951405. Previously, similar criticisms had been put forth in the OECD DAC's periodic peer-review process.

[8] Development Council Conclusions, 21 May 1999.

[9] Communication from the Commission to the Council and the European Parliament, The European Community's Development Policy, COM(2000)212 final, 26 April 2000.

[10] But, see Williams (2005: 320–1): 'the Communication seems to be perpetuating exactly the same kind of presentational faults which it earlier admitted it had been widely criticized for'.

[11] COM(2001)264 final, 15 May 2001.

external dimension (Towards a Global Partnership for Sustainable Development).[12] The former Communication contained a number of cross-cutting policy proposals relating to policy coherence and decision-making, and identified a set of headline objectives and specific measures in the four priority areas of *climate change*, transport, public health and natural resources. The latter Communication made a direct plea for stronger global governance and political institutions in order to create an effective counterweight to global market forces, and was prepared specifically with a view to negotiations for the 2002 Johannesburg World Summit on Sustainable Development. However, it should be noted that these two Commission Communications did not constitute or produce an official EU sustainable development strategy, since neither reflects an inter-institutional legislative process, and they were they not formally (fully) endorsed either by the European Council or by the Council of Ministers.[13] This ambiguous formal status of the EU sustainable development strategy was remedied in June 2006, when the Brussels European Council adopted the 'Renewed EU Sustainable Development Strategy' in a single, coherent document, which identified climate change as a cross-cutting issue and included it among the seven 'key challenges' with corresponding targets, operational objectives and actions.[14]

   A major change in the EU's development cooperation policy was heralded in December 2005 by the adoption of the European Consensus on Development, which is still in effect today.[15] This brought, for the first time, a common framework for EU development cooperation at both the Member State and the EU level, agreed jointly by the Commission, the European Parliament and Member States represented in the Council. The European Consensus confirms that the overarching objective of EU development cooperation is poverty eradication in the context of sustainable development, including the pursuit of the Millennium Development Goals (MDGs). In addition to endorsing general values such as good governance, democracy, the rule of law and multilateralism, the Consensus also stressed specific development principles, including notably DC ownership and partnership, political dialogue and the participation of civil society. In explicit acknowledgement of the

[12] COM(2002)82 final, 13 February 2002.
[13] The proposals contained in the 2001 'internal dimension' Communication were only partly endorsed by the Göteborg European Council, and mostly translated into general principles and vague objectives, excluding the Commission's headline objectives and concrete measures in four priority sectors including climate change (see Presidency Conclusions, Göteborg European Council, 15–16 June 2001, paras. 19–30). Hence, only the endorsed basic elements can be said to form an official EU sustainable development 'strategy' reflecting consensus and an inter-institutional process (Commission and Council, though not European Parliament). The 2002 'external dimension' Communication was never even submitted to, let alone endorsed by, the European Council, contrary to the Commission's claims in subsequent communications; according to Pallemaerts (2006: 10–12) it was discussed only by the Environment and Development Councils and again only partly and selectively endorsed. For further details on the EU sustainable development strategy see Pallemaerts (2006).
[14] Presidency Conclusions, Brussels European Council, 15–16 June 2006, Annexed Note 10917/06, para. 13.
[15] Joint statement of 20 December 2005 by the Council and the Representatives of the Governments of the Member States meeting within the Council, the European Parliament and the Commission on European Union Development Policy: 'The European Consensus', OJ C46/1, 24 February 2006.

link between poverty eradication and environmental issues, the 2005 Consensus reinforces the importance of the environment in development policy by upgrading it from a cross-cutting issue under the 2000 European Community Development Policy to one of nine cross-cutting *priority areas* for EU action and funding. With the Consensus, Member States committed to delivering more and better aid, in particular by, inter alia, increasing their ODA to reach the pledged 0.7% of GNI by 2015, increasing budget support, further untying aid, and improving the coordination and complementarity of their development assistance.

Building on these commitments and with a view to reinforcing and implementing the Consensus on Development, in 2007 the Commission and the Member States adopted the EU Code of Conduct on Complementarity and Division of Labour in Development Policy.[16] Despite its 'soft law' nature (it is voluntary, flexible and self-policing), the Code is intended to guide the development cooperation policies of the Commission and the Member States, and may serve as a model for international complementarity standards. With this growing focus on a more logical division of labour, the EU should be able to differentiate further the roles of donor countries in the partner countries, thereby reinforcing aid effectiveness and coherence, and enhancing aid 'ownership' by partner countries, and to avoid spreading the budget too thinly across them.[17]

Both the European Consensus on Development and the supporting Code of Conduct on Complementarity and Division of Labour in Development Policy place aid effectiveness at the centre of their shared vision, which is congruent with the OECD's 2005 Paris Declaration on Aid Effectiveness (see Section 4.3.4). In addition, in the wake of the Consensus, during 2006, the EU Commission and Council adopted a so-called 'EU Aid Effectiveness Package' consisting of three strategic policy documents reinforcing the EU's commitment to aid effectiveness, above and beyond the Paris Declaration commitments.[18] This policy package, bearing the subsequently oft-cited slogan *EU aid: Delivering more, better, and*

---

[16] The Code of Conduct is annexed to the Development Council conclusions of 15 May 2007, Document 9558/07. The Council and the Representatives of the Member States meeting with the Council added the explicit reference to 'complementarity' to the title of the code, which had originally been proposed by a Communication from the Commission to the Council and the European Parliament, EU Code of Conduct on Division of Labour in Development Policy, COM(2007)72 final, 28 February 2007, building on the guiding principles on complementarity and division of labour adopted in the GAERC Council and Member States Conclusions of 16 October 2006 on Complementarity and Division of Labour: Orientation Debate on Aid Effectiveness, Document 14029/06.

[17] At present, the EU has the largest spread of any OECD DAC donor, with ODA being distributed to some 145 partner countries.

[18] Communication from the Commission, EU Aid: Delivering More, Better, and Faster, COM(2006)87 final, 2 March 2006; Communication from the Commission to the Council and the European Parliament, Financing for Development and Aid Effectiveness: The Challenges of Scaling EU Aid 2006–2010, COM(2006)85 final, 2 March 2006; and Communication from the Commission to the Council and the European Parliament, Increasing the Impact of EU Aid: A Common Framework for Drafting Country Strategy Papers and Joint Multiannual Programming, COM(2006)88 final, 2 March 2006.

faster, was endorsed by the Member States in the GAERC in April 2006.[19] Some commentators have described the adoption of the European Consensus on Development, its supporting Code of Conduct and the subsequent Aid Effectiveness Package as indicative of 'a change of course whereby the Member States are more committed to improving the effectiveness of EU aid and the visibility of the EU in international development, including its ability to shape the global agenda' (Carbone, 2008: 340). In this respect, the EU and its Member States may be moving towards a response to what the EU Commissioner for Development and Humanitarian Aid, Louis Michel, cited as 'one of the EU's most central challenges in development cooperation', that is, 'to ensure a coherent and effective approach between 26 [*sic*] different actors, the 25 Member States [*sic*] and the European Commission, with 26 [*sic*] development policies' (CEC, 2005; cited in Holland, 2009: 349–50). However, on the other hand, as the most recent OECD DAC peer review pointed out, '[t]here is a risk that the ambitious, multiple objectives of the Consensus, including expanded political ones, could diffuse a focus on development and undermine longterm strategic priorities' (OECD DAC, 2007: 13, cited in Holland, 2009: 359).

In late 2006, the new policy of the 2005 Development Consensus was strengthened by a complementary Regulation establishing a new EU Development Cooperation Financing Instrument.[20] This new legislation streamlined the existing array of some 35 financial instruments into a more manageable set of 10, and aligned the programming cycles of the main budget lines for development cooperation, to simplify harmonization and management of funds.[21] A further consolidation may occur in 2013, when most current financial instruments expire.

6.2.2 Policy coherence for development

Evidently, the objectives of global eradication of poverty and the sustainable development of DCs cannot be achieved through development assistance alone, as long as the needs and interests of DCs are not also taken into account in non-development aid policies of donor ICs. In fact, development cooperation efforts may be undermined by incoherencies with such non-aid policies as macro-economic policies, trade, export credits, foreign direct investment, migration, the arms trade, agriculture, fisheries, environment, climate change, etc. In view of this reality, since the early 1990s, the OECD DAC has sought to develop initiatives to increase coherence across all such policies towards the attainment of development cooperation objectives, but initially

---

[19] GAERC conclusions of 10–11 April 2006, Document 7939/06, with the conclusions on the Aid Effectiveness Package as an Annex.

[20] Regulation (EC) No. 105/2006 of the European Parliament and of the Council of 18 December 2006 establishing a Financing Instrument for Development Cooperation, OJ L 378/41, 27 December 2006.

[21] The main multi-annual instruments, namely the European Development Fund (EDF) and the Development Cooperation Instrument (DCI), which are discussed in Section 6.2.5 below, will now both run until 2013.

without much success (Carbone, 2008: 324; citing Forster and Stokke, 1999).[22] In the EU, the concept of policy coherence for development had been formally institution-alized in EU law by the 1992 Maastricht Treaty, but its operationalization had been lacking.[23] According to Article 178 EC Treaty, the EU 'shall take account' of the development cooperation objectives laid down in Article 177 EC Treaty in its other policies 'which are likely to affect' DCs and, moreover, Article 3 EU Treaty stipulates that 'the Union shall in particular ensure the consistency of its external activities as a whole in the context of its external relations, security, economic and development policies. The Council and the Commission shall be responsible for ensuring such consistency and shall cooperate to this end.'[24]

However, only in 2000, following the adoption of the MDGs did the concept of policy coherence for development gain headway, as it became obvious that development aid alone would be insufficient to meet these goals. The EU became a global frontrunner in this area,[25] culminating in the formulation of a central position for policy coherence for development in the December 2005 European Consensus on Development. The groundwork for this had been laid by the Commission in its April 2005 Communication on Policy Coherence for Development.[26] The Commission identified 11 non-aid policy areas (trade; environment; security; agriculture; fisheries, the social dimension of globalization, employment and decent work; migration; research and innovation; the information society; transport; and energy), which directly or indirectly impact on the attainment of the MDGs, and established specific 'coherence for development commitments' for each of these priority areas. In May 2005, the Council endorsed this ambitious policy agenda in the context of its MDG-attainment package for the 2005 'Millennium +5' Summit, and added one further priority area for policy coherence for development: climate change.[27]

---

[22] The first discussions on policy coherence for development were held at the OECD DAC High-Level Meeting in December 1991. Since 2000, the DAC has included a separate chapter on policy coherence in its peer reviews of the development policies of its members.

[23] Within the EU, policy coherence for development emerged as a part of the Maastricht Treaty's so-called '3Cs package' of complementarity, coordination and coherence. Besides coherence in the sense of policy coherence for development, complementarity refers to the shared (parallel) competences of the EU and its Member States in development policy; coordination denotes the need for the EU and its Member States to coordinate their parallel development policies and consult on their aid programmes and international cooperation commitments and negotiations.

[24] Note that the Lisbon Treaty further strengthens the EU's commitment to policy coherence.

[25] See for a detailed discussion, including overviews of specific policy sectors, the special issue on 'Policy Coherence and EU Development Policy' of the *Journal of European Integration* (2008), **30**(3), 323–427.

[26] Communication from the Commission to the Council, the European Parliament and the European Economic and Social Committee, *Policy Coherence for Development: Accelerating Progress towards Attaining the Millennium Development Goals*, COM(2005)134 final, 12 April 2005. This document formed a package with two further Communications of the same date: *Accelerating Progress towards Achieving the Millennium Development Goals: The European Union's Contribution*, COM(2005)132; and *Financing for Development and Aid Effectiveness*, COM(2005)133.

[27] GAERC conclusions of 23 May 2005 on the Millennium Development Goals, Document 9266/05.

The progress of the policy coherence initiative is to be monitored and reviewed by the Commission in biannual reports. In September 2007, the Commission published the first such evaluation. While noting that a certain amount of progress had been made, particularly in the areas of trade, agriculture and fisheries, the Commission concluded critically that policy coherence for development is still at an early stage in the EU, including as regards the process of mainstreaming climate change into development cooperation.[28] Specifically with regard to climate change, the evaluation concluded from an analysis of the new CSPs of African, Caribbean and Pacific (ACP) countries that the recognition of climate change as an important issue for policy coherence for development is 'very low' and that it is 'one of the least frequently mentioned issues', with only six CSPs addressing it as a relevant issue for policy coherence for development.[29] Building on this evaluation and with a view to accelerating progress, a 2008 Commission report identified the policy areas of climate change and energy policies, migration and research as priority sectors with great development potential, and made specific proposals designed to strengthen synergies with EU development policies.[30]

6.2.3 Special relationships

Whereas the EU provides development funding to almost all DCs,[31] it has special relationships with some, including the ACP countries and the countries in the Mediterranean basin. The cooperation with the ACP countries dates back to the early years of the EU when the Member States expressed solidarity with their colonies and overseas countries and territories, and aimed to contribute to their prosperity through the creation of an association. The ACP–EU relations were governed initially by the Yaoundé Convention and subsequently by the Lomé Conventions (I–IV). In 2000, these were replaced by the Cotonou Agreement, which was concluded for a 20-year period.[32]

---

[28] Commission Working Paper, *EU Report on Policy Coherence for Development*, COM(2007)545, 20 September 2007; accompanied by Commission Staff Working Paper, *EU Report on Policy for Development*, SEC(2007) 1202, 20 September 2007.

[29] Commission Staff Working Paper, *EU Report on Policy for Development*, SEC(2007)1202, 20 September 2007.

[30] Commission Staff Working Paper, *Policy Coherence for Development*, SEC(2008)434, 9 April 2008; with an accompanying Communication from the Commission to the European Parliament, the Council, the European Economic and Social Committee and the Committee of the Regions, *The EU: A Global Partner for Development. Speeding up Progress towards the Millennium Development Goals*, COM(2008)177, 9 April 2008.

[31] With a total of 145 partners, the EU disburses development assistance more widely than any other OECD DAC donor.

[32] Partnership Agreement between the African, Caribbean and Pacific Group of States (ACP), and the European Community and its Members States, signed in Cotonou on 23 June 2000, OJ L 317/1, 15 December 2000. The Cotonou agreement entered into force on 1 April 2003, following its ratification by two-thirds of the ACP states, the EC and its (then) 15 Member States. Its first revision, of June 2005, entered into force on 1 July 2008.

The Euro-Mediterranean Partnership was forged between all EU Member States and the Eastern and Southern Mediterranean countries after the adoption of the Barcelona Declaration in 1995.[33] In 2008, at the Paris Summit for the Mediterranean,[34] this partnership was relaunched as the Union for the Mediterranean. For many years, the implementation of the Euro-Mediterranean bilateral and regional cooperation has taken place through the MEDA programme.[35] However, in early 2007, this programme was replaced by the European Neighbourhood Partnership Instrument (ENPI).[36]

6.2.4 EU development cooperation policy in financial terms

In 2007, the multilateral development aid of the EU, represented by the European Commission, accounted for 10% of the global total of official development assistance (ODA) (CEC, 2008a).[37] The ACP countries received the highest amount of funding, followed by the Mediterranean countries. Sectorally, more than 55% of the disbursements were spent on social and economic infrastructure projects. The specific area of the environment and sustainable management of natural resources received 3.2% of the EU's ODA (CEC, 2008a).

In recent years, the proportion of EU ODA managed by the Commission and destined for low-income countries, including the least developed, has stabilized at approximately 45%, despite the Commission's intention to continue strengthening the poverty focus of ODA (CEC, 2008a).[38] With the mid-term review of the MDGs in 2005, the Commission committed to prioritizing Africa and to increasing the volume of the EU's aid to the African continent by at least 50% by 2010. The OECD DAC's 2007 periodic peer review of the EU's multilateral development cooperation programmes and policies (OECD DAC, 2007) recommended, inter alia, that '[t]he Commission needs to maintain the integrity of its development agenda and to emphasise results. To achieve this, more coherent operation strategies would help assure that poverty eradication, the MDGs and cross-cutting issues, including gender, environment and HIV/AIDS are fully addressed.'

---

[33] Final Declaration of the Barcelona Euro-Mediterranean Ministerial Conference, 27–8 November 1995.

[34] Joint Declaration of the Paris Summit for the Mediterranean, 13 July 2008.

[35] Council Regulation (EC) No. 1488/96 of 23 July 1996 on financial and technical measures to accompany (MEDA) the reform of economic and social structures in the framework of the Euro-Mediterranean partnership, OJ L 189, 30 July 1996.

[36] Major policy documents and legal instruments include Communication from the Commission: European Neighbourhood Policy – Strategy Paper, COM(2004)373 final, 12 May 2004; Regulation (EC) No. 1638/2006 of the European Parliament and of the Council of 24 October 2006 laying down general provisions establishing a European Neighbourhood and Partnership Instrument, OJ L 310/1, 9 November 2006; and Communication from the Commission to the Council and the European Parliament on Strengthening the European Neighbourhood Policy, COM(2006)726 final, 4 December 2006.

[37] This is additional to the bilateral ODA disbursed by the EU Member States, which amounts annually to over half of the global ODA flows.

[38] For a critical assessment of this trend, see Williams (2005).

6.2.5 *The process of delivering aid*

Although the Commission has reformed and simplified its aid administration in recent years, the process of delivering aid is still quite complex (OECD DAC, 2007).[39] At the EU level, the primary services for development cooperation include the Directorate-General for Development (DG DEV), the Directorate-General for External Relations (DG RELEX), the EuropeAid Cooperation Office (EuropeAid), which was established in 2000, and the Directorate-General for Humanitarian Aid (DG ECHO).

The implementation of the European Consensus on Development is expected to make the DG DEV and DG RELEX work more closely together and challenge their historically separate organizational responsibilities. In the words of OECD DAC (2007: 33): 'A more unified Commission organisation for development could minimise institutional redundancy, promote a development policy informed by the full geographic range of development experience and ensure that aid leadership is unambiguously dedicated to the priority issues of development.'

The European Consensus pushes the Commission to work on the principles of engagement, delivery and policy to include objective criteria for resource allocation, improvements in managing harmonization with external donors, headquarters reorganization, devolution of authority to the field, streamlining of internal procedures, reinforcement of quality control and better use of results.

The substantive devolution of management responsibility away from Brussels to the Commission's 81 empowered field delegations has been a key component of the reform process. According to the OECD DAC (2007), this devolution is highly appreciated by the EU's partners in the field and has played a major role in improving the efficiency of its operations. Furthermore, the EU has committed itself to reducing the number of parallel implementation units. However, despite these successful reforms, in its most recent peer review the DAC warned against complacency and suggestions that the reform has been 'completed'. It recommended further enhancing the Commission delegations' responsibilities in prioritizing and applying the thematic programmes, project approval and reporting of results in partner countries.

The main instruments of EU development cooperation include general budget support and thematic and geographically based programmes. Both the European Development Fund (EDF) and the Development Cooperation Instrument (DCI) are implemented by EuropeAid, although the EDF and DCI have different sets of rules and regulations. The EDF is a multi-annual programme for which DG DEV is

---

[39] These complexities are well illustrated by comments made by former Development Commissioner (and Danish development minister) Poul Nielsen, shortly after taking up his post in 2000: 'The Commission machine was never constructed to deliver development assistance. It was designed for producing directives, regulations, trade negotiations, to facilitate political relations between EU states. For development assistance it doesn't work.' Quoted in Olsen (2005: 598).

responsible, and is funded by voluntary contributions from Member States outside the EU budget. It provides support to the 77 ACP countries. The DCI is a major budget stream under the auspices of DG RELEX and is financed directly from the EU's annual budget. It supports development programmes principally in Asia and Latin America.

However, one special aspect of the EU's cooperation with ACP countries, which is the responsibility of DG DEV, is the designation of a National Authorising Officer to ensure that programmes are consistent with the EU's rules and regulations. The National Authorising Officer is often assisted by an 'EDF Cell' composed of both local government staff and special contractors, which in some cases resembles a parallel implementation unit. Such mandatory National Authorising Officers do not exist in the countries under the auspices of DG RELEX, although the latter has increasingly been calling for the involvement of non-state actors in the programming process.[40]

6.3 The linkage of development cooperation and climate change

6.3.1 The evolution of climate change incorporation

The debate about including climate change concerns in EU development cooperation policy originates in earlier discussions about the integration of environmental policy considerations into other EU policies and measures. More precisely, the integration principle, as introduced by the European Commission in the early 1980s, in its third Environmental Action Programme,[41] states that the environmental dimension should be taken into account in the development and implementation of the EU's sectoral policies, including development cooperation policy.

During the 1980s and 1990s, the legal and policy status of this environmental integration principle has been strengthened, first by including it in the environmental title and later by its 'upgrading' to a general principle of European law in Article 6 EC by the Amsterdam Treaty in 1997 (see Dhondt, 2003). On this basis, the Cardiff summit in June 1998 launched a process to promote the integration of the environment into all EU policy areas with a view to promoting sustainable development.[42] At the December 1998 Vienna summit, environmental policy integration was officially extended to include development cooperation.[43] Notably, at both these summits the European Heads of State and Government highlighted climate

---

[40] See e.g. http://www.acp-programming.eu.
[41] Resolution of the Council of the European Communities and of the Representatives of the Governments of the Member States, meeting within the Council, of 7 February 1983 on the continuation and implementation of a European Community policy and action programme on the environment (1982–6), OJ C 46, 17 February 1983.
[42] European Council Conclusions Cardiff Summit, June 1998.
[43] European Council Conclusions Vienna Summit, December 1998.

change as the most obvious example of the need for integration of environmental concerns into other policy areas.

In 1999, the Commission working document *EC Economic and Development Co-operation: Responding to the New Challenges of Climate Change*[44] served as input for a Development Council meeting in November. The Council reaffirmed that the problem of global climate change should be prioritized and invited the Commission to report on the progress made in integrating climate change considerations into EU economic and development cooperation policies, in a document that was also to include an action programme.[45]

In response, in 2003, the Commission published a Communication on Climate Change in the Context of Development Cooperation.[46] It proposed an integrated strategy for addressing climate change and poverty reduction concerns, also to strengthen the ongoing process of integrating the environment into EU development cooperation and the sustainability dimension of EU external policies.[47] With the Communication, the Commission invited the Member States, European Parliament, civil society and other stakeholders to contribute to the formulation and implementation of a coherent and coordinated EU climate change strategy and an action plan for support to partner countries.

In the conclusions from the December 2003 meeting, the Council – in its General Affairs and External Relations Council (GAERC) configuration[48] – welcomed the Communication but requested that further work be undertaken by a Council Expert Group to elaborate the Action Plan.[49] A year later, the GAERC adopted the elaborated Strategy on Climate Change in the Context of Development Cooperation and the accompanying Action Plan 2004–2008 (see Section 6.3.2), reiterating that 'Climate change is a risk to development. Adaptation strategies should seek to manage the risk, thereby supporting developing countries in building resilience to climate change impacts, and protecting national and EU efforts to eradicate poverty.'[50] It further considered that 'mainstreaming of responses to climate change into poverty reduction strategies and/or national strategies for sustainable development is the main avenue to address both adaptation to the adverse effects and mitigation of the causes of climate change'.

---

[44] C(1999)3532, 3 November 1999.   [45] Development Council Conclusions, 11 November 1999.

[46] COM(2003)85 final, 11 March 2003.

[47] See the Commission Staff Working Paper *Integrating the Environment into EC Economic and Development Co-operation*, SEC(2001)609, 10 April 2001; Commission Communication: Towards a Global Partnership for Sustainable Development, COM(2002)82 final, 13 February 2002; and GAC meeting doc. 6927/02.

[48] The GAERC configuration has as a key task the coordination of the work of the nine sectoral Council formations.

[49] GAERC Conclusions on Climate Change in the Context of Development Policy, 5 December 2003.

[50] GAERC Conclusions on the subject of climate change in the context of development cooperation (15164/04), including an action plan to accompany the EU Strategy on Climate Change in the Context of Development Cooperation – Action Plan 2004–2008, 11 November 2004.

In 2007, in order to strengthen its commitment to the Action Plan and as a contribution to the Conference of the Parties to the Climate Convention in Bali, the Commission proposed building a Global Climate Change Alliance (GCCA) between the EU and the poor, most vulnerable developing countries.[51] According to the External Relations Council, this initiative needs to be seen in the broader context of the objective of policy coherence for development, which aims at enhancing synergies between development cooperation and other policies (see Section 6.2.2).[52] Notably, the Council framed the incorporation of climate change into development cooperation as a security issue.[53]

In June 2009, the Environment Council adopted conclusions on integrating the environment into development cooperation.[54] The EU Environment Ministers invited the Commission to 'prepare an ambitious EU-wide environment integration strategy, to be presented to the Council by late 2011' building on the 2005 European Consensus on Development and to 'set up an appropriate framework, consisting of the Commission and the Member States, to prepare and monitor the implementation of the EU approach to environment integration'. Among the priority elements of such a strategy are the enhancement and further development of the 'quality, relevance and use of environmental integration tools' (see Section 6.3.3), taking into account the climate change dimension.

6.3.2 The policy instruments

Moving away from the broader picture of EU law and policy, this section elaborates on the main policy instruments on climate change and development, including the Strategy on Climate Change in the Context of Development Cooperation, its accompanying Action Plan 2004–2008, the GCCA and the Thematic Strategy for the Environment and Sustainable Management of Natural Resources, including Energy (ENRTP).

---

[51] Communication from the Commission, *Building a Global Climate Change Alliance between the European Union and Poor Developing Countries Most Vulnerable to Climate Change*, COM(2007)540 final, 18 September 2007.

[52] External Relations Council Conclusions on a Global Climate Change Alliance, 20 November 2007.

[53] See also Climate Change and International Security, Paper from the High Representative and the European Commission to the European Council, S113/08, 14 March 2008. This paper warned that climate change is intensifying security risks for the EU, threatening to overburden states and regions of the world that are already fragile and prone to conflict, and undermining the efforts to attain the MDGs.

[54] Environment Council Conclusions on Integrating Environment in Development Cooperation, 25 June 2009, with Annex. See also, on incorporating climate change risks in the deployment of such environment integration tools, GAERC Conclusions on climate change ('Contribution to the Spring European Council: Taking into Account the Development Dimension for a Comprehensive post-2012 Climate Change Agreement in Copenhagen'), 16 March 2009, Document 7645/09.

Climate strategy in the context of development cooperation

The objective of the EU Strategy on Climate Change in the Context of Development Cooperation is to assist partner countries in meeting the challenges posed by climate change, in particular by helping them implement the Climate Convention and the Kyoto Protocol.[55] To this end, the Commission considers it imperative fully to mainstream climate change concerns into EU development cooperation in complete coherence with the objective of poverty reduction.

Furthermore, the implementation of the strategy is to be guided by two main principles. The first is the primacy of national ownership, meaning that DCs are the owners and drivers of development strategies and processes, and hence are themselves primarily responsible for identifying and responding to environmental issues. The second principle refers to broad stakeholder participation in the implementation process.

Although the Commission claims to adhere to the principle of national ownership of development strategies, this has not prevented it from presenting suggestions for possible response strategies to EU partner countries. At the same time, it stresses that these strategies merely provide guidance and do not exclude any country from identifying any additional priority and obtaining concomitant development assistance. Whether and to what extent the Commission in practice succeeds in finding a balance between creating a climate and/or environment 'conditionality' on the one hand and respecting the principle of national ownership on the other remains to be seen.

The Commission identified four strategic priorities in its strategy, including *raising the policy profile* of climate change, support for *adaptation*, support for *mitigation* and *capacity development*. These priorities are elaborated below. Interestingly, the strategy document observed that some overlap may exist between these priorities and that, although on the face of it measures for adaptation and mitigation seem to be mutually exclusive, in practice they can have synergistic effects.

The Commission noted in its strategy that climate concerns were absent from almost all of the EU and partner countries' development strategies. It therefore concluded that it was important to *raise the policy profile of climate change*, both in dialogue and cooperation with partner countries and within the EU itself. In this respect, the EU should take advantage of existing institutional frameworks (such as the Cotonou Agreement, the Euro-Mediterranean Partnership, or other bilateral agreements) and its extensive network of delegations and representations, in order

---

[55] GAERC Conclusions on the subject of climate change in the context of development cooperation (15164/04), including an Action Plan to accompany the EU Strategy on Climate Change in the Context of Development Cooperation – Action Plan 2004–2008, 11 November 2004.

to enhance the dialogue on climate change and to support the preparation of national strategies for sustainable development, with climate change as a cross-cutting component.

In addition, the Commission announced that it would start incorporating climate change concerns into all strategic programmes and sectors of EU development cooperation and into other EU internal and external policies having possible impacts on partner countries. Furthermore, the Commission intended to examine the synergies with other international initiatives, inter alia in the areas of desertification, biodiversity, forests and water.[56] Moreover, it aimed to persuade the European Investment Bank (EIB) and the European Bank for Reconstruction and Development (EBRD) to take climate change into consideration. In this respect, it should be noted that in particular the EBRD has had an environmental mandate ever since its inception, and has a policy of making adherence to international environmental standards in addition to national provisions part of the requirements for providing development funding.[57] The EIB is also increasingly incorporating the environmental dimension into development funding decisions, including through the recent establishment of an Environmental and Social Office, with environmental assessment, climate change and social working groups.[58] Furthermore, the EBRD and the EIB have jointly established the Multilateral Carbon Credit Fund (MCCF) as a key instrument in their strategies for combating climate change. Fully subscribed, with Euros 190 million in commitments, the MCCF is one of the few carbon funds dedicated specifically to countries from Central Europe to Central Asia.[59]

The strategy defines *adaptation* as responses that may reduce vulnerability to climate change, involving individual and collective coping and risk management strategies, including adjustment in practices, processes, or structures of systems. The EU aims to promote 'mainstreaming' of adaptation concerns and national action plans for climate change reported in national communications or national adaptation programmes of action (NAPAs), where they exist, into strategic frameworks such as national strategies for sustainable development. In addition, it intends to support the development of tools and capacities for the integration of adaptation concerns into national and sectoral planning. As can be seen, the EU tends to use the terms mainstreaming and integration interchangeably (see Section 6.5 and Chapter 3).

The strategy defines *mitigation* as all interventions to reduce anthropogenic emissions of greenhouse gases, either by reduction at source level or by

---

[56] Some sectors, including trade, export credits, etc., fall under the Policy Coherence for Development initiative (see Section 6.2.2).
[57] Further information is available online at http://www.ebrd.com/enviro.
[58] The Environment and Social Office was established in early 2009, following the approval of the EIB's new EIB Environmental and Social Statement on 3 February 2009. Further information available online at: http://www.eib.org/projects/news/eibs-environment-and-social-office.htm.
[59] Further information is available online at http://www.ebrd.org/mccf.

sequestration through 'sinks'. The Commission emphasized measures in the energy supply, energy use and transport sectors. However, agriculture is remarkably absent from this list of sectoral priorities. Taking also into account the Commission Communication on Energy Co-operation with Developing Countries[60] and the Energy Initiative for Poverty Eradication and Sustainable Development (EUEI) presented by the EU and its Member States at the World Summit on Sustainable Development of 2002, the Commission proposed the full menu of technical and institutional options, including energy efficiency and renewable energies. This is to be achieved through partnerships with DC governments for providing assistance with the development of sustainable energy policies as well as advice on financing opportunities from various sources. In addition, the Commission proposed to support research into alternative fuels.

Interestingly, the strategy stated that the Clean Development Mechanism (CDM) is in principle a good vehicle for the transfer of clean and modern technologies to developing countries while delivering real development benefits, and can provide an economic incentive for the greening of foreign direct investment (FDI). However, since most CDM activities are primarily driven by the private sector and fare poorly in terms of the sustainable development component, geographical spread, the additionality criteria and technology transfer (see Section 5.5), the Commission argued that some public funding might be necessary to ensure that these activities fulfil all sustainability requirements and are geographically well balanced. Using ODA for project-preparation activities, including capacity building of the host developing country, could be an important step in that direction. However, the Commission *'warns that this should not result in a diversion of ODA'*.[61] The strategy therefore recommended the development of an EU code of conduct for the use of ODA to finance CDM activities (cf. Section 5.5).

The strategy defines *capacity development* as improving the overall performance and functional capabilities of organizations, enhancing the ability to adapt, and improving the participation of DCs in future international negotiations. The 2003 Communication stated that the EU's Sixth Framework Programme for Research and Technological Development (FP6)[62] should be a vehicle for partner countries to develop knowledge, tools and methodologies that are pertinent to climate change and planned adaptation. Therefore, the programme provided opportunities for partner countries to participate in energy, transport and climate change research projects.

---

[60] COM(2002)408 final, 17 July 2002.   [61] Emphasis in the original (p. 18).
[62] Decision No. 1513/2002/EC of the European Parliament and of the Council of 27 June 2002 concerning the Sixth Framework Programme of the European Community for Research, Technological Development and Demonstration Activities, contributing to the Creation of the European Research Area and to Innovation (2002 to 2006), OJ L 232/1, 29 August 2002.

Table 6.1 *Strategic objectives of the EU Action Plan in combination with specific aims*

Strategic objective	Specific aim
1. Raising the policy profile of climate change	1.1 Dialogue and cooperation with partner countries 1.2 Dialogue and cooperation within the Community and with other donors
2. Support for adaptation	2.1 Support DCs in integrating climate risk management into planning processes 2.2 Partner countries develop research on impacts, vulnerability and adaptation
3. Support for mitigation and low-GHG development paths	3.1 Support DCs to integrate the pursuit of low-GHG development paths into the planning process 3.2 Support partner countries to benefit from thediffusion of environmentally sound technologies 3.3 Encourage the private sector to invest in mitigation and low-GHG development
4. Capacity development	4.1 Raise public awareness in partner countries 4.2 Development of human and institutional capacities in partner countries for the implementation of the Climate Convention and the Kyoto Protocol
5. Monitoring and evaluation of the action plan	5.1 Ongoing monitoring of the action plan 5.2 Carry out biannual evaluation activities

Source: GAERC, Action Plan 2004–2008.

Action Plan 2004–2008 on Climate Change and Development

With a view to implementing the above four strategic objectives of the 2003 EU Strategy on Climate Change in the Context of Development Cooperation, in November 2004 the Council adopted an Action Plan 2004–2008 on Climate Change and Development.[63] The Action Plan lists specific aims, actions and sub-actions in relation to the Strategy's four objectives (see Table 6.1). It should be noted that the Action Plan does not establish a new funding line or package with which to implement the specified actions. Rather, funding will come primarily from the ENRTP programme (discussed below) and through geographical funds raised at country and regional level.

---

[63] GAERC Conclusions on the subject of climate change in the context of development cooperation (15164/04), including an Action Plan to accompany the EU Strategy on Climate Change in the Context of Development Cooperation – Action Plan 2004–2008, 11 November 2004.

The Global Climate Change Alliance (GCCA)

As mentioned above (see Section 6.3.1), the Commission established a GCCA between the EU and the Least Developed Countries (LDCs) and the Small Island Developing States (SIDSs).[64] The Commission budgeted Euros 60 million for the period 2008–10 through its established channels for political dialogue and cooperation at national level and the ENRTP programme (discussed below), rather than setting up a new fund or governance structure specifically for the GCCA.[65] Despite the Commission's expectation that the elaboration of an implementation framework for the GCCA, in 2008, would inspire additional financing by Member States (CEC, 2008b), to date only Sweden and the Czech Republic have pledged funding for the GCCA (Euros 5.5 million and Euros 1.2 million, respectively).[66] In response to the Commission's proposal to launch the GCCA, the European Parliament adopted in 2008 a resolution welcoming the initiative but criticizing the current budget as 'woefully inadequate' (EP, 2008). Evidently, the Parliament propagated a much stronger European commitment to the mainstreaming of climate change into development cooperation. It called on the Commission to place climate change at the core of its development cooperation policy, pointing out that the funding allocated should be complementary to the ongoing process in the Climate Convention, and urged the Commission to establish a long-term financing goal of at least Euros 2 billion annually by 2010 and Euros 5–10 billion annually by 2020. To this end, the Commission and the Member States are called upon to earmark at least 25% of expected revenues from auctioning within the EU Emissions Trading Scheme (EU ETS) during the next trading period to funding of the GCCA and other climate change measures in developing countries (EP, 2008: paras. 10 and 19). Moreover, the EP promoted the involvement of the private sector as a close partner to the GCCA, recognizing that public money could play a catalysing role by incentivizing investments and delivering access to markets and technology (EP, 2008: para. 22). Spurred by the Council to consider innovative means of financing for the GCCA, the Commission is working in particular on plans for a Global Climate Financing Mechanism (GCFM) (CEC, 2008c). This mechanism, which would be based on the issuance of bonds, would frontload grants to support adaptation in vulnerable countries using guarantees linked to future ODA commitments and in particular

---

[64] Communication from the Commission, *Building a Global Climate Change Alliance between the European Union and Poor Developing Countries Most Vulnerable to Climate Change*, COM(2007)540 final, 18 September 2007.

[65] According to the GCCA Implementation Framework (CEC, 2008b), the Commission plans to earmark Euros 25 million annually in 2009 and 2010, following its Euros 10 million allocation for 2008. The total EU funding specifically targeting one or more of the GCCA priority areas, i.e. geographical (10th EDF), thematic (food security, environment) and research programmes, is projected to total at least Euros 300 million between 2008 and 2012.

[66] Several other Member States, including notably the UK, have considered but postponed decisions on additional GCCA funding, awaiting evidence of the 'added value' of the initiative compared with other (existing) options.

to revenues from EU ETS auctioning of emission rights as supported by the EP (CEC, 2008b: 2; EP, 2008). The Commission aims to raise about Euros 1 billion per year for the period 2010–14 through the GCFM.[67]

The GCCA initiative is complementary to, and supportive of, the ongoing process within the climate regime. According to the Commission, 'the GCCA's added value lies in its European approach and in its objective to start developing immediately an innovative way of addressing climate change and its effects in most vulnerable developing countries' (CEC, 2008b: 4). The Alliance aims to provide a platform for dialogue and exchange to help DCs link development strategies and climate change and to participate in global climate change mitigation activities that contribute to poverty reduction. Furthermore, the initiative distinguishes five priority areas for effective cooperation: adaptation, reducing emissions from deforestation (REDD), enhancing participation in the CDM, promoting disaster risk reduction and integrating climate change into poverty reduction efforts.

The GCCA became fully operational in 2008. In its first year, four countries were identified to start up activities under this initiative: Vanuatu, the Maldives, Cambodia and Tanzania. A further 11 countries have been identified for cooperation during 2009 and 2010. A GCCA Support Facility will be set up to support national and regional capacity building and technical assistance measures for DCs to improve their knowledge base on the expected impacts of climate change, to integrate climate change vulnerability effectively into development plans and budgets, to identify and prepare GCCA activities in particular sectors, and to facilitate dialogue activities.

Thematic Strategy for the Environment (ENRTP)

In 2007, the Commission launched a Thematic Strategy for the Environment and Sustainable Management of Natural Resources, including Energy (ENRTP).[68] The ENRTP *strategy* is the first multi-annual implementing instrument of the recently established ENRTP *programme*,[69] the environment-oriented funding line under the auspices of the Development Cooperation Instrument (DCI).[70] The ENRTP programme, with a global budget of Euros 804 million, is currently the main financial instrument for climate-change-related funding in EU development cooperation, including notably the abovementioned Action Plan 2004–2008, and the GCCA.

---

[67] See Communication from the Commission to the European Parliament, the Council, the European Economic and Social Committee and the Committee of the Regions, *Towards a Comprehensive Climate Change Agreement in Copenhagen*, COM(2009)39 final, 28 January 2009.

[68] Commission Decision of 20 June 2007 establishing the ENRTP Thematic Strategy (2007–10).

[69] Communication from the Commission to the Council and the European Parliament, *External Action: Thematic Programme for Environment and Sustainable Management of Natural Resources including Energy*, COM(2006) 20 final, 25 January 2006.

[70] Article 13 DCI. The ENRTP programme has been allocated an (indicative) total amount of Euros 469.7 million of DCI funding for the period 2007–10. According to Behrens (2008: 10), roughly half of this funding is (at least partly) related to climate-change-related initiatives.

The ENRTP strategy for the period 2007–10 prioritizes (1) assisting DCs in achieving the MDG on environmental sustainability; (2) promoting implementation of EU initiatives and helping DCs to meet internationally agreed environmental commitments; (3) promoting coherence between environmental and other policies and enhancing environmental expertise; (4) strengthening international environmental governance and policy development; and (5) supporting sustainable energy options in partner countries.

Hence, the four-year ENRTP strategy addresses environmental challenges that affect poor people, including rapidly degrading key ecosystems, climate change, poor global environmental governance and inadequate access to and security of energy supplies. The multi-annual strategy is implemented through annual action programmes.[71] The policy context for this strategy has changed significantly since it was first proposed, owing to the tremendous increase in attention and funding for climate change issues. Funding has surged from Euros 3 million in 2007 to more than Euros 16 million in 2008, of which some Euros 10 million are reserved for the GCCA (CEC, 2008a).[72] From this amount, Euros 3 million will be used to set up a GCCA Support Facility, hold events, and conduct analyses and studies, whereas Euros 7 million will be used to support pilot actions in a limited number of pilot countries addressing adaptation and/or mitigation measures. In addition, the ENRTP focuses on forestry initiatives and sustainable energy options in partner countries, building on previous funding lines that it replaced in these sectors.[73]

6.3.3 Environmental integration tools in development cooperation

A range of tools have been developed with potential use for incorporating the environment and climate change into policy making and programming for development cooperation. This section outlines the main environmental integration tools designed for use in the crucial programming phase and in the EU's aid delivery approaches, including the Country and Regional Strategy Papers (CSPs and RSPs), the Country Environmental Profiles (CEPs), Strategic Environmental Assessments (SEAs) and Environmental Impact Assessments (EIAs). This section closes with a

---

[71] The Annual Action Programme 2007 was adopted by the Commission on 3 December 2007 and amended on 19 December 2007. The 2008 Annual Action Programme was adopted on 7 August 2008 and amended on 23 December 2008. The 2009 Annual Action Programme was adopted on 14 May 2009.

[72] In addition, the Commission has allocated Euros 25 million for the GCCA in both 2009 and 2010. The Swedish government pledged an additional Euros 5.5 million to support actions in partner countries in 2008.

[73] LIFE Third Countries and Intelligence Energy Europe; Cooperation on Energy in Developing Countries IEE – COOPENER, which expired at the end of 2006; and Regulation (EC) No. 2494/2000 of the European Parliament and of the Council of 7 November 2000 on Measures to Promote the Conservation and Sustainable Management of Tropical Forests and Other Forests in Developing Countries, OJ L 288/6, 15 November 2000; and Regulation (EC) No. 2493/2000 of the European Parliament and of the Council of 7 November 2000 on Measures to Promote the Full Integration of the Environmental Dimension in the Development Process of Developing Countries, OJ L 288/1, 15 November 2000.

discussion of the Commission's *Environmental Integration Advisory Services*, which were set up to support and promote the deployment of these integration tools in EU development cooperation; and finally a discussion of the most recent work on tailoring existing environmental integration tools to the specific issue of climate change in development cooperation.

Country and Regional Strategy Papers (CSPs and RSPs)

The programming priorities for EU development assistance are set out in strategic frameworks at both national and regional levels in Country and Regional Strategy Papers (CSPs and RSPs). They set up the political guidelines on the implementation of cooperation policies and are instrumental in guiding, managing and reviewing EU assistance programmes. These tools are intended to contribute to the better planning of cooperation activities, to improve donor coordination and complementarity, and to ensure the overall coherence of external assistance policy with other EU policies, in part through extensive policy dialogue with partner countries and other donors. The following description of the strategy-paper formulation process applies equally to the regional level.

In order to qualify for development aid, partner countries are obliged to draw up CSPs, in collaboration with the Commission. Following the Commission's format,[74] the CSP outlines a partner country's circumstances, needs and development priorities, examining political, economic, social and environmental dimensions. On this basis, the Commission formulates its chosen response strategy, for which the implementation and resource-management approach is elaborated in National Indicative Programmes (NIPs). Hence, the adoption of the strategy for development cooperation between the EU and partner countries, laid down in CSPs/RSPs and NIPs/RIPs, results from an interactive programming and decision-making process organized and directed by the Commission. The active involvement of partner countries in this programming process is key to the EU's development cooperation policy, so as to 'ensure sufficient ownership to facilitate successful implementation', since, according to the Commission, 'strategies which are not owned by the country are generally less effective in the long-term than those developed with the full participation of the country itself (CEC, 2001). Thus, the programming process involves consultation with state and non-state actors of the partner countries, as well

---

[74] Commission Staff Working Paper, *Community Cooperation: Framework for Country Strategy Papers*, SEC (2000)1049. In 2001, the Commission adopted (non-binding) guidelines for the implementation of this common framework (CEC, 2001). In 2006, the Commission revised the CSP format to reflect better the political developments concerning aid effectiveness and donor coordination and harmonization, as well as the EU commitment to sustainable development and environmental integration: *Communication from the Commission to the Council and the European Parliament, Increasing the Impact of EU Aid: A Common Framework for Drafting Country Strategy Papers and Joint Multiannual Programming*, COM(2006)88, 2 March 2006. This revised common framework was subsequently endorsed by the General Affairs and External Relations Council, at its April 2006 meeting.

as with EU Member States and other donors in order to ensure that the EU's overall objectives are in line with the strategies of the countries concerned and to look for possible synergies between the EU action and the interventions of other donors. The programming process must be guided by certain underlying principles including, inter alia, adherence to the regional strategies laid down in RSPs adopted by the regional partner countries and the EU following a similar process of consultation, incorporating cross-cutting issues such as the environment; alignment with a partner country's own policy agenda and a broad stakeholder involvement in the consultation process.

The drafting of a CSP takes between 12 and 18 months and passes through three obligatory stages: first, the drafting of an initial version of the CSP and NIP; second, the assessment of the paper by EU services ('quality control'), including revisions, if necessary; and third, formal approval of the CSP. In order to guide the process, the Commission has established a Common Framework for Strategy Papers that provides minimum requirements for CSPs and a standard format (CEC, 2001). Each CSP/RSP is adopted for a five-year period, with a mid-term review to keep the strategy up to date with regard to current developments in the countries and regions concerned and to adjust the originally formulated priorities where necessary. Close to the expiry period, overall evaluations are undertaken with a view to possible revisions for the succeeding strategy papers.

The first generation of CSPs/RSPs covered the period from 2001 up to 2006, whereas the second generation of CSPs/RSPs is intended to cover the period from 2007 to 2013. The expectation is that there will be a follow-up to these papers after 2013.

Country and Regional Environmental Profiles (CEPs and REPs)

One major innovation for this second generation of CSPs/RSPs, following the Commission's 2006 revision of the Common Framework, is the structural incorporation of an analysis of the environmental situation of the country or region, in the form of Country or Regional Environmental Profiles (CEPs or REPs). According to the Commission, Country Environmental Profiles (CEPs) are 'the key tool' for integrating environmental issues into development programming from the start of the cooperation process (CEC, 2007: 10). These profiles provide an analytical overview of the availability and use of environmental resources, links with poverty and food security, institutional capacity, the policy and regulatory framework, including international agreements and environmental cooperation, and issues arising from climate change. By annexing (the summary of) an environmental profile of the country or region to each CSP or RSP, the environmental dimensions are made to form an integral component of the programming process, by their being integrated into the country/region analysis, response strategies and underlying policy dialogue for multi-annual programming decisions. Regional Environmental Profiles (REPs) focus on (transboundary) environmental issues common to a group of neighbouring countries that can be more effectively addressed through cooperation at the regional level. REPs should inform, but do not replace, CEPs for each country within a region. Draft terms of reference for CEPs and REPs are available from the Commission's Environmental Integration online portal.[75]

Strategic Environmental Assessment (SEA)

Beyond programming, environmental concerns must also be integrated into aid-delivery approaches. Strategic Environmental Assessments (SEAs) are used at the early formulation stage of aid-delivery decision-making and planning, by systematically evaluating the environmental consequences of a partner country's proposed policy, plan or programme (PPP) prior to its adoption. The SEA is a relatively new and still evolving tool. According to the Commission's Environmental Integration online portal, SEA is more of a generic term than a specific tool, in that the form of an SEA can vary as needed, from a broad-brush and more qualitative approach for policy analysis to a more detailed and quantitative approach for the analysis of specific programmes of projects.[76]

The 2005 European Consensus on Development, building on the 2001 Environmental Integration Strategy and Council Conclusions,[77] established a commitment to prepare SEAs for budget and sectoral aid.[78] In accordance with the 2005 Consensus and the Paris Declaration on Aid Effectiveness, current EU policy is promoting a shift away from the project approach towards sectoral and budget aid, thereby enhancing the importance and systematic use of SEAs. Moreover, under the Paris Declaration donors have committed to applying common approaches for SEA at the sector and national level.[79]

An SEA may be directed by the EU, by partner governments or by other donors. Whereas SEAs for national policies and strategies are better directed by those partners, the EU typically drives the SEA procedures to identify and assess linkages between the environment and the Sector Policy Support Programmes (SPSPs) which it supports.[80] In all cases, the Commission stresses the importance of 'full ownership' of the SEA by partner governments and coordination with other donors, and considers the involvement of stakeholders a key success factor.

In contrast to CEPs, the SEA is not a generally mandatory instrument; a structural obligation to undertake an SEA exists only for environmentally sensisitve sectors

---

[75] See the online portal at http://www.environment-integration.eu.

[76] See the online portal at http://www.environment-integration.eu.

[77] Presidency conclusions of the Göteborg European Council Summit, 15–16 June 2001; Development Council conclusions of 31 May 2001 on a strategy on the integration of environmental concerns into EC economic and development cooperation to promote sustainable development; both based on the Commission Staff Working Paper *Integrating the Environment into EC Economic and Development Co-operation*, SEC(2001)609, 10 April 2001.

[78] Reiterated in the 2006 Common Format for Country Strategy Papers, see above.

[79] As a follow-up to the Paris Declaration, in 2006, the OECD DAC published a *Good Practice Guidance on Applying SEA in Development Cooperation*.

[80] See EuropeAid Co-operation Office, *Guidelines for European Commission Support to Sector Programmes, 2003.*

and for certain other specified sectors and strategies. For all other sectors the Commission's *Environmental Integration Handbook* (CEC, 2007) provides general environmental integration guidance.

Environmental Impact Assessment (EIA)

An important tool for environmental integration at individual-project level is the Environmental Impact Assessment (EIA). Through *ex ante* systematic evaluations of the potential environmental impacts of a project, or by comparing alternative project designs, an EIA leads to an Environmental Management Plan (EMP) with recommendations for measures to mitigate a project's negative impacts and to optimize its positive effects, prior to its being carried out. The EU emphasizes the importance of stakeholder participation and consultation in EIA procedures, within the local institutional framework.

In comparison with SEAs, owing to the more specific project level of assessment, EIAs can be more precise and detailed, whereas SEAs may provide a better perspective on interactions or cumulative effects. As with SEAs, a structural obligation to undertake an EIA exists only for projects of a certain type, category, or scale, or where the partner country's national legislation so requires. For the remaining projects the Commission's *Environmental Integration Handbook* (CEC, 2007) provides general environmental integration guidance, in lieu of a full EIA.

The Commission's Environmental Integration Advisory Services

The use of the aforementioned tools in the EU's environmental integration effort in development cooperation is generally supported by the Commission's *Environmental Integration Advisory Services.*[81] The purpose of these services, which were initially established in 2004 as the *Helpdesk for Environmental Integration into EC Development Cooperation,*[82] is to raise awareness and build the capacities of staff and stakeholders, and to provide technical advice on the integration of the environmental dimension generally and climate change specifically into EU development cooperation and into partner countries' sector policies and programmes. Because of limited in-house expertise, the Commission has contracted environmental experts to carry out this work (see also ECA, 2006: 8). One of their main tasks was to finalize the Commission's *Environmental Integration Handbook* (CEC, 2007).[83] This Handbook (also described as the EU 'environmental

---

[81] See the online portal at http://www.environment-integration.eu.

[82] The Helpdesk was renamed and redefined in January 2009 as the 'Advisory services, methodological support and seminars on integrating the environment in development co-operation'. A previous environment helpdesk had been in place between 1999 and 2002.

[83] The 2007 Handbook replaces the 2001 draft edition of the *EC Environmental Manual* and preceding environmental manuals. The final version was originally planned for 2003, with preparations having started as early as 1998. See also ECA (2006: 9).

mainstreaming manual'), which was published in 2007, discusses in detail the tools for integrating the environment into EU development cooperation, and their intended implementation to assist EU staff and partner organizations so as to enable an early consideration of environmental sustainability concerns in country programming and in the preparation of planned operations. Climate change, however, is not a specific focus area of the Handbook.

Draft Climate Change and Environment Integration Guidelines

With a view to the growing importance of the specific issue of climate change in the development context, the Commission's Integration Advisory Services have reviewed and updated the 2007 edition of the *Environmental Integration Handbook*. Among the key results of this revision process are 'Draft Guidelines on the Integration of Environment and Climate Change in Development Cooperation'.[84] These draft guidelines aim to define a comprehensive operational framework to enhance consideration of environment and climate change concerns in EC development cooperation, as a cross-cutting dimension alongside dedicated programmes and projects. Concretely, the guidelines propose the following amendments to the current practices and tools discussed above.

- Programming phase: a new section has been introduced in the standard format for CEPs, looking at climate vulnerabilities as well as national strategies, plans and programmes and institutional capacities to address climate change.
- Sector programme approach: consideration of climate change within SEAs is recommended, making use of the new 'climate change sector scripts' introduced by the Commission Integration Advisory Services in July 2009.[85] These scripts are designed to address possible climate impacts at sector level and concomitant adaptation and mitigation options. An additional purpose is to support political dialogue on climate change implications between the Commission and partner country governments and other national partners involved in EU development cooperation activities. Scripts have been introduced for the sectors/topics of agriculture and rural development (including forestry, fisheries and food security), ecosystems and biodiversity management, education, energy supply, governance, health, infrastructure (including transport), solid-waste management, trade and investment (including technological development, employment and private-sector development), and water supply and sanitation.
- Project approach: a questionnaire is proposed in order to screen environment and climate change, mainly concentrating on actual climate variability and taking into account not only geo-climatic exposure but also sensitivity and adaptive capacity.

---

[84] Pending the internal consultation on the draft Guidelines within the Commission, the document is not publicly available. The description in this chapter is based on personal communications and emails with Peter Brinn, Project team leader of the Commission's Environmental Integration Advisory Services.

[85] The climate change sector scripts, as well as a general introductory Information Note, can be downloaded from the Integration Services portal at www.environment-integration.eu.

In June 2009, an internal consultation was launched within the Commission on the Draft Guidelines, to be followed by a consultation of Member States regarding environmental focal points. The Guidelines are expected to be presented to EuropeAid Management by October/November 2009.[86]

In addition, as mentioned above, the June 2009 Environment Council conclusions on integrating environment into development cooperation invited the Commission to prepare an 'ambitious EU-wide environment integration strategy' to be presented to the Council by late 2011 with the enhancement and further development of the 'quality, relevance and use' of the aforementioned environmental integration tools, specifically taking into account the climate change dimension.

6.4 The current state of climate change 'mainstreaming'

Whereas the previous sections outlined the EU's intentions for climate 'mainstreaming', this section compares intentions and practice concerning the incorporation of climate change into development cooperation. This assessment partly builds on research focused on the integration of environmental concerns into development cooperation policy, which can be considered partly to overlap or run in parallel with that regarding climate change 'mainstreaming'. To date, the research and policy evaluation communities have paid less attention to the specific process of integrating climate change concerns into development cooperation policy. This section aims to fill this gap by assessing the extant literature and audit reports, including notably the special report on integration of environmental concerns into development cooperation by the European Court of Auditors (ECA, 2006), a study on the policy development process of the Action Plan on Climate Change and Development (van Schaik, 2006) and the mid-term review of the Action Plan on Climate Change and Development (Teixeira Santos and de Lopez, 2007).

6.4.1 Assessing the integration of environmental concerns into development cooperation

Both researchers and policymakers increasingly admit that the Cardiff process of integrating environmental policy requirements into other policy areas has largely failed and that the political context for environmental policy integration has changed from facilitative to disruptive (e.g. Wilkinson, 2007). Among the main reasons identified for this failure are a shift in political priorities away from environmental priorities through the mid-term review of the Lisbon strategy; a

---

[86] According to personal communications/emails from Peter Brinn, project team leader of the Commission's Environmental Integration Advisory Services, on file with the authors.

lack of institutional coherence and coordination; unclear roles and responsibilities at EU level; inconsistent leadership (EEA, 2005a); a lack of clarity about the meaning of the integration principle and how it relates to sectoral characteristics (such as differing levels of EU competence, the extent of decentralization of policy responsibilities and the nature of the actors and stakeholders to be targeted); and the variable commitments to, and results of, monitoring and review (EEA, 2005b; see also Dhondt, 2003).

The European Court of Auditors (ECA) undertook an audit in 2005 focused on the Commission's management of the environmental aspects of its development aid (ECA, 2006). The audit examined whether the Commission had a comprehensive strategy for addressing the environmental aspects of its development cooperation, whether it had made adequate management arrangements to implement the strategy, to what extent and how the environment had been incorporated (integrated, 'mainstreamed') into the Commission's development programmes and projects, and the results of the Commission's environment projects. It reviewed the Commission's systems in its headquarters and delegations, examined 65 environment projects costing Euros 560 million, and 43 non-environment programmes and projects costing Euros 1073 million, in 16 countries.

The ECA concluded that the Commission had made only limited progress since 2001 in 'mainstreaming' the environment into its development cooperation. The Commission lacked a clear, comprehensive strategy for the environmental aspects of its development aid. Moreover, whilst the Commission uses the European Consensus on Development to manifest its intention to increase the priority it attaches to funding environmental programmes and projects, 'there remains the need to make the policy operational by establishing a clear strategy' (ECA, 2006: 7). The ECA argued further that the new strategy must address, in particular, 'the question of how the Commission will seek to ensure that [the policy priority of funding environmental programmes and projects] is actually translated into environment programmes and projects in the Country Strategy Papers' (ECA, 2006: 20).

With regard to the CSPs drawn up by the Commission in conjunction with beneficiary countries for the period 2001–6, the ECA remarked that these did not take sufficient account of environmental issues. On the basis of its sample of 60 CSPs, it concluded that the environmental analyses were mostly weak and that the environment had not been satisfactorily integrated (mainstreamed) into the CSPs. Furthermore, the Commission's response strategies did not adequately address environmental considerations in priority fields to be funded. None of the 60 CSPs (mixed ACP and other) sampled mentioned the MDG on environmental sustainability, while only a quarter included references to multilateral environmental agreements.

The ECA pointed out that 'to some degree the lack of environmental mainstreaming into CSPs reflected the limited priority attached to this issue by many beneficiary countries', but that at the same time 'the lack of environmental mainstreaming into development aid also reflects shortcomings on the Commission side', because it had failed to do proper country environmental analyses and had barely made use of important environmental integration tools such as SEAs and EIAs (ECA, 2006: 10). It noted that the environment is a particularly challenging area of development cooperation and that partner countries often have other, higher political priorities. It was also acknowledged that effective 'mainstreaming' of the environment into development assistance depends not only on the Commission, but also on the acceptance of environmental protection as a priority by the governments of the beneficiary countries; and that this, unfortunately, is not always the case (cited by Williams, 2007: 37).

The Commission conceded in its reply to the ECA audit report that for the next generation of CSPs it has been systematically preparing CEPs with the aim of better integrating the environment, but that: '[t]he impact of these CEPs on the new CSPs depends ultimately on the priority given by the partner countries to environmental issues' (CEC, 2006: 38). This shows that the Commission is struggling to find a balance between the *ex ante* priorities of the EU and the priorities of the partner countries to ensure 'ownership' by the latter, while avoiding creating a 'climate conditionality'.

About the environmental projects undertaken by the Commission, the ECA observed that almost all were relevant to beneficiary country needs but that project effectiveness was problematic because of over-ambitious project design, delays in project preparation and implementation, limited progress in building institutional capacity, difficulties in addressing the needs of local communities for development while meeting conservation objectives, projects' insufficient impact on the policy and legal framework, and the unrealistic goals set for financial sustainability after the end of project funding.

Notably, Williams (2007) comes to similar conclusions on the basis of her analysis of the first generation of CSPs for 46 African countries. Interestingly, one of the specific questions in her analysis concerned whether climate change was mentioned in the CSPs. However, as Table 6.2 shows, she found that climate change was mentioned in only one of the 46 CSPs examined.

With regard to the EU response strategies in CSPs, Williams observed that it is impossible to identify a coherent approach by the Commission in its assessments of a country's environmental aspirations or in its application of criteria in order to decide whether projects qualify for funding. Williams pointed to a number of reasons for this two-sided failure to integrate the environment systematically into CSPs, including the EU's failure to appreciate the obligations arising under the integration principle of Article 6 EC, its inadequate policy guidelines, the lack of data, meagre resources or staffing levels, inadequately trained staff and insufficient analysis by the Commission of

Table 6.2 *Environmental issues covered in the first generation of African CSPs* *(n = 46)*

	Yes (%)	No (%)
Environmental analysis	44	56
Sustainable development	25	75
Environmental impact assessment	28	72
Multilateral environmental agreements	26	74
Climate change	2	98

Source: Based on Williams (2007).

the CSPs and the projects they foster. Furthermore, the partner countries might not always desire an environmental component to be part of their aid agreements.

Authors of other previous studies have reached similarly critical conclusions about the level of environmental integration in the first generation of CSPs (e.g. Dávalos, 2002;[87] FERN, 2002; ACP-EU Joint Parliamentary Assembly, 2003). Yet, some hopeful signs of (early) progress with the Commission's enhanced, systematic use of CEPs in the second generation of CSPs may be found in a recent study executed by Commission staff (Palerm *et al.*, 2007a; 2007b). Albeit with a very small sample of only six CSPs for the 2007–13 period, the authors of this study note a significant improvement in the level of environmental integration in CSPs, even for non-environment sectors. In addition to the more systematic use of CEPs, resulting in increased quantity *and* quality, Palerm *et al.* also credit this improvement to the establishment of the Commission Environment Helpdesk (see Section 6.3.3), of which the study's authors were members, and enhanced training of Commission staff on environmental 'mainstreaming'.

Authors of subsequent studies based on a larger sample of new-generation CSPs have also found some signs of improvement, while reiterating the critical findings of earlier reports regarding the deployment of environmental integration tools in EU development cooperation practice. In particular, a 2007 report commissioned by the environmental NGOs FERN, the WWF and Birdlife International, analysed the 44 CEPs and 3 REPs available at that time for the presence and public availability of tools that are necessary to facilitate public consultation and participation in environmental aspects of development cooperation (Nicholson and Leal Riesco, 2007). Whilst noting significant improvements in the Commission delegations' approach to carrying out CEPs and attitudes to involvement of civil society and public availability of documentation, on the critical side the report pointed to the lack of

---

[87] For example, having identified CEPs in only 6 of the 60 CSPs assessed, Dávalos (2002: 30) concluded that '[m]any countries do not include the slightest description of the state of the environment or of the actions at national level to tackle environmental concerns'.

a systematic approach to carrying out EIAs and SEAs in accordance with commitments to this effect, as well as inconsistency in facilitating public access to crucial documents such as CEPs, EIAs and SEAs. Notably, incongruently with the emphasis placed by, inter alia, the 2005 European Consensus on Development on SEAs as an important environmental integration tool, only 4 of the 70 Commission delegations and regional desks contacted had carried out an SEA, with a further 4 foreseen. Similarly, only 24% of the delegations reported having carried out EIAs in the context of EU development funding since 1996. Perhaps indicative of the gaps in training, knowledge and communication on the issue of integration of the environment dimension in EU development cooperation among Commission representatives in the field is the fact that the report noted that, of the 70 Commission delegations contacted, only 5 indicated being aware of any evaluations of the practice in this field, despite the publication of the ECA's special audit on this topic only two years before. As a follow-up to this report, the same three NGOs published a study in 2009 analysing 19 geographically diverse CEPs and two REPs in order to assess the quality and added value of the information contained in these documents with a view to integration of environmental concerns into development cooperation (Mofolo *et al.*, 2009). Once again, significant gaps were identified across all of the documents studied, including notably, in the present context, insufficiencies in the environmental data and indicators generally, and insufficient coverage of climate change issues in partner countries, including impact data, short- and long-term adaptation priorities, and opportunities to support low-carbon development.

In sum, early hints of slight improvements notwithstanding, by and large the existing assessments concur in their conclusion that the Commission has yet to address the environmental aspects of its development cooperation sufficiently and has still to establish a comprehensive environment strategy for its development cooperation. Curiously, none of these studies assessed whether the Commission has actually fared well in the light of the Paris Declaration on Aid Effectiveness and its recommendations on conditionality.

6.4.2 *Assessing the integration of climate change concerns into development cooperation*

In 2006, the Centre for European Policy Studies (CEPS) studied, inter alia, the policy development process of the EU Strategy and Action Plan for Climate Change in the Context of Development Cooperation (Hudson, 2006). This process was evaluated as being successful, as a result of the central position of development concerns in the plan; the close cooperation between DG Development and DG Environment, in particular, through inter-service consultation; the involvement of working parties and expert groups from development and non-development streams that cooperated constructively; and the supportive stance of the presidencies involved.[88]

However, the study recognized the limitations of the EU Strategy and Action Plan, observing that the EU realizes that climate change impacts severely on DCs, but when it comes to actual financial commitments the EU relies on the Member States as its financing counterparts and is hesitant to come forward (van Schaik, 2006).

This study listed several bilateral agreements between the EU and China as well as India focusing on clean energy technologies. It stated that the EU and/or its Member States contribute to mitigation and adaptation through the Global Environment Facility (GEF), the Least Developed Countries Fund and the Adaptation Fund (see Section 5.4.3).[89] Furthermore, the EU has provided funding for some EU tenders and capacity-building efforts to enable (least) developing countries to participate in the CDM. However, van Schaik (2006) emphasized that most resources are provided by national development agencies that have decided to take sustainability into account and by the international financial institutions.

In 2007, a mid-term review of the EU Action Plan on Climate Change and Development based on a desk analysis of official communications and reports, quantitative analysis of development cooperation statistics and some 120 stakeholder interviews was published (Teixeira Santos and de Lopez, 2007). It shows that the implementation of the Plan provides a mixed picture of areas of significant achievements and areas where gaps remain and where opportunities for incorporating climate change issues into development cooperation need to be explored further.

On the positive side, the awareness of climate change issues within EU development agencies is generally high, as shown by the fact that climate change has been put on the agenda for dialogue and cooperation with partner countries, assessments of vulnerability and adaptation needs have been undertaken, and efforts to stimulate capacity building have been made. On the negative side, these activities have not yet led to any integration into operational practices. Concerning adaptation, funds are in short supply for adaptation needs of the DCs (see Tables 5.3 and 5.4), including the implementation of NAPAs, and international research projects have insufficiently addressed the challenge of delivering meaningful and useful information to DCs. Concerning mitigation, partner countries, in particular LDCs, do not emit sufficient greenhouse gases to warrant its incorporation into national development strategies.

---

[88] Hudson (2006) explains that the Developing Countries Expert Group that advised the EU Council's Working Party on International Environmental Issues/Climate Change contained strong representations from Member States' development agencies, including those of the UK, Sweden, the Netherlands, Denmark, Portugal and France. With respect to the presidencies, he describes how the Presidency played an important and constructive role, with the Irish Presidency pushing the issue forwards during the first half of 2004, and the Netherlands' Presidency bringing the initiative to a conclusion. Under the Irish Presidency, the Action Plan was revised through a process involving consultation with developing countries and civil society organizations.

[89] The author of the same study noted that 'most resources however are provided by national development agencies that have decided to take sustainability into account' (Hudson, 2006: 83).

Furthermore, CSPs guiding development with partner countries that are major gross but not per capita emitters of greenhouse gases have yet to include climate change as a priority issue.

The review of the CSPs for the period 2002–6 shows that climate change was identified as a priority by three countries only: China (energy), Brazil (forestry) and Argentina (forestry) (Teixeira Santos and de Lopez, 2007: 55). None of the other CSPs mentioned climate change as a development cooperation priority. In the CSPs for the period 2007–13, climate change issues are more explicitly discussed in accompanying CEPs. These profiles consist of an assessment of a country's environmental situation, policy and regulatory framework, institutional capacities and environmental cooperation, and aim to facilitate the integration of the environmental dimension into the CSP. The mid-term review of the 2007–13 CSPs is expected to provide further opportunities for more comprehensively incorporating climate change into development cooperation.

The review observes that the efforts to include climate change in CSPs may conflict with the principle of partner country 'ownership', especially when partners have other immediate priorities (Teixeira Santos and de Lopez, 2007: 23). They argue that, following the Paris Declaration on Aid Effectiveness, supporting DCs in low-GHG development paths is warranted only when there is a national need and request for such support (Teixeira Santos and de Lopez, 2007: 53–5). Through policy dialogue and awareness-raising, donors may signal their interest in climate change issues. However, the incorporation of climate change into national development planning ultimately rests on the willingness of DCs to prioritize these issues. Therefore, the pursuit of low-carbon paths is expected to remain peripheral to DC concerns unless it yields demonstrable immediate local benefits.

The review states that the challenge for the development community is to move from general awareness about the need for mainstreaming to operational working knowledge, namely to incorporate climate change considerations into the daily routines and processes of development agency staff (Teixeira Santos and de Lopez, 2007: 24). This requires more insight into the operational implications of the various stages of incorporation, such as resource allocation, staff training and institutional reform (Teixeira Santos and de Lopez, 2007: 28). There is, for example, a need for systematic and comprehensive climate change training for EU staff involved in development cooperation, and the development of appropriate procedures, tools and skills (Teixeira Santos and de Lopez, 2007: 24 and 34).

Another assessment of the progress with the integration, or mainstreaming, of climate change issues in development cooperation formed part of the Commission's September 2007 evaluation report on the EU Policy Coherence for Development initiative (see Section 6.2.2). In line with the general conclusions about the status of the EU's policy coherence for development, the report noted that the process of mainstreaming climate change into development cooperation through policy coherence was 'still at an early stage'.[90] More specifically, on the basis of a finding of only six references to climate change as an important issue for policy coherence for development in all the current CSPs of African, Caribbean and Pacific (ACP) countries, the evaluation report concluded that the recognition of climate change in this area is 'very low' and that in fact it is 'one of the least frequently mentioned issues' in the context of policy coherence for development.

6.5 Conclusions

This chapter's investigation of the way in which the EU has been incorporating climate change into its development cooperation process reveals (so far) a discrepancy between the EU's *policy intentions* and their implementation in *practice*.

The EU is convinced that climate change and development are substantively linked and that climate change should be incorporated into development cooperation. This is also in line with the principle of environmental policy integration as established in the EC Treaty. Thus, from the perspective of coherence and consistency, the EU has articulated a clear policy *intention* towards integration of climate change concerns.

With respect to adaptation, the EU argues that climate change and development cooperation are linked because climate change can negatively influence development prospects and the implementation of the MDGs in the DCs. Furthermore, it sees adaptation to climate change as critical to preventing security problems.

In its policy, the EU focuses not only on adaptation but also on mitigation by promoting low-GHG development paths through the establishment of sustainable energy partnerships. Moreover, the EU is not averse to using ODA resources to fund capacity-building exercises to enable DCs to participate in CDM activities, while it cautions that this should not lead to a diversion of resources from the primary goal of development cooperation: poverty eradication (cf. Section 5.5.3).

Turning to *practice*, the EU is clearly struggling to balance its own objective and mandate to incorporate climate change into development cooperation with its recognition of the critical importance of partner country 'ownership' for successful aid implementation – a lesson learned from development cooperation. Thus far, at this early juncture, the EU seems intent on striking a delicate equilibrium between its own priorities and those of partner countries by engaging in dialogue with partners, thereby avoiding the danger that its own environmental integration and climate mainstreaming goals create 'conditionality'.

---

[90] Commission Working Paper, *EU Report on Policy Coherence for Development*, COM(2007)545, 20 September 2007; accompanied by Commission Staff Working Paper, *EU Report on Policy for Development*, SEC(2007) 1202, 20 September 2007.

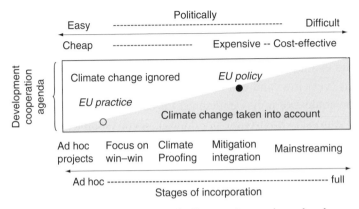

Figure 6.1. Stages of incorporation of climate change into development and development cooperation.

In relation to the stages of incorporation of climate change as defined in this book (see Figure 6.1 and in detail Section 3.3), the EU's policy intentions appear to be more in the nature of moving from ad hoc approaches towards climate proofing and integration, albeit not 'full-fledged' mainstreaming. Confusingly, the EU policy discourse blurs the distinction between 'integration' and 'mainstreaming' by using these terms synonymously. However, the assessment in this chapter has shown that, incongruently with its policy intentions, in practice the EU is still more at the stage of ad hoc and win–win approaches. Overall it appears that the EU is just at the beginning stages of incorporation of the environment, let alone climate change, into its development cooperation, although the practice of drawing up country environmental profiles in second-generation CSPs is certainly contributing to a more systematic approach. More generally, the systematic use of environmental integration tools such as SEAs and EIAs will need to be stepped up since, besides the political dimension, technical difficulties including availability and analysis of data appear to be among the key factors limiting the incorporation of climate change into development cooperation.

In conclusion, the experiences of the coming years will be decisive for a more definitive assessment of whether EU development cooperation policy can make a meaningful difference in the mitigation of, and adaptation to, climate change by developing countries. One of the main determining factors for success in this respect will be the extent to which the EU provides long-term, specific financing to achieve its ambitious policy goals in this area. Evidently, the funding allocated will need to be complementary to the ongoing processes under the Climate Convention. Given its exceptional position as the largest development assistance donor, acting both multilaterally and bilaterally, combined with its image as a global frontrunner in the response to climate change, the EU could have a major demonstrative effect on other countries. If it can find the right approach to incorporating climate change into development cooperation, it can be a global leader in this field.

Acknowledgements

The authors would like to thank Marc Pallemaerts (University of Amsterdam and Institute for European Environmental Policy), Wybe Douma (TMC Asser Institute) and Peter Brinn (project team leader of the European Commission's Environmental Integration Advisory Services) for their helpful comments on earlier versions of this chapter.

References

ACP–EU Joint Parliamentary Assembly (2003). *Report on Sustainable Management and Conservation of Natural Resources in ACP Countries in the Context of the 9th EDF Programme*. Committee on Social Affairs and the Environment, 11 October 2003, ACP-EU 3590/03/fin.

Behrens, A. (2008). *Financial Impacts of Climate Change: An Overview of Climate Change-Related Actions in the European Commission's Development Cooperation*. CEPS Working Document No. 305/September 2008. Briefing note prepared for the European Parliament. Brussels: Centre for European Policy Studies.

Carbone, M. (2008). Mission impossible: the European Union and policy coherence for development. *Journal of European Integration*, **30**(3), 323–42.

CEC (2001). *Guidelines of 4 May 2001 for Implementation of the Common Framework for Country Strategy Papers*. Brussels: European Commission.

(2005). *European Commission Adopts "European Union Strategy for Africa"*. Commission Press Release IP/05/1260, 12 October 2005.

(2006). *The Commission's Reply to the Court of Auditors Special Report No. 6/2006 Concerning the Environmental Aspects of the Commission's Development Cooperation*, OJ C 235/31, 29 September 2006.

(2007). *Environmental Integration Handbook for EC Development Cooperation*. Brussels: European Commission EuropeAid Cooperation Office.

(2008a). *Annual Report 2008 on the European Community's Development and External Assistance Policies and their Implementation in 2007*. Luxembourg: Office for Official Publications of the European Communities.

(2008b). *Commission Staff Working Document on the Implementation Framework of the Global Climate Change Alliance*, SEC(2008)2319, 15 July 2008.

(2008c). *Global Climate Financing Mechanism (GCFM): International Finance Facility for Climate Change*. Draft Issues Paper, 15 April 2008. Brussels/Washington, D.C.: European Commission/World Bank.

Dávalos, E. (2002). *Mainstreaming Environment in the Country Strategy Papers: A Review of 60 Countries*. Report to the EC Directorate on General Development. Brussels: Commission of the European Communities.

Dhondt, N. (2003). *Integration of Environmental Protection into Other EC Policies*. Groningen: Europa Law Publishing.

ECA (2006). *Court of Auditors Special Report No. 6/2006 Concerning the Environmental Aspects of the Commission's Development Cooperation (Pursuant to Article 248(4), Second Subparagraph, EC)*. OJ C 235/1, 29 September 2009.

EEA (2005a). *Environmental Policy Integration in Europe. State of Play and an Evaluation Framework*. EEA Technical Report No. 2/2005. Copenhagen: European Environment Agency.

   (2005b). *Environmental Policy Integration in Europe. Administrative Culture and Practices*. EEA Technical Report No. 5/2005. Copenhagen: European Environment Agency.

EP (2008). *European Parliament Resolution of 21 October 2008 on Building a Global Climate Change Alliance between the European Union and Poor Developing Countries Most Vulnerable to Climate Change*, 2008/2131(INI). Rapporteur: Anders Wijkman, A6–0366/2008.

FERN (2002). *Forests at the Edge: A Review of EC Aid Spending*. Brussels: FERN.

Forster, J. and Stokke, O., eds. (1999). *Policy Coherence in Development Co-operation*. London: Frank Cass.

Holland, M. (2009). The EU and the global development agenda. *Journal of European Integration*, **30**(3), 343–62.

Hudson, A. (2006). Case study: climate change in the context of development cooperation, in *Policy Coherence for Development in the EU Council: Strategies for the Way Forward*, ed. L. van Schaik, M. Kaeding, A. Hudson and J. N. Ferrer. Brussels: Centre for European Policy Studies, pp. 57–9.

Mofolo, M. J., Leal Riesco, I., Nemcova, T., Nicholson, S. and Phillips, J. (2009). *Environmental Tools in EC Development Cooperation: An Analysis of Country and Regional Environmental Profiles*. Brussels: WWF, FERN and Birdlife.

Nicholson, S. and Leal Riesco, I. (2007). *Environmental Tools in EC Development Cooperation. Transparency and Public Availability of Documentation. A Review*. Brussels: WWF, FERN and Birdlife.

OECD DAC (2007). *DAC Peer Review of the European Community*. Paris: Organisation for Economic Co-operation and Development.

Olsen, G. R. (2005). The European Union's development policy: shifting priorities in a rapidly changing world, in *Perspectives on European Development Co-operation: Policy and Performance of Individual Countries and the EU*, ed. P. Hoebink and O. Stokke. London: Routledge, pp. 573–608.

Palerm, J., Ledant, J.-P. and Brinn, P. (2007a). *Environmental Integration in EC Development Co-operation Programming: Experiences in the Use of Country Environmental Profiles*. Report of the European Commission Helpdesk for Environmental Integration in EC Development Cooperation. Brussels: European Commission EuropeAid Cooperation Office. Available online at https://ec.europa.eu/development/icenter/repository/env_helpdesk_CEPs_en.pdf.

   (2007b). Environmental integration in EC development co-operation multi-annual programming: experiences in the use of country environmental profiles. *Impact Assessment and Project Appraisal*, **25**(3), 163–74.

Pallemaerts, M. (2006). The EU and sustainable development: an ambiguous relationship, in *The European Union and Sustainable Development: Internal and External Dimensions*, ed. M. Pallemaerts and A. Azmanova. Brussels: VUB University Press, pp. 19–52.

Teixeira Santos, S. and de Lopez, T. (2007). *First Bi-Annual Progress Report on the Implementation by the EU of the Action Plan to Accompany the EU Strategy on Climate Change in the Context of Development Cooperation: Final Report*. Rotterdam/Lisbon: ECORYS-NEI and CESO CI.

van Schaik, L. (2006). Fiche on EU climate change policy, in *Policy Coherence for Development in the EU Council: Strategies for the Way Forward*, ed. L. van Schaik, M. Kaeding, A. Hudson and J. N. Ferrer. Brussels: Centre for European Policy Studies, p. 54.

Wilkinson, D. (2007). *Environmental Policy Integration at EU Level: State-of-the-Art Report*. London: Institute for European Environmental Policy (IEEP).

Williams, R. (2005). Community development cooperation law, sustainable development and the Convention on Europe: from dislocation to consistency?, in *The Yearbook of European Environmental Law*, Volume 4, ed. T. F. M. Etty and H. Somsen. Oxford: Oxford University Press, pp. 303–75.

  (2007). The integration of environmental protection requirements into EC development cooperation policy. Draft paper for the first EPIGOV conference 'Better Integration: Mainstreaming Environmental Concerns in European Governance', Brussels, 15 February 2007. (For information about the EPIGOV project, see http://www. ecologic.eu/projekte/epigov.)

7

Incorporating climate change into EU Member States' development cooperation

HARRO VAN ASSELT AND JOYEETA GUPTA

7.1 Introduction

Given that 80% of development cooperation resources are managed at European Union (EU) Member State level, this chapter moves from an analysis of the policy processes at EU level to the processes within individual countries. It examines the extent to which climate change adaptation and mitigation are incorporated into development cooperation policies and practices in selected EU Member States, with a view to identifying relevant best practices.

The EU's 27 Member States are not all equally active in aid provision (see Chapter 10), and not all Member States are members of the OECD DAC (see Chapter 4). Furthermore, environmental awareness in the Member States differs, implying differing degrees of commitment to environmental issues in general and in aid strategies. This chapter examines the aid policies and practices in Denmark, France, Germany, the Netherlands and the UK, thereby including some of the largest EU donors in absolute and relative terms (the UK and Germany), countries that are considered leaders (Denmark and the Netherlands) and laggards (France, Germany) in terms of aid quality (e.g. Concord, 2009) and countries that spend a large proportion of their aid on environmental and climate-change-related activities (Teixeira Santos and de Lopez, 2007: 9). This choice implies a bias in favour of countries that provide green aid, but provides good early experiences on how climate change is being incorporated into aid policies (cf. OECD, 2009b). To provide an indication of incorporation activities in other Member States, the chapter also covers Italy, a Southern Member State with a fragmented aid organization, and Poland, a relatively new Member State, which only recently turned from a partner country into a donor country (see Section 7.7).

Since each country has its own reporting systems, and it is often difficult to determine whether climate-related development cooperation is provided as part of official development assistance (ODA) or whether it is new and additional (e.g. Bird

Mainstreaming Climate Change in Development Cooperation: Theory, Practice and Implications for the European Union, ed. Joyeeta Gupta and Nicolien van der Grijp. Published by Cambridge University Press. © Cambridge University Press 2010.

Table 7.1 *Danish ODA over time*

	1991–2 (average)	1996–7 (average)	2006	2007
Total ODA budget (USD million)	1296	1705	2236	2562
Percentage of GNI	1.03	1.15	0.80	0.81

Source: OECD (2009a: 156–7).

and Peskett, 2008), it is difficult to draw clear comparisons. Although climate-related ODA can be tracked through the OECD system of 'Rio Markers', it is difficult to determine whether these self-reported figures are accurate (Roberts *et al.*, 2008). Therefore, this chapter is based on an analysis of relevant documents and information regarding how climate change is incorporated into development cooperation, supplemented by interviews with key actors.

The chapter presents the five country studies (see Sections 7.2–7.6), a comparative analysis (see Section 7.7), and concluding thoughts and indications of best practices (see Section 7.8).

7.2 Denmark

7.2.1 Policy context

Development aid enjoys widespread public support in Denmark (Danida, 2008a: 181). Denmark has met the 0.7% ODA/GNI target, and has committed to keep this above 0.8% in the future (OECD, 2007: 18); see Table 7.1. Danish aid is of high quality, ranking third on the Commitment to Development Index in 2008 (CGD, 2008), and its 'decentralised but highly integrated' system was praised by the OECD DAC (OECD, 2007: 14). However, it includes debt relief in its ODA contributions (OECD, 2009a: 236).

The 2002 Act on International Development Co-operation states that Danish development assistance aims to support developing countries' (DCs') 'endeavours aimed at promoting economic growth, thereby making contributions to ensuring social progress and political independence in accordance with the aims and principles of the United Nations Charter, and to promote mutual understanding and solidarity through cultural co-operation'. The government's key strategy paper – 'Partnership 2000' – establishes poverty reduction in support of sustainable development as the 'overriding' objective of Danish aid (Danida, 2000: 10). Human rights and democracy, environmental sustainability and gender equality constitute 'cross-cutting issues' in Danish development aid, which need to be considered in all development activities (Danida, 2007c). The government releases an annual policy paper, outlining its development policy priorities and budget for the subsequent five years (OECD, 2007: 38). These priorities include children and youth, HIV/AIDS and peace operations. In 2007, the government prioritized climate change, energy and environment, migration and development, and stability and democracy (Danida, 2007a). Although the annual identification of priorities provides flexibility, its consequences for consistency and sustainability are not clear (OECD, 2007: 33).

In 2007, about 65% of Danish ODA was disbursed as bilateral assistance (OECD, 2009a: 175). Apart from assistance to 90 countries (OECD, 2007: 31–2), bilateral aid primarily focuses on long-term, binding partnerships with 16 'programme countries'[1] selected on the basis of criteria including the level of development of a country, the presence of other donors, the potential for promoting sustainable development, respect for human rights, gender aspects and previous experiences with Danish aid (Danida, 2008a: 186).

This aid is concentrated in Least Developed Countries (LDCs), particularly in Africa. In 2006–7, more than half of ODA was allocated to LDCs (OECD, 2009a: 209), going beyond the donor community commitment to provide 0.15% of GNI to this country group. Of its 16 programme countries, 10 are in Africa. In 2006–7, Uganda received the most (3.7%), followed by Tanzania (3.7%) and Nigeria (a non-programme country; 3.7%) (OECD, 2009a: 221). The government intends to increase its allocation to Africa to two-thirds of bilateral aid (Danida, 2008a: 35). Although Danish aid targets poorer countries, it focuses on the 'safer' countries, ignoring higher-risk countries (OECD, 2007: 32).

The ODA disbursed through multilateral channels consists of contributions to UN agencies (38%), the EU (26%), the World Bank (15%) and regional development banks (8%) (Danida, 2008a: 1). For each organization, strategies outlining the Danish government's objectives are prepared. The share of multilateral ODA has been debated in Denmark: after USD 25 million initially was reallocated from multilateral to bilateral ODA (OECD, 2007: 35), the government decided to reprioritize multilateral cooperation in 2008, following an analysis of how multilateral aid could be made more effective (Danida, 2008c: 5–6).

The choice of aid modality is guided by the principles of the Paris Declaration as well as cost-efficiency considerations, and 'depends on the conclusions of in-depth analyses of the national policy framework for poverty reduction, the role and strength of both the private sector and civil society, and the quality of public sector management' of partner countries (Danida, 2005b: 9). The modalities used include sector-wide approaches (including sector programmes), general and sector budget

---

[1] The countries are Bangladesh, Benin, Bhutan, Bolivia, Burkina Faso, Egypt, Ghana, Kenya, Mali, Mozambique, Nepal, Nicaragua, Tanzania, Uganda, Vietnam and Zambia. Development assistance to Bhutan, Egypt and Vietnam will be phased out in favour of Africa (Danida, 2008a).

support, and stand-alone projects. In addition, 'mixed credits' – interest-free or low-interest loans for financing equipment or services in development activities – are used in a few countries of interest to Danish trade and industries. Mixed credits are the only form of tied aid still used, although untied mixed credits are also used (Danida, 2008e). Mixed credits lead to development results, but are questionable, especially when used in LDCs (OECD, 2007: 55–6).

Increasingly, aid is provided through sector programmes replacing individual projects since the mid 1990s (OECD, 2007: 34; Danida, 2008a: 195). While general budget support may account for up to 25% of bilateral aid, it was limited to six countries in 2006 (about 6% of bilateral aid). This is unlikely to change in the near future, despite criticism from the OECD DAC peer review. The use of general budget support is contingent on the evaluation of a programme country according to ten guiding principles, which include the partner country's adherence to good governance, the existence of a solid poverty reduction strategy and sound public financial management (OECD, 2007: 54; Danida, 2008b). The use of sector budget support, which goes mainly to the health and education sectors, is likely to increase (Danida, 2008b: 60).

Sector support concentrates on two to four priority sectors per partner (Danida, 2008a: 195). Most aid (44.2%) was allocated to social infrastructure in 2007 (mainly public administration and civil society, education and health), followed by economic infrastructure (13.5%; primarily transport and energy), the productive sectors (10.2%; mainly agriculture and industry), emergency aid (9.1%), debt relief (8%) and multi-sector projects (including the environment; 6.9%) (Danida, 2008a: 158–9).

7.2.2 Operational context

The Ministry of Foreign Affairs coordinates policy development through consultation with stakeholders including the Parliament, NGOs, embassies, representatives of multilateral organizations and partner countries. The Ministry and the Minister for Development Cooperation coordinate bilateral and multilateral policies, while the Danish Parliament approves annual budgets. Within the Ministry, Danida[2] manages the bilateral and multilateral aid programmes (OECD, 2007: 18–19). The Board for International Development Cooperation (Danida Board), consisting of nine stakeholders acting in their personal capacity, provides professional and technical advice on planned development activities.

---

[2] Although Danida as a separate aid agency ceased to exist in 1991, the abbreviation is used to refer to 'Danish international development assistance', and denotes the organization of aid within the Ministry of Foreign Affairs.

Danish development policy has been heavily decentralized to embassies in programme countries. This allows one to account for local circumstances, reduce bureaucracy and increase efficiency (OECD, 2007). While strategic policy decisions are taken at Danida's department of policy development, embassies in programme countries are responsible for ensuring 'the effective and efficient identification, preparation and implementation of Danish development cooperation according to the established policies and programmes' (Danida, 2006b: 37). Embassies help draft multi-annual country strategies in consultation with local stakeholders. The strategies should include the choices about programmes to be implemented, as well as information about how progress is monitored and evaluated (Danida, 2007b). Heads of embassies draft annual business plans, which include the goals of Danish aid in a country, the expected budgets, activities and results, and an indication of how the activities are monitored and reported (OECD, 2007: 42). After approval for a programme or project has been secured, embassies enjoy considerable flexibility in daily management activities.

The main link between the embassies and headquarters is provided by Danida's quality assurance and technical advisory services departments (OECD, 2007: 41–2). The department of quality assurance oversees the financial and performance management of projects and programmes, whereas the department of technical advisory services controls the technical quality of development cooperation (Danida, 2006b). An overall Programme Committee ensures coherence between the different sector- and country-specific strategies and Denmark's overall development policy. Although this committee does not have any decision-making authority, it advises on issues such as bilateral programmes (above DKK 30 million; about USD 5.7 million[3]), strategies for multilateral organizations and thematic policies (Danida, 2008d). After the necessary documents have been completed by the embassies, the Danida Board approves or rejects proposals for programmes above DKK 30 million.

7.2.3 Incorporating adaptation and mitigation

Environmental protection, including climate change, is a cross-cutting issue in Danish development cooperation policy, and is incorporated in all development activities (Danida, 2000: 28). In recent years, climate change has been a recurring issue in the annual Danish development policy papers. Partly in the light of the decision to host the 15th Conference of the Parties to the FCCC in 2009, climate change was prioritized in Danish aid in 2007, and development assistance was to be made 'climate-proof' (Danida, 2007a). Whereas environmental objectives were

---

[3] The exchange rates used in this chapter are as of 29 July 2009.

initially the main driving forces for the inclusion of climate change concerns, climate change has become an important driver for environmental protection in Danish aid.

The policy priority assigned to climate change has resulted in pledges of increased funding. In 2009, the Danish government set aside DKK 200 million (about USD 38.2 million) for climate-related development programmes, twice the amount of 2008 (Danida, 2008c: 4). This amount will increase annually by DKK 100 million (about USD 19.1 million) during the period from 2010 to 2012.

The government's 2005 'climate and development action programme', which was drafted in response to the 2004 EU action plan (see Chapter 6; Danida, 2005a: 7), followed the structure of the EU document, focusing on raising the policy profile of climate change, adaptation, mitigation and capacity development. The relatively short time-frame within which the programme was developed forced Danida to keep it as simple and pragmatic as possible. The document outlines how these elements should be pursued in bilateral and multilateral development cooperation, assigns responsibilities to actors, and identifies 'entry points' for incorporation (Danida, 2005a). Relevant actions include requesting information from multilateral organizations about their climate-related activities and defining the scope for cooperation on adaptation and mitigation with relevant organizations. For bilateral cooperation, embassies in programme countries and countries receiving special environmental assistance[4] are responsible for including climate change considerations in national planning processes (e.g. in Poverty Reduction Strategy Papers) and the development of country strategies. Proposed actions include the introduction of climate screening tools (see below), undertaking vulnerability assessments, and pursuing mitigation options in country programmes. Embassies are also responsible for taking into account climate change in existing and planned sector programmes. In addition, the action programme provides an indication of how climate change considerations should be incorporated into aid provided through mixed credits. The Danish programme was praised because it 'has set the stage to become an exemplary showcase of the integration of climate change into development cooperation, using a comprehensive, yet ultimately practical and flexible approach' (Teixeira Santos and de Lopez, 2007: 40). A 2008 review of the programme provided the basis for an update (Danida, 2009). The review stresses the need for better links between adaptation and disaster risk reduction, and argues for a long-term approach to 'climate-proofing'. However, an updated action programme is not expected before 2010, mainly because the Ministry is focusing on the post-2012

---

[4] These countries are the programme countries Bhutan, Bolivia, Egypt, Kenya, Mozambique, Nicaragua, Tanzania, Vietnam and Zambia, and the non-programme countries Cambodia, Indonesia, Malaysia, South Africa and Thailand.

climate negotiations, but also because an updated overall development strategy following up 'Partnership 2000' as well as a new environment and development strategy are expected in that year.

Within Danida, the department of technical advisory services is responsible for implementing the objectives related to cross-cutting issues, including environmental protection and climate change (Danida, 2006b). The department has launched a climate change screening process for its development aid. First, climate change screening is part of the regular development activities through the mandatory process of drafting environmental screening notes. If such a note indicates that development activities are climate-related (both in terms of adaptation and in terms of mitigation), one of the possible tools is a climate change screening note (Danida, 2006a: 14). The notes contain basic information about a country's or sector's vulnerability to climate impacts, mitigation opportunities and existing and planned policies, and provide a first indication of the possible risks for Danish aid programmes.[5] Second, between 2005 and 2008 climate change screening reports were prepared for 17 partner countries.[6] The first three reports focused on identifying and designing climate-related projects, but later reports aimed at a comprehensive climate screening of sector programmes. Although the formats of the reports differ, most reports (e.g. for Bhutan, Cambodia, Ghana, Kenya and Nepal) explicitly include a simple climate change screening of the Danish portfolio in the form of an assessment of the climate risks and adaptation options. The reports also examined the activities of other donors in the partner countries. The emphasis of the exercises was primarily on adaptation, but mitigation options were taken into account as well in some of the reports through assessing the programme country's CDM potential. However, for some of the LDCs examined, mitigation was perceived to be linked to carbon finance, which could divert attention from the more imminent risks of climate change impacts (Linddal, 2008).

The screening reports showed that the existing Danida portfolio was not directly at risk from climate impacts, but that there was a strong case for including climate change in the design of new programmes (Linddal, 2008). They indicated the importance of communicating risks to stakeholders in partner countries in understandable terms, for instance relating directly to impacts (such as droughts and floods) experienced in those countries. Incorporation activities – such as screening – should be simple and operational for those working in the development community. The Danish government allocated DKK 5 million (about USD 0.95 million) for

[5] A standard climate change screening note is available online at http://www.danidadevforum.um.dk/en/menu/Topics/ClimateChange/ClimateAndDevelopment. For the screening reports, the note was developed into a climate change screening matrix, which focused on climate risks and adaptation options. The main reason for further developing the screening note into a matrix was to simplify the tool.

[6] Most screening reports are available online at
http://www.danidadevforum.um.dk/en/menu/Topics/ClimateChange/ClimateAndDevelopment.

implementing the recommendations of the report for each country (Danida, 2009). This funding led to new climate-related projects and started discussions on how best to spend the money in an ad hoc fashion. However, Linddal (2008) suggests that in the long term climate change should be integrated, rather than just focused on ad hoc projects. The rationale behind the funding for stand-alone projects can be found in the political need to sell 'mainstreaming' activities through explicitly allocating new and additional resources to projects labelled as climate-relevant.

Finally, the CDM has been an integral part of Danish efforts to incorporate climate change into development cooperation. The government states that '[W]hereas [ODA] cannot be used to purchase Certified Emission Reductions (CERs), [it] may be used to build capacity to develop CDM projects' (Danida, 2005a: 22; see also Danida, 2004). For instance, in 2005, Denmark allocated approximately USD 7.6 million to 'prepare CDM credits', whereas in 2007, USD 3.7 million was spent for this purpose (Danida, 2005c: 16; 2008a: 85).[7] Although some of the capacity-building efforts were aimed at generally increasing the programme countries' attractiveness for the CDM, part of the expenditure was also used with a view to purchasing credits by the Danish government, including the preparation of Project Design Documents – even though purchases were not funded through ODA. This way of using ODA for CDM purposes has been questioned by Michaelowa and Michaelowa (2007) and others (see Section 5.5.3). The importance of the CDM within Danish ODA has decreased following the creation of a new Ministry of Climate and Energy in late 2007 and the subsequent shift of responsibility for the CDM programme to this ministry and away from the Ministry of Foreign Affairs.

7.2.4 Conclusion

Danish incorporation efforts have received a boost with the explicit commitment by the government to 'climate-proof' Danish development assistance and the development of a specific strategy document on how to achieve this, which clearly assigns responsibilities and specifies the actions that need to be undertaken. One of the main practical accomplishments has been the completion of several screening reports in partner countries, resulting in an increase in awareness both in Danida and in the partner countries of the impact of climate change and variability on the effectiveness of development aid.

Danish efforts to incorporate climate change in development are justified by the argument that climate change can best be addressed by 'doing development better'

---

[7] The main countries with which Denmark cooperates on the CDM are China, Indonesia, Malaysia, South Africa, Thailand and, more recently, India and Vietnam.

and more systematically enhancing the resilience of the poor. However, implementation challenges have led mainly to ad hoc projects and a focus on raising awareness about climate impacts. Arguably, not all climate-related activities are motivated by altruism or enlightened self-interest: the government supports capacity building for CDM host countries, even though the focus has moved away from the CDM in recent years.

After an impressive start, developments seem to have slowed down somewhat. Although screening the aid portfolio in partner countries is a step towards 'climate-proofing' Danish aid, these efforts may end up as mere awareness-raising exercises. Even though Danida views climate change as more than just a fad, it clearly does not want to force its views on partner countries in line with the provisions in the Paris Declaration about ownership.

7.3 France

7.3.1 Policy context

France is a key donor in absolute terms, ranking third among all OECD countries and second among EU Member States in 2007 (OECD, 2009a: 148). However, in relative terms, French ODA has decreased, with ODA/GNI dropping from 0.62% in 1991–2 to 0.38% in 2007 (see Table 7.2). The French government enhanced its commitment to development cooperation in 2002, committing to increasing ODA to 0.5% of GNI by 2007 and to 0.7% by 2012. However, the 2007 target was missed, and the 2012 target has been postponed to 2015 (OECD, 2008: 39). Furthermore, a large and increasing percentage of the funds is focused on debt relief (OECD, 2008: 40), while refugee and student costs also form a small share of total ODA (Concord, 2009).

Although France does not have an overarching policy document setting out the objectives of development aid, it seeks to promote economic growth, protect global public goods and reduce poverty (OECD, 2008: 26). In practice, however, it is argued to be less focused on poverty reduction (Nunnenkamp and Thiele, 2006). The protection of global public goods includes the fight against transmissible and emerging diseases, as well as the issues of climate change and biodiversity loss (OECD, 2008: 26; see also France, 2002). Finally, French aid aims at building capacity in the partner countries (OECD, 2008: 67).

Most ODA is provided bilaterally (~63% in 2007), although the share of multilateral aid is slowly increasing (OECD, 2009a: 176). In 2007, most of the multilateral contributions were to the EU (~60%), followed by the World Bank (~15%) (OECD, 2009a: 187–8).

Table 7.2 *French ODA over time*

	1991–2 (average)	1996–7 (average)	2006	2007
Total ODA budget (USD million)	7828	6879	10 601	9884
Percentage of GNI	0.62	0.47	0.47	0.38

Source: OECD (2009a: 156–7).

French bilateral development cooperation policy is strongly influenced by its historical ties with partner countries, including colonial relations (Schraeder *et al.*, 1998; Alesina and Weder, 2002). This has meant that the geographical and sectoral priorities set by the French government need to be adapted to the changing international context (OECD, 2008: 23). Bilateral aid is spread over 55 countries, which comprise the 'Priority Solidarity Zone'. For these countries, the French priority is to achieve the Millennium Development Goals (MDGs) by 2015. Many (43 out of 55) of these countries are in (francophone) Africa, and they include former colonies. Although Africa is thus an important focal area, and African countries are the main partners, aid to LDCs has decreased in recent years (OECD, 2008: 43–4). However, France has announced its intention to focus on a smaller number of countries, especially in Africa and LDCs (OECD, 2008: 45). The main individual partner countries of aid in 2006–7 were Iraq and Nigeria, mainly for reasons of debt relief, whereas in the past former colonies in the Pacific such as French Polynesia and New Caledonia received the largest share of ODA (OECD, 2009a: 222). The government's 2008 action plan on aid effectiveness announced the French government's intention to identify 'core target countries', where France would be the leading donor (France, 2008: 8).

The French government has identified seven priority sectors, for which strategies are being developed: education, water and sanitation, health and the fight against AIDS, infrastructure development in sub-Saharan Africa, agriculture and food security, environmental and biodiversity protection, and development of the productive sector. There are also three cross-cutting strategies on governance, sustainable development and gender equality (OECD, 2008: 27).

The French government prefers a mix of aid instruments (OECD, 2008: 63). Partner country ownership is considered to be important, and French aid increasingly promotes the use of programme-based approaches and, where possible, general budget support (CINI, 2009: 17). General budget support is provided to countries with sound macro-economic policies, poverty reduction strategies in line with the MDGs and a good public financial management system. Project aid still plays an important role in bilateral development cooperation as a complement to budget support (OECD, 2008: 63–4).

7.3.2 Operational context

The organization of French development cooperation, even after significant reforms in 1998 and 2004, is still fragmented (OECD, 2008). The French Development Agency is responsible for implementing bilateral development activities, while strategic decisions on ODA are taken by two ministries, namely the Ministry of Foreign and European Affairs and its Directorate-General for International Cooperation and Development, and the Ministry of Economy, Finance and Employment and its Directorate-General of the Treasury and Economic Policy. The mandates of the two Directorates and the Development Agency are partly overlapping, and none of the organizations focuses exclusively on ODA. In addition, since 2007 the Ministry of Immigration, Integration, National Identity and Cooperative Development has become involved in planning development aid.

The inter-ministerial committee for international cooperation and development, which was created in 1998, provides policy guidance and sets the priorities for French aid (OECD, 2008). This committee is chaired by the French prime minister, and includes all ministers involved in development cooperation. The committee and the prime minister receive annual advice from the High Council of International Cooperation, allowing for interaction between public and private actors engaged in international cooperation. Although the committee improved the coordination in the aid organization, 'it has not succeeded in simplifying the system, and this hampers its efficiency' (OECD, 2008: 16).

Besides the French Development Agency, many other actors are involved in implementing development aid, including a great number of local authorities, several research institutes and civil society organizations. In other words, the multiplicity of actors makes the development activities very dispersed, and the great number of people working in French aid are not being used effectively (OECD, 2008).

At the partner country level, the French Development Agency and the Directorate-General for International Cooperation and Development are responsible for managing aid, under the coordination of the local ambassador. 'Partnership Framework Papers' are developed every five years in the partner country. These documents identify French priority sectors in line with a partner country's Poverty Reduction Strategy Paper. Furthermore, annual 'Strategic Guidance and Programming Conferences' are held, where French officials meet to review and plan the French aid portfolio and allocations to countries (OECD, 2008: 54).

7.3.3 Incorporating climate change into ODA policy

Addressing climate change is indirectly included in French aid policy through the importance assigned to the protection of global public goods, as well as the need to support economic growth. In this context, the French engagement with emerging economies – e.g. Brazil, China and India – has been explained as follows: 'It is thus in France's interest to step up its cooperation with the emerging countries with a view to preserve global public goods, while at the same time promoting French economic interests' (AFD, 2007: 7). France sees its commitments to DCs in the area of climate change as defined by the climate negotiations, through dialogue with other countries and through commitments taken on as a part of the G8. The inter-ministerial committee for international cooperation and development has developed a sectoral strategy for the environment, in which climate change issues are addressed (CICID, 2005). The strategy states that France pledges to spend USD 40.8 billion annually on climate-related aid. Furthermore, it indicates that the focus of projects and programmes in LDCs is primarily on adaptation, whereas the focus in the Mediterranean region, Asia, Latin America and South Africa is mainly on ensuring that countries pursue a low-carbon development path through the promotion of energy efficiency and renewable energy technologies. The activities are mainly implemented by the French Development Agency, which is responsible for the largest share of climate-related ODA, and, to a more limited extent, the French Global Environment Facility.

The French Development Agency drafted its own climate change strategy for the period 2006–8 (AFD, 2005), forming the basis for various climate-related activities. The agency reports that between 2005 and 2008 it spent USD 1.8 billion on climate-related projects (AFD, 2009a). In 2008, 30% of the agency's activities were climate-related (AFD, 2009b). The focus is primarily on climate change mitigation, including renewable energy, energy efficiency and biosequestration. This also includes investments in CDM-related activities in Africa, from general capacity building to pilot projects; however, the agency does not purchase any CERs. Some projects are related to adaptation, for example in the agriculture and water sectors. Most projects are implemented in Asia, followed by the Mediterranean region and sub-Saharan Africa (AFD, 2009a).

The agency has developed two screening tools. First, the 'carbon footprint' tool measures the greenhouse gas emissions of development projects, as well as emission reductions achieved. For the measurement of emissions, the tool has been applied systematically *ex post* to about half of the agency's existing portfolio, while it is the intention to apply it *ex ante* to new projects and programmes. For the measurement of emission reductions, the tool has been used *ex ante* since 2007. The carbon footprint sketches a picture of the overall carbon footprint of the agency's development activities, but further strategic decisions on emission reduction objectives are needed. Second, the agency has developed a vulnerability analysis tool to measure what share of its portfolio is relevant for helping DCs adapt to climate change (e.g. projects in the water sector). While tools have been developed both for adaptation and for mitigation, the agency's focus is primarily on the latter.

The French Global Environment Facility was established in 1994, and funds various climate-related projects. The amount of ODA for which the facility is responsible is much smaller than that controlled by the French Development Agency. Late in 2004, the Facility was implementing 30 climate change projects, mostly in sub-Saharan Africa, but also in the Mediterranean region and Asia. The majority of projects is focused on renewable energy and energy efficiency, while some projects concern the protection or promotion of carbon sinks, or are related to the CDM (e.g. capacity building or funding projects) (FGEF, 2005).

Finally, the French government has emphasized the importance of innovative financing mechanisms additional to ODA in the context of climate change. Although it has mentioned the idea of using revenues raised by auctioning emission allowances in this regard, no specific proposals have been made.

7.3.4 Conclusion

Despite reorganizations in recent years, the French aid system is still rather complex and non-transparent. Although France aims to increase its ODA to reach the 0.7% target by 2015, it will be challenging to achieve this target, given the large share of debt relief and the decrease of aid in recent years.

Climate change as such is not prioritized in French aid policy, but climate protection is one of the global public goods which have become a key element of French development cooperation policy. France has pledged to commit USD 40.8 billion to climate-related activities in development cooperation annually, and since 2005 climate change has been on the agenda of the French Development Agency, which drafted a specific strategy for addressing the problem and is increasingly funding climate-related development activities. The agency also developed two tools to measure climate-related development activities. In addition, the French Global Environment Facility funds climate-related projects in the areas of renewable energy and energy efficiency, as well as CDM-related activities and projects promoting carbon sinks.

7.4 Germany

7.4.1 Policy context

German development cooperation policy sees the achievement of the MDGs as critical (BMZ, 2001). The 2008 White Paper outlines four 'guiding principles' for development cooperation, namely poverty reduction, building peace and enhancing democracy, promoting equitable forms of globalization and protecting the environment and climate (BMZ, 2008a: 12–15). Poverty reduction is also a cross-cutting

Table 7.3 *German ODA over time*

	1991–2 (average)	1996–7 (average)	2006	2007
Total ODA budget (USD million)	7236	6729	10 435	12 291
Percentage of GNI	0.38	0.30	0.36	0.37

Source: OECD (2009a: 156–7).

task in German development cooperation, and, 'as a matter of principle, all development activities should make a direct or indirect contribution towards poverty reduction in developing countries' (BMZ, 2005: 4).

Even though it was the second largest OECD donor in 2007 in absolute terms, on a relative scale Germany's 0.36% ODA/GNI ranked below the OECD average of 0.45%. Furthermore, 31% of its ODA was related to debt relief (OECD, 2009a: 148 and 196). Still, German ODA has steadily increased in recent years (see Table 7.3), after a period of decline, and the government has pledged to increase its ODA to 0.51% in 2010 and 0.7% in 2015 (OECD, 2006c: 27).

Over the years, multilateral cooperation has become more important for Germany, with an increasing emphasis on ensuring an effective international system for aid delivery (OECD, 2006c: 33). The government has published strategy papers outlining its goals and recommendations for development cooperation through the EU, the World Bank and regional development banks (BMZ, 2003; BMZ, 2006; BMZ, 2007a). Multilateral ODA is channelled through the EU (56% in 2007) and the World Bank (25%), and in 2007 it was the largest donor in absolute terms to these bodies (OECD, 2009a: 189–90).

In the late 1990s, Germany decided to concentrate its bilateral relationships, reducing the number of partner countries from 120 through 70 (OECD, 2006c: 30) to 58 in 2008, with 24 of these countries located in sub-Saharan Africa. In 2006–7, most aid went to Iraq (9.6%), Nigeria (7.3%), Cameroon (3.9%) and China (3.5%) (OECD, 2009a: 222–3). Most ODA in these years was spent in lower-income countries (57.2%, and 30.6% in LDCs) (OECD, 2009a: 209).

The criteria for partner selection include a country's development needs, the quality of governance, the presence of other donors and existing ties.[8] 'Priority' partner countries receive intensive development cooperation in three priority areas and regular 'partner' countries receive ODA in one priority area (OECD, 2006c: 30). Besides these countries, aid is also provided to others through regional and sectoral programmes (BMZ, 2008c). In addition, Germany introduced in 2004

---

[8] See http://www.bmz.de/en/countries/partnercountries/laenderkonzentration/index.html.

the notion of 'anchor countries', namely countries that are economically and politically important at regional and global levels (BMZ, 2004a).[9] German ODA is not provided to all of these countries; the 'anchor' designation is rather aimed at developing a strategic bilateral partnership with these countries, reaching beyond aid.

Aid modalities used include projects, programme-based approaches and budget support (BMZ, 2008e: 12), although Germany is moving away from project-based funding to programme-based approaches (OECD, 2006c; BMZ, 2008b: 18). German actors were initially sceptical about the use of budget support, or 'programme-oriented joint financing', to support policy reform in the partner countries (BMZ, 2008b: 36–7). Although concerns about budget support still exist, the implementing agencies have made commitments to expand its use. Still, the emphasis is rather on a mix of modalities, and budget support is used primarily for those countries that have developed Poverty Reduction Strategy Papers, have good governance, sufficient administrative capacity and a high quality of public financial management, and satisfy certain criteria related to economic stability. Budget support to fragile states is available only under exceptional circumstances (BMZ, 2008d).

For each partner country one or more 'priority areas' are selected, including democracy and civil society, peace building and conflict prevention, education, health (including HIV/AIDS) and family planning, water and sanitation, food security and agriculture, environmental protection and sustainable use of natural resources, sustainable economic development, energy, and transport and communication (BMZ, 2008e: 9). A broad sectoral breakdown (excluding debt relief) shows that ODA is spent mainly on education (15.2% in 2007), economic infrastructure (13.5%), government and civil society (10.2%), and water and sanitation (6.2%) (OECD, 2009a: 196).

7.4.2 Operational context

The organization of German aid is 'pluralistic' (OECD, 2006c: 22) and complex. The Federal Ministry for Economic Cooperation and Development (BMZ) is responsible for coordinating development policy. There are several implementing agencies for bilateral cooperation. These notably include two federal enterprises: the German Agency for Technical Cooperation (GTZ), for technical assistance and capacity development; and the KfW development bank, regarding financial cooperation. Other agencies focus on human resources cooperation (e.g. the German

---

[9] These countries are Argentina, Brazil, China, Egypt, India, Indonesia, Iran, Mexico, Nigeria, Pakistan, Russia, Saudi Arabia, South Africa, Thailand and Turkey.

Development Service and Capacity Building International), while civil society organizations are also involved in implementing development activities (OECD, 2006c: 42). Although most ODA is disbursed by the BMZ, other ministries as well as the federal states provide a portion of German ODA. Among the implementing agencies, most ODA is managed by the KfW, followed by the GTZ (OECD, 2006c: 23).

The BMZ is represented in partner countries by its own development cooperation officers and sometimes regular embassy staff from the German Foreign Office (OECD, 2006c: 53). In addition, field operations are carried out by staff of the implementing agencies. The operations and responsibilities of the implementing agencies have increasingly been decentralized, and especially the GTZ has a large representation in partner countries (OECD, 2006c: 54). This structure implies that it is challenging to coordinate German development cooperation on the ground. Furthermore, the division of responsibilities can be confusing for other actors (including other donors and NGOs), especially since the implementing agencies cannot act on behalf of the German government (BMZ, 2008b: 68). Steps have been taken to address these uncertainties, such as the establishment of country teams including representatives from ministries and the implementing agencies at headquarters and in the field (BMZ, 2008b: 69; see also BMZ, 2004b), and the appointment of priority area coordinators in the partner countries (OECD, 2006c: 59).

Strategic planning of German bilateral cooperation takes place through multiannual country strategy papers, which outline how the BMZ's objectives are implemented in a partner country and identify the priority area(s). Priority area strategy papers indicate which mix of aid instruments is appropriate, how the different instruments are connected and which indicators are to be used to measure progress towards country objectives (BMZ, 2008b: 50). The partner country government has a limited role in drafting the country strategy papers, although such papers are to be in line with that country's priorities and strategies, and require consultation with stakeholders (OECD, 2006c: 58–9). The country and priority area strategies form the basis for bilateral negotiations leading to a binding international agreement on development cooperation between Germany and the partner country.

The implementing agencies put the strategies into practice. The procedures for the GTZ and KfW are provided in the 2007 'Guidelines for Bilateral Financial and Technical Assistance with Developing Countries' (BMZ, 2008e). The BMZ uses different frameworks for cooperation with its implementing agencies: for technical cooperation, it employs the standard 'Development Policy Framework for Contracts and Cooperation', which provides an overview of the key aspects of the development measures. The framework leaves the GTZ some scope for discretion (OECD, 2006c: 59). For financial cooperation, the KfW first carries out an appraisal of a proposed activity, and then, after this has been approved by the Ministry, it can enter into contracts with implementing organizations in the partner country.

7.4.3 *Incorporating adaptation and mitigation*

Although consideration of climate change in development cooperation had already received attention in the early 1990s (BMZ, 1993), this has been strengthened in recent policy documents. The 2008 White Paper identifies protection of the environment and climate as one of the four guiding principles for German development cooperation (BMZ, 2008a). Furthermore, climate change is the main focus of bilateral activities in several partner countries (e.g. Indonesia and Tunisia). Climate-related activities consist of both a specific climate-relevant portfolio and efforts to incorporate climate change considerations into the general portfolio. While the emphasis was initially on mitigation activities, adaptation has become a more prominent item in the last few years. Mitigation activities are justified by the need to induce economies like those of China and India to take stronger action on climate change, while adaptation activities are centred on LDCs (Michaelowa and Michaelowa, 2007: 18–19). Climate change concerns are to be incorporated since this issue affects the provision of global public goods such as climate stability, which is necessary in order to reach broader development goals, including poverty alleviation.

Most climate-relevant development assistance flows through the BMZ, which spent Euros 450 million on the issue in 2007 (Scholze *et al.*, 2008: 174), and intends to spend around Euros 1 billion annually from 2008 onwards. Funding for climate activities has also become available through the government's International Climate Initiative, which set aside Euros 120 million both in 2008 and in 2009 (BMU, 2009: 4), almost all of which is ODA. This initiative 'contributed substantially to increasing the ODA/GNP ratio of the German federal government' (BMU, 2009: 16). It is managed by the Federal Ministry for the Environment, Nature Conservation and Nuclear Safety, while the BMZ is responsible for ensuring coherence with development cooperation policy objectives. The initiative focuses heavily on mitigation activities in Asia, but adaptation also plays an important role (BMU, 2009: 17). Although there is ongoing coordination between the two ministries, more strategic coordination could enhance the policy coherence of German efforts to incorporate climate change.

Strategies for adaptation and mitigation are being defined in 2009. While these strategies are of a conceptual nature, they will most probably be accompanied by a strategic 'guideline' containing mandatory operational guidance for the implementing agencies on how to incorporate climate change into development activities, specifically through the country and priority area strategy papers. Prior to the guideline, operational guidance had already been developed in the implementing agencies. Indeed, the guideline itself is inspired by the 'climate check' developed by the GTZ (see below), while the KfW has included climate protection in its operations since the late 1990s. The guideline will need to be operationalized for each implementing agency.

Climate change mitigation and adaptation features in approximately one-fifth of the GTZ's development activities (GTZ, 2008: 4–5), whereas 39% of the KfW's portfolio included climate-related projects (KfW, 2008: 69). In the early 1990s, the GTZ had started the Climate Protection Programme on behalf of the BMZ, but its projects were not systematically screened for potential climate risks (Klein, 2001). By now, however, it has created a separate division on Environment and Climate Change with a Task Force on Climate Change, and is actively incorporating climate change considerations into many sectors of development cooperation. It employs several experts in Germany and in the field in various climate-relevant sectors, provides policy advice to Germany and partner countries, and screens its portfolio (see below).

Climate change has also become important in the activities of the KfW, which devoted its 2007 annual report to the issue (KfW, 2008). Following the 2004 Renewables Conference organized by the German government, the KfW and BMZ set up the Special Facility for Renewable Energies and Energy Efficiency with Euros 1.3 billion for loans to DCs. The Special Facility was followed by the Initiative for Climate and Environmental Protection, which aims to provide an additional Euros 2.4 billion in loans to climate-friendly projects in DCs between 2008 and 2011 (KfW, 2008). In 2009, the KfW shifted the operation of its carbon fund into the development bank, facilitating the integration of development aspects into carbon finance.

Mitigation activities include forest conservation and biodiversity protection, CDM capacity building, promoting the use of renewable energy in partner countries and enhancing energy efficiency (BMZ, 2007b). They include Euros 500 million for renewable energy programmes and energy efficiency measures between 2003 and 2007, funded through the 'Sustainable Energy for Development' programme (BMZ, 2007c). This programme has sought to reduce the CO_2 emissions from development activities and to provide advice to partner country governments on policies promoting renewable energy and energy efficiency. Germany also supported capacity building for CDM projects in many countries, but most prominently in China and India (GTZ, 2008), and more recently focused on Africa.

Adaptation activities include support for incorporation of climate change concerns into national policymaking, for example in Tunisia for the agricultural sector and in Indonesia for the water sector (GTZ, 2007a; GTZ, 2007b). Furthermore, close cooperation on adaptation research with research institutes has been established by the GTZ. In general, Germany aims to incorporate climate risks systematically into existing and new programmes and portfolios (Scholze *et al.*, 2008: 174).

Germany's climate assessment, which is based on the 'climate check' developed by the GTZ and the Potsdam Institute for Climate Impact Research and implemented together with the KfW, screens for climate risks (through a climate-proofing tool) and mitigation opportunities (through an emission-saving tool). The climate-proofing tool aims at increasing adaptive capacity and undertaking adaptation measures to reduce these risks. The emission-saving tool seeks to ensure that development activities result in the maximization of reductions in greenhouse gas emissions by the development programmes (Scholze *et al.*, 2008: 176). Projects or programmes are first screened with regard to climate risks and reduction potentials, followed by an in-depth analysis where necessary. Furthermore, measures are applied to guarantee the incorporation of the recommendations into the project or programme design (from proposal to monitoring).

This methodology is accompanied by training of staff, and the tools are to be integrated with project and programme appraisals that are carried out anyway. The climate check has been designed in a flexible way, in order to avoid 'mainstreaming fatigue'. The tool has been piloted in partner countries, including Bolivia, Brazil, India, Morocco and Vietnam, and in the future up to 400 projects might be screened annually (Schemmel and Scholze, 2009). The climate check will be systematically applied at project and programme level. Apart from the climate check, the GTZ has also developed a non-mandatory portfolio screening tool at national level. This is used to relate the development activities to the mitigation potential and adaptation needs in a partner country (or sector in a country), and to provide recommendations for adjustment of the portfolio (Scholze *et al.*, 2008).

7.4.4 Conclusion

Since the early 1990s, Germany has been actively exploring the links between climate change and development cooperation, and these activities significantly increased in the early 2000s. The organizational complexity partly obscures the extent to which climate change is being incorporated, but clearly there are many ongoing efforts to include adaptation and mitigation in development activities at various levels. At the policy level, the BMZ has prioritized the issue, by linking its role to the achievement of broader development objectives, commissioning strategies for adaptation and mitigation, allocating financial resources and providing operational guidance for various levels of the aid organization. At the operational level, the GTZ has set up a separate Division on Environment and Climate Change as well as a Task Force on Climate Change and developed the climate check methodology, while the KfW has established a number of financial products to support climate change mitigation and adaptation.

Table 7.4 *Dutch ODA over time*

	1991–2 (average)	1996–7 (average)	2006	2007
Total ODA budget (USD million)	2635	3097	5452	6224
Percentage of GNI	0.87	0.81	0.81	0.81

Source: OECD (2009a: 156–7).

A remaining challenge is that the BMZ's manpower is limited compared with that available in other countries, yet decision-making processes are still highly central-ized. Another obstacle is the perception that climate change is yet another 'main-streaming' issue that needs to be addressed in all development activities. This could be overcome through the climate check methodology which ensures a focus on projects with clear links to climate change.

7.5 The Netherlands

7.5.1 Policy context

The Netherlands has a high commitment to providing ODA, and is regarded as one of the leaders in aid supply (OECD, 2006a: 21). It provided 0.81% of its GNI in 2007 (OECD, 2009a: 124), making it one of the few countries consistently to exceed the 0.7% UN target (Table 7.4). Furthermore, the country ranks second on the Commitment to Development Index on aid (CGD, 2008).

Development cooperation policy focuses on achieving sustainable poverty reduction in the light of the MDGs (MFA, 2007b). The ODA resources are allocated to a number of activities, including humanitarian aid, civil–military operations, political reform, improving the investment climate, education, health, gender, environmental and water, immigration policy, contributions to European and multi-lateral funds and banks, cancellation of export credit debts and creating support for Dutch foreign policy. Most funding is directed at poverty reduction and human and social development, followed by peace and security. In 2007, the government expressed its intention to focus on fragile states, women's rights, sustainable economic growth and equity, and environment and energy (including climate change) (MFA, 2007a).

Aid is provided through multilateral cooperation (29% in 2005), bilateral coop-eration (46% in 2005 – embassies 19%, humanitarian aid 6%, debt relief 8%, other bilateral aid 13%) and cooperation with non-governmental actors (25% in 2005 – civil society 21%, business 4%) (MFA, 2007b: 10).

Multilateral cooperation takes place through contributions to the EU and UN agencies, as well as multilateral and regional development banks, such as the World Bank. Dutch policy aims at ensuring that multilateral financing matches its thematic and sectoral foci (MFA, 2003; OECD, 2006a), and bilateral funds are channelled through the multilateral institutions through earmarked contributions.

Since 1997, the bilateral assistance programme has been influenced by World Bank research (e.g. Burnside and Dollar, 1997; World Bank, 1998) arguing that aid is most effective if provided to countries that adhere to good governance principles. In 2003, 'full cooperation' and 'thematic cooperation' strategies were merged in favour of long-term partnerships with fewer countries (MFA, 2003). After 2003, bilateral cooperation initially focused on 36 partner countries[10] selected for reasons including the level of poverty, adherence to good governance standards, a discernible added value of a bilateral relationship, the capacity of a country to use aid and pre-existing relations (MFA, 2006a: 39–44). Although this aid policy was influenced by the discourse on good governance, other criteria underlying country selection significantly weakened its effect (Hout and Koch, 2006). However, few of these countries were LDCs (MFA, 2006a: 54). The focus on good governance thus 'led to a situation whereby the countries that need help the most do not receive it because they do not meet the criteria for programme support' (MFA, 2007a: 17); see also Chapter 2. The government then amended its policy: it phased out cooperation with seven countries because of their relatively high income, the emergence of new partnerships between the EU and these countries, and/or political difficulties; it added 'fragile states' such as Burundi, the Democratic Republic of Congo and Sudan to the list of partner countries; and it classified partner countries in terms of three 'profiles' to guide aid efforts. This new categorization led to a new list of 42 partner countries. The profiles for the partner countries are as follows.

- Accelerated achievement of the MDGs: aimed at the poorest countries that are lagging behind in the achievement of the MDGs, but which have 'a reasonable level of stability' (MFA, 2007a: 38).
- Security and development: focused on 'fragile states' where achievement of the MDGs has become difficult, with a view to enhancing 'the legitimacy and capacity of government to enable it to maintain security and provide essential social services' (MFA, 2007a: 42).

---

[10] These are Afghanistan, Albania, Armenia, Bangladesh, Benin, Bolivia, Bosnia-Herzegovina, Burkina Faso, Cape Verde, Colombia, Egypt, Eritrea, Ethiopia, Georgia, Ghana, Guatemala, Indonesia, Kenya, Macedonia, Mali, Moldova, Mongolia, Mozambique, Nicaragua, Pakistan, Palestine, Rwanda, Senegal, South Africa, Sri Lanka, Suriname, Tanzania, Uganda, Vietnam, Yemen and Zambia. Dutch ODA reaches other countries too. In fact, the top recipient in 2006–7, Nigeria (OECD, 2009a: 227), is not on this list. OECD (2006a) notes that in 2005 disbursements were made to 125 countries in total.

- A broad-based relationship: targeted at relatively high-income countries that have made progress towards achieving the MDGs, but which could use support to achieve certain MDGs as well as broader economic development objectives (MFA, 2007a: 44).

Partnerships with non-governmental actors, including civil society organizations and business, constitute the third channel of aid provision. The Netherlands uses a co-financing system, under which NGOs applying for funding of development projects need to show that they can secure additional funding (25%) from other sources (OECD, 2006a). In addition, some NGOs receive funding from the Ministry independently.

Dutch bilateral aid has moved from project-based to programmatic or sector-wide approaches (including budget support), emphasizing partner country ownership. In addition to reducing the number of partner countries, the Netherlands focuses on two or three sectors chosen on the basis of the partner country's national strategies, institutions and processes. The focal sectors are in general education, health, environment, water and private-sector development, as well as the themes of sexual and reproductive health and rights, and HIV/AIDS. Good governance and human rights are cross-cutting themes throughout development cooperation policy (MFA, 2007b: 8).

Sector-wide approaches have become an 'organising principle' (MFA, 2006a: 26) making programme aid and, where possible, budget support the preferred aid modalities (OECD, 2006a: 66). According to the OECD (2006a: 17), the Ministry of Foreign Affairs 'considers budget support as the most effective form of aid since it ensures that partner countries assume responsibility for implementing their own development agenda and contributes to a better alignment of aid with policy and systems of partner countries'. However, the transition from project-based to sector-wide approaches has not been smooth, because of a lack of capacity in government agencies and partner countries, and the inevitable trade-offs between the advantages of more structural support and the risks of not reaching marginalized communities (MFA, 2006a). Government-to-government aid has meant that donors lost touch with the poorest communities. In response, the Dutch government proposed to seek more active participation of local stakeholders in the planning and monitoring of development policy (MFA, 2007a).

7.5.2 *Operational context*

The primary responsibility for the coordination of ODA resources lies with the Directorate-General for International Cooperation in the Ministry of Foreign Affairs, which is responsible for about 80% of the ODA resources (OECD, 2006a). The agency has several thematic departments, including the Environment

and Water Department. Other departments indirectly involved (e.g. through monitoring and evaluation) include the Effectiveness and Quality Department, which is
responsible for overseeing the implementation of the Paris Declaration. Country
teams for the partner countries have been established to support the embassies
(MFA, 2008a).

The Ministry of Foreign Affairs relies on cooperation with civil society organizations, and has to a very large extent decentralized the ODA delivery process to
embassies in partner countries (van Steenbergen, 2008: 10). The embassies maintain
contacts with partner country governments, coordinate with other donors, formulate
support to country and sector policies, manage finances, and assess, approve and
monitor the implementation of activities, including budget support (OECD, 2006a;
MFA, 2008a).

To align Dutch policy priorities with the partner country context, various instruments are used (MFA, 2008a: 24–5). Embassies need to develop four-year multi-
annual strategic plans in which they specify the concrete goals and development
activities in pursuance of the MDGs (OECD, 2006a: 55–6). The Ministry reviews
the plans and needs to approve them (EuropeAid, 2008: 20). Embassies use a 'track
record' tool to choose aid modalities (budget support, programmatic or project-
based aid, or a mix thereof) by annually assessing the partner countries' performance on their commitment to poverty reduction, macro-economic policy and business climate, good governance and dialogue and harmonization (OECD, 2006a:
57). Finally, embassies prepare a 'sector track record' analysing and monitoring the
priority sectors in partner countries, and selecting the appropriate aid modalities.
The Department for Effectiveness and Quality coordinates the assessment and
approval process of the general track records, whereas the thematic departments
are responsible for ensuring that the strategic decisions in the multi-annual strategic
plans are in line with the sector track records (MFA, 2008a: 24–5).

Procedural guidelines based on the policy document on management and supervision (MFA, 1998) are used to select, assess and implement specific development
activities (MFA, 2008a: 26–7). To ensure funding for a specific activity, an internal
appraisal document needs to be prepared by embassies. The procedures for sector or
project aid are slightly more stringent than those for budget support, since an
organizational capacity assessment of the counterpart is required (van
Steenbergen, 2008: 11).

Central governments in partner countries play a role in the implementation of
Dutch development cooperation policy, with capacity-building support from the
embassies where needed. In several countries, civil society and the private sector are
also actively involved. However, the level of ownership among citizens in some
countries is still limited (MFA, 2008a: 27–8).

Table 7.5 *Distribution of Dutch environmental ODA by channel (Euros 1000)*

Channel	2005	2006	2007	2008
Bilateral	212 012	241 047	261 623	286 670
Multilateral	131 969	103 172	117 568	103 682
Non-governmental	133 037	167 001	210 333	215 955
Other	24 465	26 591	25 834	29 085
Total	501 482	537 811	615 358	635 392
Percentage of GDP	0.100	0.100	0.109	0.105
Percentage of ODA	11.85	11.46	12.70	12.60

Source: MFA (2006b; 2007d; 2009).

7.5.3 Incorporating adaptation and mitigation

Although climate change is not a focal area of Dutch aid, it received attention in 2007 (MFA, 2007a), while several sectors and themes deal with it also in relation to the MDGs, including MDG 7 on environmental sustainability. The Netherlands dedicates 0.1% of GDP – or about 12.5% of total ODA – to environmental protection, including water (van Steenbergen, 2008); see Table 7.5.

Table 7.5 shows that between 2005 and 2008 most green aid was delivered bilaterally. The main recipients of green aid were the partner countries where the environment or water is a priority sector.[11] Bilateral assistance in the environment sector was mainly through projects due to 'the varied and sometimes fragmented nature of the environmental sector, and the fact that responsibility is often shared by more than one ministry' (MFA, 2007b: 66), although this category includes activities that could also be regarded as sector support (van Steenbergen, 2008: 76). Implementing a sector-wide approach for the environment has been challenging, since the definition of the sector is unclear, environmental management is often carried out in a decentralized fashion, and the relation between aid input and environmental impacts is often not straightforward (van Steenbergen, 2008). Nevertheless, sector-wide approaches are used in several partner countries, including Cape Verde, Colombia, Senegal and Vietnam, where embassies employ specialized staff and engage in high-level discussions with partner countries (van Steenbergen, 2008: 37). Environmental activities include the incorporation of environmental concerns into other activities, e.g. promoting good governance, poverty reduction and macro-economic policy (MFA, 2007b), but also activities in other sectors (e.g. water).[12]

---

[11] Albania, Cape Verde, Colombia, Ghana, Guatemala, Mali, Mongolia, Pakistan, Senegal, Sri Lanka, Suriname and Vietnam. However, environmental ODA was not limited to these countries.

[12] For instance, the top three recipients of green aid in 2004 according to the OECD are countries for which environment is not one of the priority sectors (van Steenbergen, 2008: 21).

Within the environment sector, climate change, energy, forests, biodiversity and desertification are key themes. In 2007, the government indicated that '[m]ore attention needs to be given to the effects of climate change and the need for a more responsible energy policy' (MFA, 2007b: 77; see also MFA, 2007a: 34–6). To enhance access by the poor to sustainable energy by 2015, the government pledged Euros 500 million for the period 2008–13. Whereas Euros 125 million will be generated from existing ODA, the remainder will be new and additional. This funding is for direct investments for the production, access to and efficient use of renewable energy, particularly in Africa (Euros 470 million); promoting sustainable biomass production, including biofuels (Euros 30 million); influencing renewable energy policies of important partners, including the G8, the EU and the World Bank (no specific funding); and capacity development on renewable energy (no specific funding) (MFA, 2007c). Other mitigation activities include support to partner countries (e.g. in Indonesia, the Congo Basin and the Amazon) concerning emissions from deforestation (MFA, 2007a). Furthermore, CDM capacity building.

Adaptation activities are becoming more prominent in Dutch aid (e.g. MFA, 2007a). In 2008, Euros 18 million was spent on bilateral adaptation activities, whereas for 2009 the government projected to spend Euros 34 million.[13] The implications of climate change were analysed in a 'quick scan' of three partner countries (Bangladesh, Bolivia and Ethiopia). The studies indicated that 'climate change poses a real and immediate risk to poverty reduction' and that climate risk management should be integrated from the start of strategic planning at various levels (van Aalst *et al.*, 2007: 23). Furthermore, the quick scans identified and helped prioritize climate risks in development activities. Still, the quick scans were organized in an ad hoc fashion, and systematic screening tools are not under development.

Climate-related activities should be seen in the context of an increased focus on sector-wide approaches and partner country 'ownership'. The Netherlands provides advice and builds capacity and knowledge in partner countries, for example in the water sector (MFA, 2008b). Although the Netherlands seeks to raise the issue of climate change in the partner country, and provides expertise in this regard, the partner country ultimately decides how climate change fits into its development strategy. The sector-wide approach has yielded results at the institutional level in the partner country through connecting the often-marginalized environmental departments to other, more powerful departments, while budget support addressed more systemic problems. The remaining challenges include the shift of development activities to the central government level in partner countries and the reduced role

---

[13] Letter from the Ministers of Housing, Spatial Planning and the Environment and of Development Cooperation to the Dutch Parliament (16 February 2009).

Table 7.6 *British ODA over time*

	1991–2 (average)	1996–7 (average)	2006	2007
Total ODA budget (USD million)	3222	3316	12 459	9849
Percentage of GNI	0.32	0.27	0.51	0.36

Source: OECD (2009a: 156–7).

of sub-national authorities at 'the risk of losing touch with reality and implementation issues at the lowest level' (van Steenbergen, 2008: 74).

7.5.4 Conclusion

Although environmental concerns, including climate change, are important and at least 0.1% of GDP is budgeted for it, the focus on partner country ownership through the sector-wide and budget support approaches implies that partners have a stronger role in defining the priorities. However, the focus on central governments risks neglecting the implementation on the ground, which is particularly important in environmental management and poverty reduction.

The various policy documents show that the Netherlands focuses on increasing the use of renewable energy in partner countries, reducing the rate of deforestation and reducing the climate risks in partner countries, especially in the water sector. Furthermore, what is clear is that the Dutch government seeks to provide resources for climate-related development cooperation in addition to the 0.7% ODA.

7.6 The United Kingdom

7.6.1 Policy context

The UK's ODA policy aims at poverty reduction through achieving the MDGs. The UK has not met the 0.7% target, and its ODA/GNI decreased from 0.51% to 0.36% in 2006–7 (OECD, 2009a: 132), mainly due to the provision of lower amounts of debt relief. Still, in absolute terms it was the fourth largest OECD donor in 2007, while total ODA was more than three times the ODA in 1991–2 (Table 7.6), and its aid is steadily increasing with a view to reaching the UN target by 2013 (Concord, 2009: 8). On aid delivery it ranked sixth (out of 22) on the Commitment to Development Index (CGD, 2008).

The 2002 International Development Act provides a legal mandate for the Department for International Development (DFID) to provide development assistance with the objective of reducing poverty. Government-wide 'Public Service Agreements' provide specific goals, against which the Department's performance can be measured and evaluated. The most important performance indicators in this regard relate to the MDGs (HM Treasury, 2007a). Furthermore, the Department has seven strategic objectives against which progress is measured (DFID, 2008a: 14–15):

- promoting good governance, economic growth, trade and access to basic services;
- promoting climate change mitigation and adaptation measures and ensuring environmental sustainability;
- responding effectively to conflict and humanitarian crises and supporting peace in order to reduce poverty;
- developing a global partnership for development (beyond aid);
- making all bilateral and multilateral donors more effective;
- delivering high-quality and effective bilateral development assistance; and
- improving the efficiency and effectiveness of the organization.

Most aid is provided bilaterally (57% in 2007), with the remainder flowing through multilateral institutions, in particular the EU, the World Bank and, to a lesser extent, UN institutions and regional development banks (OECD, 2009a: 188–9). The UK's contribution to multilateral institutions has increased in the past few years, with a view to increasing the effectiveness of the international system at reducing poverty. The activities and goals of multilateral institutions are coordinated with domestic objectives through institutional strategies, which are developed together with the multilateral institutions and civil society (OECD, 2006a: 34–5).

Bilateral aid flows to a wide range of partner countries, with India (9% of bilateral ODA in 2007), Ethiopia and Nigeria (both 5%) being the top recipients (DFID, 2008b: 30). In the future, the government intends to focus on fewer countries (DFID, 2009). The UK considers aid partnerships to be effective if a partner country is committed to poverty reduction and achieving the MDGs, respects human rights and international obligations, improves financial management, promotes good governance and transparency, and addresses corruption (DFID, 2006a: 23). The focus has mainly been on the poorest countries, particularly in (sub-Saharan) Africa and South Asia (DFID, 2006a). The UK was the largest OECD donor to low-income countries in 2006–7, allocating more than 85% of ODA to this group (OECD, 2009a: 209); its target is at least 90% (HM Treasury, 2007b: 4). To determine the allocations of aid among the low-income countries, a resource-allocation model is used, which takes into account the level of poverty and the likelihood of UK aid contributing to poverty alleviation (OECD, 2006b: 31). The UK also seeks to support fragile states that might not have a good track record in terms of good governance criteria, but where aid is much needed (DFID, 2006a; DFID, 2009). Finally, the historical context also plays a role, with some of the major recipients, such as Nigeria and India, being members of the British Commonwealth (OECD, 2006b: 21).

British aid follows thematic rather than sectoral priorities, which are strongly aligned with the MDGs through the Public Service Agreements (OECD, 2006b: 38). British ODA goes to economic infrastructure (15.4% in 2007), government and civil society (14.3%), programmatic aid (10%), health (8.1%) and education (12.1%) (OECD, 2009a: 196–7).

British bilateral aid focuses increasingly on 'poverty reduction budget support', which encompasses both general and earmarked sector budget support, if countries' circumstances are 'appropriate' (OECD, 2006b: 71). The appropriateness is assessed by looking at a partner country's budget priorities for poverty reduction, the administrative, technical and financial capacity, and the costs and benefits compared with other aid modalities (DFID, 2004: 4–5). Although budget support has become the preferred modality (NAO, 2008), the UK relies on a mix of modalities, including non-budget sector-wide approaches, projects and programmes, as well as technical cooperation (DFID, 2008b: 13). Project aid is used for policy experiments, demonstration purposes, transferring skills and, increasingly, for supporting non-state actors (DFID, 2004: 5).

7.6.2 Operational context

Since 1997, the DFID has coordinated aid policy. Strategic decisions are made by the management board (consisting of a permanent secretary, four directors-general and two non-executive directors). The directors-general supervise divisions on country programmes, international affairs, policy and research, and corporate performance. The country programmes division supervises country offices, while the international division controls multilateral activities. The policy and research division supports policy development, whereas the corporate performance division is responsible for financial and corporate performance, human resources and evaluation, among other things (DFID, 2008a: 183–4).

The operational activities have been decentralized, with about half of the department's staff being stationed in country offices abroad (OECD, 2006b: 55). Offices in countries where the DFID spends more than £20 million (about USD 33 million) per year are responsible for developing and implementing multi-annual country business plans. These plans outline how ODA can contribute to the MDGs in a partner country, and provide a framework for monitoring and evaluating the Department's performance. The plans include an explanation of the chosen aid modalities as well as a risk assessment, and need to be in line with a partner country's poverty reduction strategy (OECD, 2006b: 59). External and internal consultations are compulsory, although the intensity varies according to the circumstances of a particular country. Aside from consultations with other departments within the UK government, external consultations are conducted with NGOs and other national organizations (DFID, 2007a).

Proposals for aid spending need to be accompanied by project documents, with enhanced stringency for activities over £1 million (about USD 1.6 million) (DFID, 2008c). Country offices have considerable discretion when it comes to the distribution of financial resources for decisions up to £5 million or £7.5 million (about USD 8.2 million or USD 12.4 million) (EuropeAid, 2008: 39). Above those amounts, the director for the regional programme has to authorize expenditure, while spending over £20 million (about USD 33 million) needs to be approved at the highest level by Ministers (Thornton and Cox, 2008: 24).

7.6.3 Incorporating adaptation and mitigation

Various policy documents on development cooperation have emphasized climate change (e.g. DFID, 2000; 2005; 2006a; 2009). The 2006 White Paper acknowledged that '[c]limate change poses the most serious long term threat to development and the [MDGs]' (DFID, 2006a: 91), and made climate change one of the four strategic priorities. The heightened attention to climate change in the DFID results in part from an increase in overall attention in the British government since 2005, including hosting the G8 Summit in Gleneagles in that year. The most recent White Paper assigns even more importance to climate change, highlighting the fact that the response to climate change can be either a development disaster or a success. In the White Paper, the government pledged to pursue an 'ambitious, comprehensive and equitable' global climate deal in Copenhagen in December 2009 and to 'provide new and additional public finance for climate change over and above existing development assistance commitments' (DFID, 2009: 50). In addition, the government has acknowledged that some climate-related spending is also poverty-related, but limits this to at most 10% of ODA (DFID, 2009: 53). The White Paper outlines three main activities for the UK: building knowledge and capacity on climate change; increasing investments in low-carbon development in the energy and forestry sectors; and helping DCs adapt (DFID, 2009: 56).

The UK has made aid commitments for several climate-related activities (Table 7.7). In 2007, the government announced a new £800 million (about USD 1318 million) international window for its Environmental Transformation Fund in order to 'support adaptation to climate change, provide access to clean energy and help tackle unsustainable deforestation' during the period 2008–11 (HM Treasury, 2007b: 240). The fund supports the World Bank-administered Climate Investment Funds, comprising the Strategic Climate Fund and the Clean Technology Fund, as well as the Forest Carbon Partnership Facility and the Congo Basin Forest Fund. Of

Table 7.7 *Specific climate change commitments in British development cooperation*

Purpose	Commitment	ODA
Environmental Transformation Fund, including	£800 million	Yes
protection of rainforests in the Congo Basin	£50 million	
low-carbon development in China	£50 million	
Forest Carbon Partnership Facility	£15 million	
Research and capacity building on impacts and enhancing climate resilience, including	£100 million	Yes
focus on Asia and Latin America	£50 million	
focus on Africa (CCAA programme)	£24 million	
Chars Livelihoods Programme (Bangladesh)	£50 million	Yes
Special Climate Change Fund/Least Developed Countries Fund	£20 million	77% is ODA

Source: DFID (2008a; 2008d).

the £800 million, a £50 million (about USD 82.4 million) grant supports rainforest conservation in the Congo Basin, with the same amount for energy efficiency and renewable energy in China, if China elects to access the Clean Technology Fund. In addition, £15 million (about USD 24.7 million) has been allocated to the World Bank's Forest Carbon Partnership Facility, which aims to assist DCs tackling deforestation. British NGOs have expressed reservations about the Environmental Transformation Fund, for the following reasons: aid will partly be provided in the form of concessional loans rather than grants;[14] the World Bank is the trustee, and arguably does not take the interests of partner countries sufficiently into account; and ODA is used for climate funding.[15] The UK has responded to these concerns. While 10% of the UK money is grant funding and 90% consists of loans, the loans are mainly concessional and equivalent to a grant element of about 75%. Furthermore, the trust fund committees that govern the World Bank funds are made up of equal numbers of DCs and ICs. Furthermore, the UK's contribution can be seen as new funding that is additional to the UK's agreed ODA development spending.

The DFID also supports disaster-risk reduction and adaptation activities in partner countries. For instance, the contribution to the Environmental Transformation Fund partly goes to the World Bank's Pilot Program for Climate Resilience, which aims to integrate climate resilience into national planning

---

[14] See the *Guardian*, 'UK demands repayment of climate aid to poor nations', available online at http://www. guardian.co.uk/environment/2008/may/16/climatechange.internationalaidanddevelopment.

[15] See, for instance, the letter from British development NGOs to the Secretary of State, available online at http://www.bond.org.uk/data/files/resources/190/feb08_deg_letter_ministers.pdf.

activities (DFID, 2009). In terms of bilateral aid, the Chars Livelihoods programme in Bangladesh seeks to help people adapt to flooding by rebuilding homes above the flood line and by diversifying livelihood options. Furthermore, up to 10% of humanitarian aid following a natural disaster is spent on reducing the vulnerability to new disasters (DFID, 2006b).

Furthermore, the DFID funds various other climate-change-related research programmes (DFID, 2008e), and has agreed to increase its spending for climate research and capacity building to £100 million (about USD 165 million) over the next five years (DFID, 2009: 56). For instance, together with the Canadian International Development Research Centre, it has started the Climate Change Adaptation in Africa programme, which aims to support research and capacity building on adaptation in Africa. Future research will include programmes on climate science and mapping vulnerabilities at various levels, and with a focus on Africa; the incorporation of climate change into policy frameworks; strengthening decision-making on adaptation options; and investigating mitigation options in DCs (DFID, 2008e: 11).

Although the British government sees an important role for itself in the funding of climate-related activities in developing countries, it also emphasizes recipient countries' ownership, and has suggested a 'compact approach', whereby the main decisions on how to spend financial resources on mitigation and adaptation are made by the recipient countries (DFID, 2009: 54). In this context, the DFID intends to establish a Climate and Development Knowledge Network to build capacity through improving developing countries' access to world-class expertise, helping them build climate-resilient economies and societies. The DFID will allocate £50 million over 5 years (2009/10 to 2014/15) to establish and run the network, and is hoping that other agencies will join in co-funding.

The UK has screened the risks of climate impacts and identified adaptation options. In 2005, it argued that '[n]ew investments should take account of current climate variability and future climate change' (DFID, 2005: 24). Subsequent to the 2006 White Paper, multiple studies screened the climate risks of development activities. The first few studies applied the 'Opportunities and Risks of Climate Change and Disasters' (ORCHID) methodology to screen aid portfolios in Bangladesh, India and China in order to find measures integrating disaster risk reduction and adaptation into development activities (Tanner *et al.*, 2007a; 2007b; 2008). A slightly different methodology, 'Climate Risk Impacts on Sectors and Programmes' (CRISP), was piloted in Kenya (Downing *et al.*, 2008). Whereas ORCHID assessed the risks at project level, the CRISP methodology also identifies risks and adaptation options at higher levels, including programmes and sectors.

The CRISP methodology reflects the Department's ambition to move towards incorporating climate change at the programmatic level.[16] This ambition was also expressed in the 2009 White Paper, which calls for a strategic review of the DFID's activities on climate change, to be piloted in eight countries, expanding to all partner countries by 2013 (DFID, 2009: 66). Through the strategic review, the DFID aims to identify how it could best contribute to climate change mitigation and adaptation in partner countries, taking into account local circumstances and activities of other stakeholders. In addition, a climate assessment will be included in the mandatory environmental screening process (DFID, 2009), which is part of normal practice for projects over £1 million (about USD 1.6 million) (DFID, 2003; Paterson, 2006).

Following the 2006 White Paper, an internal implementation plan has been developed. The plan maps the DFID's current engagement in climate-related activities and provides several options to increase this engagement. The plan itself does not set internal targets or present indicators. However, measuring progress towards climate change objectives is part of the regular performance assessments in the context of the department's strategic objectives and the Public Service Agreements.

7.6.4 Conclusion

The UK has been active in incorporating climate change considerations into development cooperation, especially after 2005. Its activities received a strong impetus following the 2006 White Paper, whereas the 2009 White Paper promises to attach even more importance to the issue, arguing that 'climate change is so critical to the prospects of development that it will take centre stage in the UK's international development efforts' (DFID, 2009: 54). Although the DFID has no public strategic document covering all its climate-related activities, the 2009 White Paper provides a coherent policy framework, while regular performance assessment processes make it possible to measure progress. The priority assigned to climate change at the policy level was followed by concrete financial commitments, several activities aimed at screening and reducing the climate risks for its development portfolio, an increase in the budget for climate change research and capacity building, and an increase in the human resource capacity to engage in this area.

While the screening exercises using the ORCHID and CRISP methodologies amounted to a rather ad hoc approach towards screening the British development assistance portfolio, changes are under way so that it will be possible to assess systematically the climate risks and opportunities for the DFID's activities, taking into account lessons learned from these first screening activities.

---

[16] See the minutes of the HMG and NGO meeting on the Climate Investment Funds: Adaptation, 18 February 2008.

7.7 Comparative analysis

7.7.1 General comparison

A comparative analysis of the five member states reveals that the biggest EU donors (France, Germany and the UK) have not yet achieved the UN target, while Denmark and the Netherlands have consistently exceeded it. However, the donors which are lagging behind have made commitments to increase their contributions, while the others have promised to keep theirs above the 0.7% level during the next few years (Table 7.8).

Table 7.9 provides a snapshot comparison of the countries' total ODA, as well as the distribution of bilateral and multilateral aid in 2007. The share of multilateral ODA is relatively small in the Netherlands (~25% in 2007) and quite large in the UK (~43%), while for the other countries the share is about 35%. Furthermore, contributions to the EU are an important part of multilateral aid for France, Germany and the UK. These observations indicate the administrative levels at which the incorporation of climate change into development assistance has to be dealt with.

The effectiveness of aid differs according to the country concerned and for different reasons. Generally, the Danish and Dutch systems are praised for their level of decentralization, while the French and German systems have been criticized for their high levels of fragmentation and complexity. Furthermore, development NGOs consider the British, Dutch and Danish systems to be more transparent than the French and German systems (Concord, 2009: 12). French aid is also seen as rather unpredictable (Concord, 2009: 24). The quality of aid depends also on whether it is used for debt relief. France and Germany devote a large share of their aid to debt relief, whereas British aid is hardly delivered in this form (Table 7.9). The small share of debt cancellation in aid from Denmark and the Netherlands means that, even if this is taken into account, the 0.7% mark would still be passed (Concord, 2009: 8). Debt relief also limits the potential for incorporating climate change. The same holds for aid provided through costs for refugees in the donor countries, or funding foreign students, which occurs in France and Germany (Concord, 2009: 11).

In terms of the geographical distribution of aid, each donor has its own motivations for selecting partner countries. While France, the Netherlands and the UK focus on low-income countries and past colonies, countries like Nigeria also received aid for debt relief. Denmark, followed by the Netherlands and the UK, has the strongest focus on the poorest countries, spending more than half of its ODA on LDCs, mainly in sub-Saharan Africa. The new emerging economies are important recipients of aid from the five Member States, including China (from France and Germany) and India (from Germany and the UK).

Table 7.8 *Aid as a percentage of GNI, and future commitments*

Country	% of GNI in 1996–7	% of GNI in 2007	Commitment to increase/ stabilize ODA/GNI
Denmark	1.01	0.81	>0.8%
France	0.46	0.38	0.51% in 2010; 0.7% in 2015
Germany	0.30	0.37	0.51% in 2010
Netherlands	0.81	0.81	>0.8%
UK	0.27	0.36	0.56% in 2010–11; 0.7% in 2013

Source: OECD (2009a: 157).

Table 7.9 *ODA breakdown in 2007 (USD million)*

Country	Denmark	France	Germany	Netherlands	UK
Bilateral ODA	1 651	6 258	7 950	4 664	5 602
Multilateral ODA, including	912	3 625	4 341	1 580	4 247
EU contributions	238	2 156	2 452	569	2 143
UN agencies contributions	347	235	274	528	576
World Bank contributions	137	541	1 097	185	987
Total ODA	2 563	9 884	12 291	6 224	9 849
Of which: debt relief	123 (5%)	1 485 (15%)	2 867 (23%)	392 (6%)	70 (0.7%)

Source: OECD (2009a).

7.7.2 Comparison of climate change incorporation activities

All five countries view climate change mitigation and adaptation as relevant to their development cooperation activities. Denmark is the only one to have adopted a comprehensive strategy addressing the relation between climate and development cooperation, and to have stated its intention to make all its activities 'climate proof'. Germany and the UK also prioritize climate change, but follow-up strategy documents are still under development or internal. Climate-related activities in the Netherlands' development cooperation rely more on the principle of partner country 'ownership', and there is no overarching strategy for systematically incorporating mitigation and adaptation. Finally, climate change also plays an increasingly important role in French development cooperation through the emphasis on global public goods (see Table 7.10).

Table 7.10 *Overview of activities to incorporate climate change*

	General			Mitigation	Adaptation
	Environment and/or climate as priority area	Climate change strategy	Climate screening tools	CDM capacity building	Adaptation projects
Denmark	Yes	Yes	Yes	Yes, mainly in Africa	Yes
France	Yes (environment)	Yes (for the French Development Agency)	Yes	Yes, mainly in Africa	Yes, mainly in the water sector
Germany	Yes	Adaptation/ mitigation strategies under development	Yes	Yes, first in Asia, now more in Africa	Yes
Netherlands	Yes (environment)	No	Yes	Yes	Yes, mainly in the water sector
UK	Yes	Internal strategy	Yes	Unclear	Yes

Table 7.11 *Climate change Rio Markers 2005–7 (USD million)*

Country	2005	2006	2007
Denmark	82	38.9	83.7
France	199.8	178.6	441.8
Germany	622.2	830.2	...
Netherlands	7.3	0.6	...
UK	0.0	57.2	47.8

Source: OECD Creditor Reporting System database.

Table 7.12 *Overview of screening tools*

Country	Screening tool	Adaptation/ Mitigation	Ad hoc/ systematic	Follow-up
Denmark	Climate change screening reports	Focus on adaptation, but mitigation included	Partly systematic	Yes; DKK 5 million (~USD 0.95 million) per country
France	'Carbon footprint' tool	Mitigation	Partly systematic	Unclear
	Vulnerability analysis tool	Adaptation	Ad hoc	No
Germany	Climate check	Adaptation and mitigation	Systematic	Yes; but no specific funding
Netherlands	Quick scans	Only adaptation	Ad hoc	No
UK	ORCHID	Only adaptation	Ad hoc	No
	CRISP	Only adaptation	Ad hoc	No

On the mitigation side, most of these five countries invest in energy efficiency, renewable energy and combating deforestation. While all are engaged in purchasing CDM credits, the extent to which ODA is used to support these activities is not always clear. Only Denmark has stated explicitly that it uses ODA for capacity building and project development documents, but not for the purchase of credits. Germany and the Netherlands set aside a (small) part of ODA for CDM preparation. Finally, all five countries have developed tools to screen their development portfolio in the context of climate change.

All five Member States have pledged to set aside funding for climate change in the context of development cooperation, as is indicated by the (self-reported) Rio Markers (see Table 7.11). It is not always clear whether publicly made pledges are part of a country's ODA or not, although the countries examined have made this

---

Box 7.1 Aid and climate change incorporation in Italy

Italy, a founding member of the EU in 1958, provided only 0.19% of its GNI as assistance in 2007, making it one of the least generous OECD DAC donors on a relative scale (after Greece, the USA and Japan) (OECD, 2009a: 148). Furthermore, a significant part of this amount was provided as debt relief, and Italy has the highest share of tied aid of all EU donors (ActionAid, 2008; Concord, 2009). The country also failed to meet its 2002 Barcelona commitment to increase assistance to 0.33% in 2006, achieving only 0.2%. In 2007, most aid was channelled through multilateral organizations, while bilateral aid primarily flows to a large number of countries, in particular sub-Saharan Africa (ActionAid, 2008).

The organizational structure of Italian aid has been criticized in recent years for being too centralized, fragmented and non-transparent (e.g. OECD, 2004; ActionAid, 2008; Concord, 2009). In 2007, radical changes were proposed to unite all Italian development cooperation activities under the Ministry of Foreign Affairs. A separate implementing agency was created within the ministry, but political differences have hindered significant reforms of the existing structure (ActionAid, 2008).

Although poverty reduction is the main objective of Italy's development cooperation policy, Italy has struggled to take a coherent approach towards this goal (OECD, 2004). The main priority sectors are health, gender equality and, since 2008, rural and agricultural development and environmental protection. Given the structure of Italian development cooperation, it is difficult to assess whether its development activities are actively taking into account the impacts of climate change. Some bilateral projects, especially in the Mediterranean region, aim at mitigation. In the context of multilateral aid, Italy is also one of the main donors to the United Nations Convention to Combat Desertification (OECD, 2009a: 120).

---

explicit in some cases: the UK's pledge to allocate £800 million (about USD 1318 million) is clearly part of its official aid, although in the long run it has pledged to provide climate funding over and above ODA commitments; Germany's Euros 120 million allocated to the International Climate Initiative is also part of ODA; whereas the Dutch government's (later reduced) pledge to spend Euros 500 million on climate and energy is explicitly *not* part of the country's ODA contribution. Furthermore, according to Concord (2009: 22), Denmark has spent 8% of its 2008 ODA budget on climate change, including preparations for hosting COP-15 in Copenhagen.

The screening tools developed and applied differ to some extent, with three countries (Denmark, France and Germany) including mitigation in their tools. Furthermore, only the German tool is being used systematically for specific types of aid (at project and programme levels), and is mandatory for all implementing agencies. Although the screening reports in Denmark followed a common format,

---

Box 7.2 Aid and climate change incorporation in Poland

Poland, a country with an economy in transition, joined the OECD in 1996 and the EU in 2004. In 1998, it developed its first aid programmes; and, in 2003, it adopted a development cooperation strategy (MFA Poland, 2003). Although its aid provision is increasing, it devoted only 0.09% of its GNI to ODA in 2006. However, it has pledged to provide 0.17% ODA/GNI by 2010 and 0.33% in 2015. Most aid is provided through multilateral channels, although the share of bilateral aid is increasing (PolishAid, 2007: 10). Most bilateral aid flowed towards Eastern European countries, including Montenegro, Ukraine and Belarus (OECD, 2009a: 138).

The Polish development aid strategy aims to contribute to sustainable development, including poverty reduction, thereby contributing to the achievement of the MDGs and the EU's development policy (MFA Poland, 2003). In the 1990s, Poland was still a major recipient of assistance, which contributed significantly to its transformation and the adoption of social, political and economic reforms. This is a motivating factor to assist others. Furthermore, as an EU Member State, the country both receives assistance and is expected to assist others.

Polish ODA is used for environmental protection and access to drinking water (PolishAid, 2007), but no significant efforts at incorporating environmental issues or climate change have taken place. Poland prefers to channel ODA resources to support countries in transition, possibly because it has good experience in this field and there is a likelihood of visibility and success. Furthermore, given its large military, action in Iraq and Afghanistan also has been deemed to be more effective. Still, several projects with a climate-related component have been initiated, such as a reforestation programme in the Kyrgyz Republic (PolishAid, 2007).

---

they ended up looking different because of the various teams involved in the screening exercises in different countries. To some extent, continuity was sought by pledging DKK 5 million (about USD 0.95 million) for each country where screening took place, and follow-up processes are in place. Although recommendations following from the climate check are being put into practice, this is connected to specific funding pledges. In contrast, the British and Dutch screening exercises were mainly conducted on an ad hoc basis, meaning that development cooperation activities are not continuously checked in terms of climate risks and opportunities (see Table 7.12).

Not all EU Member States have the same level of engagement (yet). Boxes 7.1 and 7.2 give a brief description of the (lack of) climate-change-related development activities in two other Member States, Italy and Poland. Whereas climate change can be integrated more easily in countries with a long-standing tradition of aid provision and previous experience of mainstreaming other activities (e.g. gender and environment), countries like Italy and Poland are still in the process of establishing a clear direction for their general development cooperation policies.

7.8 Conclusions

This chapter has provided an overview of the 'mainstreaming' activities in several important aid-providing Member States of the EU. Given the recent emergence of climate change on the development cooperation agenda, this chapter can only draw some provisional conclusions.

In contrast to the strategic approach to climate change taken at EU level (see Chapter 6), Member States take different approaches to incorporating the climate change agenda into their development cooperation strategies. The use of terminology in the documents in various countries is not always consistent (e.g. 'climate-proofing' means something different in Denmark and Germany), and the situation in most Member States is still in a state of flux. These differences primarily concern bilateral aid. However, a substantial percentage of Member State ODA also flows through the EU and the multilateral agencies.

Denmark and the Netherlands are already meeting the 0.7% target. The Netherlands seems to have accepted the idea of 'new and additional' resources (see also Chapter 5), and provides climate change assistance above and beyond the commitment of 0.7% of GNI (see Chapter 4). In contrast, Denmark, the UK, France, Italy and Germany are increasing their ODA via increases in expenditure on climate change activities.

Whereas some countries are still developing ad hoc climate-relevant projects, others are engaging in climate screening their development cooperation portfolio. Of the Member States examined in this chapter, incorporation of climate change is taken most seriously in Denmark and Germany. Denmark's climate change strategy provides a simple but clear overview of the objectives to be reached, specific actions to meet these objectives and indicators through which progress can be measured, as well as 'entry points' identifying the relevant actors in the supply chain for undertaking the action. Germany has advanced most on the operational level, with operational guidance for incorporating climate change existing or under way throughout the aid organization. Although other countries, such as the UK, have also made climate change a priority issue in their development policies, they have not yet made clear how specific actions should achieve those objectives and which tools should be applied. Figure 7.1 illustrates this point, using the categorization introduced in Chapter 3.

Several countries have developed and applied tools for incorporating climate change. Whereas some tools, such as the 'quick scans' introduced in the Netherlands, have been applied in an ad hoc fashion, the more systematic types of

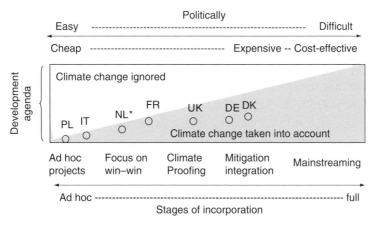

Figure 7.1 Mapping countries in the stages of incorporation of climate change. N.B. The asterisk next to NL denotes that the Netherlands has opted to allow partner countries to develop their own strategies.

screening of new programmes and projects, such as the German 'climate check', better reflect the enduring nature of climate change. The Danish experience shows that the screening process should not be overly complex, and that one should be able to communicate the problem of climate change in understandable development terms.

The structure of the development cooperation bureaucracy in the donor countries ranges from quite complex in Germany and France to relatively streamlined in Denmark, the Netherlands and the UK. It is difficult to point to the exact implications of the bureaucratic framework for incorporating climate change into development assistance; however, the German experience shows that, in complex organizations, guidance needs to be provided at different levels and to different actors.

Some of these countries rank high in terms of the quality of their development assistance. This may be an indicator that, if they try to incorporate climate change into their activities, they may be more successful in the future. However, there is a clear tension here. While demand-driven and partner country approaches may be more successful than supply-driven approaches (see Chapter 2), a strong focus on incorporating climate change into development policies may be seen as a new conditionality, unless the partner country also agrees to this. The Netherlands seems to have chosen to avoid the conditionality approach through extensive dialogue with partner countries, while others use a hybrid approach.

Finally, although various Member States have started to mainstream climate change in their development aid, there are still a few laggards. The brief examples of Italy and Poland cited above showed that including climate change in development aid may need to wait until the development aid organization has reached a certain level of maturity and effectiveness.

Acknowledgements

This chapter has benefited from research carried out by Ramona Banut, Adeline Bichet, Noortje Dijkstra, Arcella Francesco, Anna Meczkowska and Marta Miros. We also wish to thank the various interviewees (Maria Arce Moreira, Eleanor Briers, Aart van der Horst, Michael Linddal, Bertrand Loiseau, Ulf Moslener, Imme Scholz, Michael Scholze, Matthias Seiche and Mike Speirs) for kindly sharing their expertise.

References

ActionAid (2008). *Italy and the Fight against World Poverty: 2008 Report. Ready for the G8 Presidency?* Milan: ActionAid.

AFD (2005). *CIS "Climat" 2006–2008*. Paris: Agence Française de Développement.
   (2007). *Strategic Plan of the Agence Française de Développement: 2007–2011*. Paris: Agence Française de Développement.
   (2009a). *AFD and Climate Change: Balancing Development and Climate*. Paris: Agence Française de Développement.
   (2009b). *Le groupe de l'AFD et la lutte contre le changement climatique. Note de synthèse des résultats de l'Exercice 2008*. Paris: Agence Française de Développement.

Alesina, A. and Weder, B. (2002). Do corrupt governments receive less foreign aid? *American Economic Review*, **92**, 1126–37.

Bird, N. and Peskett, L. (2008). *Recent Bilateral Initiatives for Climate Financing: Are They Moving in the Right Direction?* ODI Opinion No. 112. London: Overseas Development Institute.

BMU (2009). *The International Climate Initiative of the Federal Republic of Germany*. Berlin: German Federal Ministry for the Environment, Nature Conservation and Nuclear Safety.

BMZ (1993). *Climate Protection in Development Cooperation*. Bonn: German Federal Ministry for Economic Cooperation and Development.
   (2001). *Poverty Reduction: A Global Responsibility. Programme of Action 2015. The German Government's Contribution toward Halving Extreme Poverty Worldwide*. Bonn: German Federal Ministry for Economic Cooperation and Development.
   (2003). *Combating Poverty: Our Goals at the Regional Development Banks*. Bonn: German Federal Ministry for Economic Cooperation and Development.
   (2004a). *Anchor Countries: Partners for Global Development*. Bonn: German Federal Ministry for Economic Cooperation and Development.
   (2004b). *Harmonisation and Coordination of Donor Practices in German Development Cooperation*. Bonn: German Federal Ministry for Economic Cooperation and Development.
   (2005). *Enhanced Aid Effectiveness: Focusing German Development Cooperation on the Millennium Development Goals. Implementing the Paris Declaration on Aid Effectiveness*. Bonn: German Federal Ministry for Economic Cooperation and Development.

(2006). *The Development Policy of the European Union*. Bonn: German Federal Ministry for Economic Cooperation and Development.

(2007a). *World Bank Group: Key Messages of German Development Cooperation*. Bonn: German Federal Ministry for Economic Cooperation and Development.

(2007b). *Climate Change and Development: Setting Development Policy Priorities*. Bonn: German Federal Ministry for Economic Cooperation and Development.

(2007c). *Sustainable Energy for Development: Sector Strategy Paper*. Bonn: German Federal Ministry for Economic Cooperation and Development.

(2008a). *Development Policy White Paper: The German Government's 13th Development Policy Report*. Bonn: German Federal Ministry for Economic Cooperation and Development.

(2008b). *Evaluation of the Implementation of the Paris Declaration: Case Study of Germany*. Bonn: German Federal Ministry for Economic Cooperation and Development.

(2008c). *German Bilateral Development Cooperation*. Bonn: German Federal Ministry for Economic Cooperation and Development. Available online at http://www.bmz.de/en/zentrales_downloadarchiv/laender/laenderliste_mit_weltkarte_en_0803111.pdf.

(2008d). *Budget Support in German Development Policy: A Contribution to Increased Aid Effectiveness and Ownership*. Bonn: German Federal Ministry for Economic Cooperation and Development.

(2008e). *Leitlinien für die bilaterale finanzielle und technische Zusammenarbeit mit Kooperationspartnern der deutschen Entwicklungszusammenarbeit*. Bonn: German Federal Ministry for Economic Cooperation and Development.

Burnside, C. and Dollar, D. (1997). *Aid, Policies and Growth*. World Bank Working Paper No. 1777. Washington, D.C.: World Bank.

CGD (2008). *Commitment to Development Index 2008*. Washington, D.C.: Center for Global Development (CGD). Available online at http://www.cgdev.org/doc/2008/CDI%20Main%20Report_2008.pdf.

CICID (2005). *Environmental Protection Sectorial Strategy*. Paris: Le Comité Interministériel de la Coopération Internationale et du Développement. Available online at http://www.diplomatie.gouv.fr/en/france-priorities_1/development_2108/french-policy_2589/governmental-strategies_2670/sectorial-strategies-cicid_2590/environmental-protection-may-2005_3079.html?artsuite=.

CINI (2009). *Planning for Good: Lessons Learnt on Aid Effectiveness from 6 EU Donors*. Rome: Coordinamento Italiano dei Network Internazionali.

Concord (2009). *Lighten the Load: In a Time of Crisis, European Aid Has Never Been More Important*. Brussels: Concord.

Danida (2000). *Partnership 2000: Denmark's Development Policy. Strategy*. Copenhagen: Ministry of Foreign Affairs of Denmark.

(2004). *Environmental Strategy. Strategy for Denmark's Environmental Assistance to Developing Countries 2004–2008*. Copenhagen: Ministry of Foreign Affairs of Denmark.

(2005a). *Danish Climate and Development Action Programme. A Tool Kit for Climate Proofing Danish Development Cooperation*. Copenhagen: Ministry of Foreign Affairs of Denmark.

(2005b). *Modalities for the Management of Danish Bilateral Development Cooperation*. Copenhagen: Ministry of Foreign Affairs of Denmark.

(2005c). *Globalisation: Progress Through Partnership. Priorities of the Danish Government for Danish Development Assistance 2006–2010*. Copenhagen: Ministry of Foreign Affairs of Denmark.

(2006a). *Danida Environment Guide: Environmental Assessment for Sustainable Development*. Copenhagen: Ministry of Foreign Affairs of Denmark.

(2006b). *Organisation Manual for the Management of Danish Development Cooperation*. Copenhagen: Ministry of Foreign Affairs of Denmark.

(2007a). *A World for All: Priorities of the Danish Government for Danish Development Assistance 2008–2012*. Copenhagen: Ministry of Foreign Affairs of Denmark.

(2007b). *Guidelines for Country Strategy Processes (Joint and Bilateral)*. Copenhagen: Ministry of Foreign Affairs of Denmark.

(2007c). *Guidelines for Programme Management*. Copenhagen: Ministry of Foreign Affairs of Denmark.

(2008a). *Denmark's Participation in International Development Cooperation. Danida's Annual Report 2007*. Copenhagen: Ministry of Foreign Affairs of Denmark.

(2008b). *Phase One of the Evaluation of the Implementation of the Paris Declaration: Case-Study Denmark*. Copenhagen: Ministry of Foreign Affairs of Denmark.

(2008c). *Priorities for the Danish Government for Danish Development Assistance: Overview of the Development Assistance Budget 2009–2013*. Copenhagen: Ministry of Foreign Affairs of Denmark.

(2008d). *Guidelines Programme Committee for Bilateral and Multilateral Cooperation*. Copenhagen: Ministry of Foreign Affairs of Denmark.

(2008e). *Mixed Credits: Promoting Growth. Brief Guidelines*. Copenhagen: Ministry of Foreign Affairs of Denmark.

(2009). *Review of 2005 Climate Change and Development Action Programme: Final Draft Report*. Copenhagen: Ministry of Foreign Affairs of Denmark.

DFID (2000). *Achieving Sustainability: Poverty Elimination and the Environment*. London: UK Department for International Development.

(2003). *DFID Environment Guide: A Guide to Environmental Screening*. London: UK Department for International Development.

(2004). *Poverty Reduction Budget Support*. London: UK Department for International Development.

(2005). *Climate Proofing Africa: Climate and Africa's Development Challenge*. London: UK Department for International Development.

(2006a). *White Paper on International Development: Eliminating World Poverty. Making Governance Work for the Poor*. London: UK Department for International Development.

(2006b). *Reducing the Risk of Disasters: Helping to Achieve Sustainable Poverty Reduction in a Vulnerable World*. London: UK Department for International Development.

(2007). *Country Assistance Plan Guidance*. London: UK Department for International Development.

(2008a). *Annual Report 2008*. London: UK Department for International Development.

(2008b). *Statistics on International Development 2003/04–2007/08*. London: UK Department for International Development.

(2008c). *Blue Book: Essential Guide to Tools and Rules*. London: UK Department for International Development (DFID). Available online at http://www.dfid.gov.uk/aboutDFID/blue-book/bluebook.asp.

(2008d). *Degrees of Separation: Climate Change. Shared Challenges, Shared Opportunities*. London: UK Department for International Development.

(2008e). *DFID Research Strategy 2008–2013*. London: UK Department for International Development.

(2009). *Eliminating World Poverty: Building Our Common Future*. London: UK Department for International Development.

Downing, C., Preston, F., Parusheva, D. *et al.* (2008). *Final Report, Kenya: Climate Screening and Information Exchange*. Oxfordshire: AEA Group.

EuropeAid (2008). *EuropeAid Comparative Study of External Aid Implementation Process*. Herts/Hamburg: HTSPE Ltd and GFA Consulting Group GmbH.

FGEF (2005). *French Global Environment Facility (FGEF): Climate Change*. Paris: FGEF secretariat.

France (2002). *Global Public Goods*. Paris: Directorate-General for Development and International Cooperation, Ministry of Foreign Affairs/Treasury Directorate, Ministry of the Economy, Finance and Industry.

(2008). *French Action Plan on Aid Effectiveness*. Paris: Directorate-General for Development and International Cooperation, Ministry of Foreign Affairs/Treasury Directorate, Ministry of the Economy, Finance and Industry. Available online at http://www.diplomatie.gouv.fr/en/IMG/pdf/12p_anglais.pdf.

GTZ (2007a). *Factsheet: Development of a Strategy for Adaptation to Climate Change in the Tunisian Agricultural Sector*. Eschborn: Deutsche Gesellschaft für Technische Zusammenarbeit.

(2007b). *Factsheet: Adapting to Climate Change in Indonesia*. Eschborn: Deutsche Gesellschaft für Technische Zusammenarbeit.

(2008). *Tackling Climate Change. Contributions of Capacity Development*. Eschborn: Deutsche Gesellschaft für Technische Zusammenarbeit.

HM Treasury (2007a). *PSA Delivery Agreement 29: Reduce Poverty in Poorer Countries through Quicker Progress towards the Millennium Development Goals*. London: HM Treasury.

(2007b). *Meeting the Aspirations of the British People: 2007 Pre-Budget Report and Comprehensive Spending Review*. London: HM Treasury.

Hout, W. and Koch, D. J. (2006). *Selectiviteit in het Nederlandse hulpbeleid, 1998–2004: Voorstudie voor de evaluatie van het landenbeleid en de sectorale benadering in de structurele bilaterale hulp*. The Hague: Ministry of Foreign Affairs of the Netherlands.

KfW (2008). *Working Together: For Our Climate. Annual Report 2007 on Cooperation with Developing Countries*. Frankfurt am Main: KfW Bankengruppe.

Klein, R. J. T. (2001). *Measures to Implement the UNFCCC. Adaptation to Climate Change in German Official Development Assistance. An Inventory of Activities and Opportunities, with a Special Focus on Africa*. Eschborn: Deutsche Gesellschaft für Technische Zusammenarbeit.

Linddal, M. (2008). *Completion Note: Climate Change Screening of the Danish Development Cooperation with Nepal, Bhutan and Cambodia, 24 June 2008*. Copenhagen: Ministry of Foreign Affairs of Denmark. Available online at http://ccs-asia.linddal.net.

MFA (1998). *Nota beheer en toezicht*. The Hague: Ministry of Foreign Affairs of the Netherlands.

(2003). *Mutual Interests, Mutual Responsibilities. Dutch Development Cooperation en Route to 2015*. The Hague: Ministry of Foreign Affairs of the Netherlands.

(2006a). *From Project Aid Towards Sector Support: An Evaluation of the Sector-Wide Approach in Dutch Bilateral Aid 1998–2005*. The Hague: Ministry of Foreign Affairs of the Netherlands.

(2006b). *Homogene Groep Internationale Samenwerking (HGIS) Nota 2007*. The Hague: Ministry of Foreign Affairs of the Netherlands.

(2007a). *Our Common Concern: Investing in Development in a Changing World. Policy Note Dutch Development Cooperation 2007–2011*. The Hague: Ministry of Foreign Affairs of the Netherlands.

(2007b). *Results in Development*. The Hague: Ministry of Foreign Affairs of the Netherlands.

(2007c). *Beleidsnotitie milieu en hernieuwbare energie in ontwikkelingssamenwerking*. The Hague: Ministry of Foreign Affairs of the Netherlands.

(2007d). *Homogene Groep Internationale Samenwerking (HGIS) Nota 2008*. The Hague: Ministry of Foreign Affairs of the Netherlands.

(2008a). *"Ahead of the Crowd?" The Process of Implementing the Paris Declaration. Case Study: The Netherlands*. The Hague: Ministry of Foreign Affairs of the Netherlands.

(2008b). *Vaststelling van de begrotingsstaten van het Ministerie van Buitenlandse Zaken (V) voor het Jaar 2009. Memorie van toelichting*. The Hague: Ministry of Foreign Affairs of the Netherlands.

(2009). *HGIS jaarverslag 2008*. The Hague: Ministry of Foreign Affairs of the Netherlands.

MFA Poland (2003). *Strategy for Poland's Development Co-operation*. Warsaw: Ministry of Foreign Affairs of the Republic of Poland.

Michaelowa, A. and Michaelowa, K. (2007). Climate or development: is ODA diverted from its original purpose? *Climatic Change*, **84**, 5–21.

NAO (2008). *Department for International Development: Providing Budget Support to Developing Countries*. London: National Audit Office.

Nunnenkamp, P. and Thiele, R. (2006). Targeting aid to the needy and deserving: nothing but promises? *World Economy*, **29**(9), 1177–201.

OECD (2004). *Italy Development Assistance Committee (DAC) Peer Review*. Paris: Organisation for Economic Co-operation and Development.

(2006a). *Netherlands Development Assistance Committee (DAC) Peer Review*. Paris: Organisation for Economic Co-operation and Development.

(2006b). *United Kingdom Development Assistance Committee (DAC) Peer Review*. Paris: Organisation for Economic Co-operation and Development.

(2006c). *Germany Development Assistance Committee (DAC) Peer Review*. Paris: Organisation for Economic Co-operation and Development.

(2007). *Denmark Development Assistance Committee (DAC) Peer Review*. Paris: Organisation for Economic Co-operation and Development.

(2008). *France Development Assistance Committee (DAC) Peer Review*. Paris: Organisation for Economic Co-operation and Development.

(2009a). *Development Co-operation Report 2009*. Paris: Organisation for Economic Co-operation and Development.

(2009b). *Policy Guidance on Integrating Climate Change Adaptation into Development Co-operation*. Paris: Organisation for Economic Co-operation and Development.

Paterson, C. (2006). *Review of DFID Environmental Screening*. Evaluation Working Paper No. 21. London: UK Department for International Development.

PolishAid (2007). *Development Co-operation Poland. Annual Report 2006*. Warsaw: Ministry of Foreign Affairs of the Republic of Poland.

Roberts, J. T., Starr, K., Jones, T. and Abdel-Fattah, D. (2008). *The Reality of Official Climate Aid*. Oxford Energy and Environment Comment November 2008. Oxford: Oxford Institute for Energy Studies.

Schemmel, J.-P. and Scholze, M. (2009). *The GTZ Climate Check: More than a "Tick Box"*. Presentation, 19 June 2006, on file with the authors.

Scholze, M., Schemmel, J.-P. and Fröde, A. (2008). Climate proofing in development cooperation: up-to-date practical experiences, in *Sustainability: The Unfinished Business Challenges in International Cooperation*, ed. G. Bachmann, S. Paulus and S. Giwer-Marschall. Eschborn/Berlin: GTZ and German Council for Sustainable Development, pp. 173–82.

Schraeder, P. J., Hook, S. W. and Taylor, B. (1998). Clarifying the foreign aid puzzle: a comparison of American, Japanese, French, and Swedish aid flows. *World Politics*, **50**(2), 294–323.

Tanner, T. M., Hassan, A., Islam, K. M. N. *et al.* (2007a). *ORCHID: Piloting Climate Risk Screening in DFID Bangladesh*. Sussex: Institute of Development Studies.

Tanner, T. M., Nair, S., Bhattacharjya, S. *et al.* (2007b). *ORCHID: Climate Risk Screening in DFID India*. Sussex: Institute of Development Studies.

Tanner, T. M., Xia, J. and Holman, I. (2008). *Screening for Climate Change Adaptation: A Process to Assess and Manage the Potential Impact of Climate Change on Development Projects and Programmes in China*. Beijing: Institute of Geographical Sciences and Natural Resources Research, Chinese Academy of Sciences.

Teixeira Santos, S. and de Lopez, T. (2007). *First Bi-Annual Progress Report on the Implementation by the EU of the Action Plan to Accompany the EU Strategy on Climate Change in the Context of Development Cooperation: Final Report*. Rotterdam/Lisbon: ECORYS-NEI and CESO CI.

Thornton, N. and Cox, M. (2008). *Evaluation of the Paris Declaration. DFID Donor HQ Case Study*. London: Agulhas Applied Knowledge.

van Aalst, M., Hirsch, D. and Tellam, I. (2007). *Poverty Reduction at Risk. Managing the Impacts of Climate Change on Poverty Alleviation Activities*. Leusden: Netherlands Climate Assistance Programme.

van Steenbergen, F. (2008). *Managing Environment, Managing Aid: Evaluation of the Sector-Wide Approach in Environment in Dutch Bilateral Aid 1998–2007*. London: Meta Meta, ODI and AidEnvironment.

World Bank (1998). *Assessing Aid: What Works, What Doesn't and Why*. New York: Oxford University Press.

Part IV

Case Studies

8

The need for climate assistance

NICOLIEN VAN DER GRIJP AND JOYEETA GUPTA

8.1 Introduction

This book has argued thus far that climate change, development and development cooperation are closely linked (see Chapter 1). It has further argued that there are various ways in which climate change can be incorporated into the development process and that mainstreaming is the most comprehensive of these approaches. Against this background, this part focuses on the implications of the type of assistance needed by partner countries and provided by donors for the challenge of linking climate change and development cooperation.

This chapter examines the assistance that developing countries (DCs) claim to need in the climate change arena. Evidently, this specific subset must be seen in the context of the general assistance needs expressed by DCs. For example, the closing declaration of the G-77 and China at their summit in 2005 scarcely mentioned any environmental issue (apart from biodiversity) and made no reference to climate change (G-77 and China, Summit Declaration, 2005). Rather, it focused mostly on the Millennium Development Goals (MDGs) and additional issues such as debt relief and intellectual property rights. It also highlighted that development assistance should not be accompanied by any conditionality.

Within the context of the climate change regime, the DCs are preparing National Communications (NCs), National Adaptation Programmes of Action (NAPAs) and Technology Needs Assessments (TNAs) for the Secretariat of the Climate Convention. From these documents, a number of climate change needs can be derived. This chapter highlights these expressed needs of DCs. It does so by undertaking four sectoral case studies, concerning energy, forestry, biodiversity and agriculture, and focusing on a selection of countries.

This chapter first explains the approach (see Section 8.2) and then presents a cross-country, sectoral analysis of the materials (see Sections 8.3–8.6), before drawing conclusions (see Section 8.7).

Mainstreaming Climate Change in Development Cooperation: Theory, Practice and Implications for the European Union, ed. Joyeeta Gupta and Nicolien van der Grijp. Published by Cambridge University Press. © Cambridge University Press 2010.

Table 8.1 *Financial data on case-study countries*

Country	ODA received 2002–6 (annual average) USD million	Rank	ODA as percentage of 2006 GDP	ODA per capita (USD)	Country DAC class
Brazil	166	81	0.0	0.9	Low Middle Income
China	1507	9	0.1	1.1	Low Middle Income
Comoros	28	124	6.8	45.1	Least Developed
Egypt	1109	18	1.0	14.7	Low Middle Income
India	1228	15	0.1	1.1	Other Low Income
Kenya	655	30	3.1	18.7	Other Low Income
Malawi	527	38	23.6	40.1	Least Developed
Nepal	438	43	5.4	15.8	Least Developed
South Africa	634	31	0.2	13.4	Upper Middle Income
Tanzania	1600	8	12.5	40.5	Least Developed

Source: Compiled from OECD DAC statistics.

8.2 The case-study approach

Since within the climate change regime DCs form a large group of approximately 150 states (non-Annex 1 countries), this chapter focuses on 10 countries that are important to the European Union (EU) and its Member States both per se and specifically in terms of climate change. It recognizes that the EU distinguishes among Overseas Countries and Territories, African, Caribbean and Pacific (ACP) countries, Asian and Latin American (ALA) countries, countries in the Mediterranean region, and countries in Eastern Europe and Central Asia. Reflecting these various development assistance programmes of the EU, our selection of 10 case-study countries consists of India (ALA), China (ALA), South Africa and Brazil (ALA) as newly emerging economies that are vulnerable to the impacts of climate change and receive funding as part of bilateral programmes; Kenya, Malawi and Tanzania as ACP countries that fall under the Cotonou Agreement with the EU (see Chapter 6); Nepal as an ALA country that receives aid based on bilateral programmes; Egypt as a Union for the Mediterranean country; and the Comoros as an ACP country and small island state. Table 8.1 provides some key financial data on the case-study countries.

All Parties to the Climate Convention are obliged to prepare a National Communication (NC) following the prescribed format. These NCs provide information about a country's overall situation and specific information relevant to climate change. They specify, inter alia, the needs of developing countries, in explicit and more implicit terms. In addition, the non-Annex 1 countries are encouraged to undertake assessments of country-specific technology needs, called Technology Needs Assessments (TNAs). These TNAs follow a country-driven approach bringing together stakeholders to identify needs, methodologies, areas and sectors to be covered, and to develop plans to meet those needs. To facilitate the assessment of technology needs, the UNDP in collaboration with the Climate Technology Initiative (CTI) developed a *Handbook for Conducting Technology Needs Assessment for Climate Change*. Thus far, the main focus of TNAs has been on energy generation and use, agriculture and forestry, and transport in relation to mitigation; and on agriculture and forestry, water management, and systematic observation and monitoring in relation to adaptation (FCCC Secretariat, 2009).

The National Adaptation Programmes of Action (NAPAs) have emerged from the recognition of the special needs of the 49 Least Developed Countries (LDCs). They are meant to assist them to prepare an analysis of the impacts of climate change and to learn how they can adapt to these potential impacts. It is assumed that the poorest countries will be relatively vulnerable to climate change, while probably the least able to identify their own needs and deal with them without assistance. The NAPAs identify the strategies of local populations to deal with climatic variation and seek to find ways to turn these into national priorities. Hence, community-based information is seen as vital. The NAPAs contain a ranked list of priority adaptation activities and projects, as well as short profiles of each activity or project designed to facilitate the development of proposals for implementation. Table 8.2 presents the key aspects of these three reporting frameworks under the Climate Convention.

The NCs and the NAPAs illustrate how closely intertwined development and climate change policy are. At the same time, they also show that there are major trade-offs. For the emerging economies, these trade-offs are more obvious and they are often defensive in their approach. Brazil argues, for example, that

> Therefore, the developed countries, which have emissions of greenhouse gases dating back to the industrial revolution, bear most of the responsibility for causing climate change. In addition to that, historical emission data indicate that these countries will continue to be the main contributors for another century.
>
> *(Brazil, Government of, 2004: 2)*

It goes on to state in its Executive Summary that this implies that Brazil should not have to make commitments to reduce or limit its emissions. Although clearly there are overlaps between Brazilian development policy and climate change policy, for the purpose of climate change negotiations the Brazilian government tends to see these as distinct. China's NC takes a constructive and yet clear line, arguing that economic growth will result in greater emissions, although aspirations concerning sustainable development will keep such emissions within limits, and that 'China can

Table 8.2 *Main features of the reporting frameworks under the Climate Convention*

	NC	NAPA	TNA
Established in/by	FCCC, (Art. 12)	COP decisions (5/CP.7 and 28/CP.7, 2001)	COP decision (4/CP.7, 2001)
Purpose	To report on action taken on climate change	To assess impacts of climate change and make proposals on adaptation	To assist in identifying and analysing priority technology needs
Who compiles it	Multi-stakeholder effort with guidance of the FCCC	Multi-stakeholder effort with guidance of the FCCC	Multi-stakeholder effort under guidance of the UNDP and FCCC
Who must do it	All parties to the FCCC	Least developed countries	Developing countries
Who finances it	National governments	LDC Fund	GEF and UNEP
Proposed content	Must cover national conditions, inventories of greenhouse gases, impacts, policies and gaps	Must focus on national conditions, impacts and adaptation measures	Must focus on technologies needed for mitigation and adaptation

make positive contributions to mitigating global climate change while emissions have to be necessarily increased' (China, Government of the People's Republic of, 2004: 4). It is also explicit about its status as a developing country and the need for ICs to provide assistance on mitigation, adaptation and research (China, Government of the People's Republic of, 2004: 16).

Like China, India specifies that 'notwithstanding the climate friendly orientation of national policies, the development to meet the basic needs and aspirations of a vast and growing population will lead to increased GHG emissions in the future' (India, Government of, 2004: iii) but that 'by consciously factoring in India's commitment to the FCCC', it has realigned economic development to a more climate-friendly and sustainable path (India, Government of, 2004: xiv). South Africa is more clear in seeing links between its own climate change policy and sustainable development. It states that 'policies relating to climate change are considered as an integral part of the National Sustainable Development Strategy' (South Africa, Government of the Republic of, 2000: 57).

The smaller and less developed countries are less defensive in their approach. Nepal's NC links climate change with development and states that 'Hence, the country is serious in efforts to address climate change issues and integrate it into the

Table 8.3 *Case-study countries and documents submitted (year of publication)*

	NC	TNA	NAPA
Brazil	2004	–	–
China	2004	1998, 2000	–
Comoros	2002	2006	2006
Egypt	1999	2001	–
India	2004	–	–
Kenya	2002	2005	–
Malawi	2002	2003	2006
Nepal	2004	–	–
South Africa	2000	–	–
Tanzania	2003	Not dated	2007

country's economic and social development plans' (Nepal, Government of, 2004: 141). Nepal argues in favour of more active technology transfer and capacity building from the developed countries and hopes to enhance its participation in the CDM. Egypt's NC claims that its climate policies are being integrated into national planning processes (Egypt, Government of the Arab Republic of, 1999: 83). Kenya's NC points out that Kenya's contribution on a global scale is negligible but that it will be affected by climatic impacts (Kenya, Government of, 2002: xxii). Although Kenya is inclined towards accepting the norms of sustainable development, it makes clear that financial assistance and technology transfer from the ICs are necessary.

The following sections compare the case-study countries' demand for climate assistance in relation to the energy, forestry, biodiversity and agriculture sectors. To structure the analysis, the demands for assistance have been categorized into the following types of measures: land use and management; research and technology; institution building; capacity building; social development; awareness and education; and other. Table 8.3 provides details about the documents examined for this study. Note that it is not the case that each of the 10 case-study countries has submitted a NAPA or a TNA.

8.3 Energy sector needs: a comparative assessment

8.3.1 Energy and climate change

The energy sector is a major source of greenhouse gas emissions due to its fossil-fuel dependence. According to the IPCC, the energy supply sector accounted for 25.9% of anthropogenic greenhouse gas emissions in terms of CO_2 equivalence globally in 2004 (IPCC, 2007). On the basis of their energy portfolios and

Table 8.4 *Energy portfolio in case-study countries in 2006*

	Fossil fuel based (%)	Non-fossil fuel based (%)
Brazil	54.6	46.4
China	82.8	17.2
Comoros	–	–
Egypt	95.8	4.2
India	69.1	30.9
Kenya	20.6	79.4
Malawi	–	–
Nepal	11.3	88.7
South Africa	86.8	13.2
Tanzania	8.3	92.7

Source: Based on 2008 data from the OECD/IEA.

dependence on fossil fuels, the 10 case-study countries can be divided into two major groups (see Table 8.4). Brazil, China, Egypt, India and South Africa are primarily fossil fuel dependent, whereas the Comoros, Kenya, Malawi, Nepal and Tanzania rely mostly on non-fossil energy sources, mainly renewables and waste. For the fossil fuel dependent countries, except Brazil, the energy sector is the main source of greenhouse gases.

8.3.2 Mitigation needs

Most of the case-study countries have identified mitigation needs in relation to the energy sector, but Brazil and South Africa have not yet done so (see Table 8.5). Brazil aims to develop other renewable energy sources besides hydropower, whereas South Africa is still considering mitigation options. Overall, choices on energy (development) are directly linked to further economic development. Significantly, Brazil, China, Egypt, India and South Africa emphasize that, at this stage of (economic) development, emission reduction should not hamper their economic growth.

In concrete terms, several case-study countries aim to promote the development and use of renewable energies, a fuel switch to natural gas, energy efficiency and clean coal technologies. Their mitigation needs relate mostly to research and technology, and to a lesser extent to capacity and institution building. Most countries need help in assessing their renewable resources (e.g. the Comoros, Egypt and Tanzania) or nuclear energy (e.g. Egypt), and/or making standards for solar cells and wind generators (e.g. Kenya). China and India have formulated sophisticated technology needs to make fossil fuel use cleaner and more efficient. Some have

Table 8.5 *Energy: mitigation needs*

Country	Type of measure	Need
Brazil	–	
China	Research and technology	High-efficiency coal production technology; technology for exploring and utilizing coal-bed methane; oil hydrogenation technology; high-efficiency, low-pollution coal-burning power-generation technology; large-capacity supercritical thermal-power generation unit technology; technology for recovery and use of natural gas from oilfield rim areas; technology for control of volatile hydrocarbons from oil–gas fields; CFBC; HEEM; technology to increase efficiency in industrial boilers; wind resource assessment; wind turbine design and wind power generation
	Capacity building	CDM project negotiation capacity; CDM project management entities; training programmes on emission-mitigation technologies; preparing project proposals for CFBC, HEEM and increasing thermal efficiency in industrial boilers, wind power and CBM power generation
	Institution building	Consultation and advice on policy implementation to overcome technological barriers; CDM project identification, CDM project procedures; CDM implementation regulations; CDM project monitoring, validation, verification and certification; CDM project standards and indicator system; CDM baseline studies and demonstration projects; information service centre
	Other	Private investors for clean energy technologies
Comoros	Research and technology	Resistant brick fireplaces for distilleries; hydroelectric power technology and plants; wind energy assessment; geothermal assessment

Table 8.5 (*cont.*)

Country	Type of measure	Need
Egypt	Research and technology	Feasibility studies on wind power and nuclear power
India	Research and technology	MSH technology to convert coal to oil; production of fuels and chemicals from CO_2 and CH_4 emissions; soft coke technology; fuel-cell technology; CCS technology; fuel-switch studies; improvement of solar-cell efficiency; emission abatement via *in situ* infusion of fly ash with CO_2 in thermal power plants; studies on technological mitigation options in the energy sector; IGCC technology; biomass gasifier system
Kenya	Research and technology	Wind power technology; wind resource assessment; feasibility studies on mini/micro hydropower; studies on geothermal and solar power, and biogas
Malawi	Institution building	Equipment and installation standards for PV, solar cells and wind generators.
	Research and technology	PV technology; mini/micro hydropower technology; biogas technology; improved cooking stoves; energy efficiency technology; biofuel technology
Nepal	Research and technology	Improved stoves; energy efficiency; solar thermal energy
South Africa	–	–
Tanzania	Research and technology	Energy audits; mini/micro hydropower technology; energy-conservation systems; economic analysis of renewable technology; solar (thermal) technology; wind resource assessment; feasibility studies on energy crops

Notes: CFBC, circulating fluidized bed combustion; HEEM, high-efficiency electric motor; MSH, multi-stage hydrogenation; CCS, carbon capture and storage; IGCC, integrated gasification combined cycle.

Table 8.6 *Energy: adaptation needs*

Country	Type of measure	Need
Brazil	Research and technology	Climatic projections to understand how water resources and hydroelectricity will be affected by climate change
China	Research and technology	Technology for observation and pre-warning against floods and droughts
Comoros	Research and technology	Assessment methodology for climate change risks for forests and fuel wood
Egypt	–	–
India	Research and technology	Software modules for impact assessment of climate change on the energy sector; impact studies of climate change on energy availability
Kenya	Research and technology	Hydrological systems assessments
Malawi	Land use and management	Afforestation and reforestation
	Research and technology	Improved cooking stoves; ethanol use; stand-by power-generation facilities
Nepal	Research and technology	Technology to assess vulnerability to climate change and adaptation measures
South Africa	–	–
Tanzania	Research and technology	Early-warning system for droughts and floods

simpler but focused needs (e.g. the Comoros, heatproof bricks in distilleries; Malawi and Nepal, improved cooking stoves). In terms of capacity building, China requests assistance especially in negotiating, developing and implementing CDM projects. With regard to institution building, China seeks assistance in overcoming technological barriers, setting up institutions for CDM management, and establishing an information service centre. Kenya seeks assistance for establishing standards for renewable energy.

8.3.3 Adaptation needs

With the exclusion of Egypt and South Africa, all of the case-study countries have identified adaptation needs in relation to the energy sector, mostly concerning research and technology (see Table 8.6). Several countries expect that climate impacts may threaten hydroelectricity production (e.g. Brazil and Kenya) and the cultivation of energy crops and fuel wood (e.g. the Comoros and Malawi). Overall, most countries do not seem fully to have assessed the potential impacts of climate change on the energy sector.

Table 8.7 *Forestry data on case-study countries (2005)*

	Forest		Plantations		Change (%)	
Country	Area (1000 ha)	Fraction of land area (%)	Area (1000 ha)	Fraction of forest area (%)	1990–2000	2000–2005
Brazil	477 698	57.2	5 384	1.1	−0.5	−0.6
China	197 290	21.2	31 369	15.9	1.2	2.2
Comoros	5	2.9	1	20.0	−4.0	−7.4
Egypt	67	0.1	67	100.0	3.0	2.6
India	67 701	22.8	3 226	4.8	0.6	NA
Kenya	3 522	6.2	202	5.7	−0.3	−0.3
Malawi	3 402	36.2	204	6.0	−0.9	−0.9
Nepal	3 656	25.4	53	1.4	−2.1	−1.4
South Africa	9 203	7.6	1 426	15.5	0.0	0.0
Tanzania	35 257	39.9	150	0.4	−1.0	−1.1
Total	797 801		42 082	5.3		
World	3 952 025	30.3			−0.2	−0.2

Note: NA, data not available.
Source: Compiled from FAO (2007).

8.4 Forestry sector needs: a comparative assessment

8.4.1 Forestry and climate change

The forestry sector accounts for approximately 20% of global greenhouse gas emissions, mainly through land use change and deforestation. In order to reduce these emissions, there are potentially two major strategies: first, to maintain or increase forest area, density and/or carbon; and second, to increase carbon stored off-site. On the basis of their reforestation and deforestation rates, the case-study countries can be divided into a group with stable or growing forest areas (China, Egypt, India and South Africa) and a group in which net deforestation is occurring (Brazil, the Comoros, Kenya, Malawi, Nepal and Tanzania) (see Table 8.7).

Apart from being a source of emissions, the forestry sector is also expected to be affected by climate change due to increases in temperature, changes to rainfall regimes, fertilization by CO_2 and the increasing frequency and intensity of disturbances such as droughts, floods, pests and fires. On balance, the evidence suggests that there are likely to be more positive than negative effects in economic terms (Easterling *et al.*, 2007), but more negative than positive impacts in terms of ecosystem health (Nabuurs *et al.*, 2007). The major strategies indicated to reduce vulnerability to climate change impacts are first, improving forest management; and second, establishing better institutional structures.

8.4.2 Mitigation needs

Despite differences in levels of forest resources, all of the case-study countries have identified mitigation needs in relation to the forestry sector, with the emphasis on land use and management, and institution building measures (see Table 8.8). Not surprisingly, the countries in which net deforestation is occurring (Brazil, the Comoros, Kenya, Malawi, Nepal and Tanzania) have identified more mitigation needs than have those with stable or growing forest areas (China, Egypt, India and South Africa), although the Comoros lags behind. Needs in the category of land use and management largely refer to reforestation and afforestation and sustainable management practices. Reforestation through plantations can sequester carbon (e.g. in China, Kenya, Malawi and South Africa); can help conserve the environment (e.g. in the Comoros); and can contribute to development (e.g. in Tanzania). Brazil is unique in supporting expansion of plantation forests (for fuel wood). However, with no overall reforestation strategy, it is likely that Brazil sees a certain amount of deforestation as being necessary for its development. The stated needs in the category of institution building mostly relate to joint and community forest management, reform of property and land rights systems, improved land use planning, and strengthening of monitoring and enforcement. Needs in relation to capacity building are focused on education and training for forestry management and afforestation expertise.

Although many countries see the CDM as a vehicle to secure funding, it is significant that global carbon markets have not been listed as a need. This is probably because the national reports preceded the current discussions on reducing emissions from deforestation and forest degradation, and countries may avoid being drawn into discussions on binding emissions targets. What is clear from the reports is that many consider the CDM important and an expansion of eligible forestry activities as a way to access funding.

8.4.3 Adaptation needs

All of the case-study countries have formulated adaptation needs in relation to the forestry sector, although their needs are less numerous than with respect to mitigation. Their adaptation needs are mostly within the categories of land use and management, and institution building (see Table 8.9). Interestingly, there is some overlap with the forestry mitigation needs insofar as the proposed measures include the need for improved forest management policies and practices,

Table 8.8 *Forestry: mitigation needs*

Country	Measure	Need
Brazil	Land use and management	Best practice and techniques for sustainable forest management and agro-forestry; utilization of already-degraded areas for agro-forestry or agriculture; plantation forests for charcoal production
	Institution building	Monitoring and enforcement capabilities of forestry institutions; reforming property rights system to protect national forests, indigenous peoples and traditional forest users; removal of incentives for deforestation from rules and policies
	Social development	Social development and poverty reduction in forest frontier states
China	Land use and management	Recovery and reconstruction of mangroves; eco-protection of forestry and grassland; reforestation and afforestation
	Institution building	Legal framework for forest protection
	Capacity building	Education and training for forestry management; training for protection of mangrove ecosystems
Comoros	Land use and management	Reconstituting forests on basin slopes; restoring degraded soils and reducing erosion
	Research and technology	New building materials substituting wood
	Capacity building	Improved management of forests
Egypt	Land use and management	Plantation management in dry areas; afforestation projects
India	Land use and management	Reforestation and afforestation
	Institution building	Monitoring and enforcement of forestry laws; joint forest management
	Capacity building	Afforestation expertise
	Awareness and education	Local awareness about climate change and forestry
Kenya	Land use and management	Sustainable forestry management and forest conservation; improved use of indigenous knowledge of forest conservation; promotion of agro-forestry, planting of medicinal trees and plantation forestry; afforestation and reforestation projects
	Research and technology	Improved forestry technologies
	Capacity building	Improved silvicultural practices
	Awareness and education	Increase participation and improve education for communities that use forests

Country	Category	Measures
Malawi	Land use and management	Afforestation and reforestation; good land use practices around wetlands; reforestation for fuel wood; agro-forestry on unproductive agricultural land; designated areas suitable for silviculture
	Research and technology	Improved technologies for household heating and cooking (to reduce demand for fuel wood)
	Institution building	Improved monitoring and enforcement of laws on tree protection and ownership; improved forest-fire management at community level
	Social development	Alternative incomes for fuel wood gatherers
Nepal	Land use and management	Adoption of agro-forestry in unproductive croplands; increasing and conserving vegetation coverage; discouraging slash-and-burn agriculture; sustainable forest management
	Research and technology	Technology to reduce fuel wood consumption; accounting framework for measuring potential changes in forest biomass stocks
	Institution building	Community ownership of forests; more protected forest areas; carbon sequestration, trading opportunities and eligibility through the CDM
South Africa	Land use and management	Improved management of vegetation; afforestation; reducing forest-fire frequency; savannah thickening
	Institution building	Land use planning; marketing of afforestation as carbon sinks
Tanzania	Land use and management	Afforestation, reforestation and enhanced natural regeneration and agro-forestry practices; control of alien or invasive species to minimize disturbances; sustainable supply of forest products and services; uptake of diverse management practices on plantations; reducing slash-and-burn agriculture; reducing tree felling by application of alternative materials
	Institution building	Forest protection and conservation; participatory forest management at community level with benefit sharing; sustainable management of forest plantations; enhanced fire control
	Capacity building	Managing and developing the forestry sector in collaboration with other sectors and the international community

Table 8.9 *Forestry: adaptation needs*

Country	Measure	Need
Brazil	Research and technology	Enhanced knowledge of biotechnology and silviculture
China	Land use and management	Afforestation; eco-forests; speedy and lush growth of forests; forests for high-efficiency coke and charcoal
	Capacity building	Prevention and treatment of forest pests
Comoros	Land use and management	Adapting traditional agro-forestry systems
Egypt	Land use and management	Forestry plantation management and species selection
India	Land use and management	Forest adaptation practices; silvicultural practices; enhanced forest resilience
	Institution building	Improved forest policies
	Social development	Reduced vulnerability
Kenya	–	–
Malawi	Land use and management	Improve forest management to reduce climate change impacts; promotion of traditional species
	Institution building	Seed bank to identify and store drought-resistant varieties; joint forest management
Nepal	Land use and management	Extensive planting of species least vulnerable to climate change; habitat management for protected wild animals and plants, with particular focus on buffer zones
	Research and technology	Integrated approach to adaptation research efforts; vulnerability and adaptation assessments, country-specific forestry models and factors
South Africa	Land use and management	Shift location of tree-planting areas
	Research and technology	Genetically engineered hybrid tree species with higher drought and heat tolerance
	Institution building	Seed banks for domestic species
Tanzania	Land use and management	Reduction of habitat fragmentation; promoting migration corridors and buffer zones
	Research and technology	New plant varieties
	Institution building	Forest seed banks

such as afforestation (e.g. China); agro-forestry (e.g. the Comoros); selection of fast-growing, heat- and drought-tolerant varieties for plantations (e.g. China, Egypt, Malawi and Nepal); and the establishment of forest seed banks for species threatened by extinction (e.g. Malawi, South Africa and Tanzania). Other options include fire management and pest/disease control, and forest habitat corridors (e.g. Nepal and Tanzania); and improved forest research and modelling of climate impacts (for all except China, which may have domestic expertise, and Egypt, which has scarcely any forests). Several countries have also recognized that successful action on climate change and forestry would require education and an increased awareness of climate change among their population, though among the nations in which net deforestation is taking place only Kenya and Malawi noted this.

8.5 Biodiversity sector needs: a comparative assessment

8.5.1 Biodiversity and climate change

It is widely held that changes in climatic conditions of regions will have various important effects on organisms and ecosystems, especially for species with limited and/or restricted climatic ranges and habitat requirements (Gitay *et al.*, 2002). It is expected, for example, that many species will experience behavioural changes, reduction in abundance (possibly extinction) and changes in phenology (which refers to the natural cycle of organisms, including migration, time of flowering, hatching, etc.). Many of these changes may lead to a type of chain reaction with cascading effects. In order to minimize loss of ecosystem services arising from the degradation and loss of biological diversity, it is therefore necessary not only to undertake activities to mitigate and adapt to climate change, but also to select strategies that will have the least negative consequences on biodiversity and ecosystems. Several of the case-study countries, are already suffering from considerable habitat degradation and biodiversity loss, including Brazil, the Comoros, Malawi, Nepal and Tanzania. The other selected countries – China, Egypt, India, Kenya and South Africa – seem to be experiencing relatively lesser problems with biodiversity loss.

8.5.2 Mitigation needs

All of the case-study countries, except Egypt, have listed mitigation needs with respect to biodiversity (see Table 8.10). Most of these needs fall under the category of land use and management. They typically entail reforestation (e.g. China, India, Kenya, Nepal and Tanzania), restoration of degraded soils (e.g. the Comoros and

Table 8.10 *Biodiversity: mitigation needs*

Country	Measure	Need
Brazil	Land use and management	Responsible forest resource exploitation
China	Land use and management	Natural resource conservation; tree planting and grass growing; reforestation; responsible use of forest resources
	Institution building	Protection of ecological values
Comoros	Land use and management	Restoring degraded soils; reconstruction of slopes
	Awareness and education	Reducing fuel wood use
Egypt	–	–
India	Land use and management	Regeneration of degraded forest; mangrove ecosystem rehabilitation; degraded arid/semi-arid lands; carbon sequestration and biodiversity conservation in village agro-ecosystems; afforestation and reforestation programmes to reduce pressure on primary forest from timber and fuel wood use
Kenya	Land use and management	Protection of forests and reforestation; medicinal tree-planting programmes; restoration and sustainable management of riparian forests; agro-forestry
Malawi	Land use and management	Improving wetland and conservation measures
	Research and technology	Use of efficient stoves and alternative energy for cooking and lighting
Nepal	Land use and management	Afforestation and reforestation to rehabilitate degraded lands; extensive tree planting; reforesting sensitive areas with tolerant varieties
	Research and technology	Technology to reduce use of fuel wood
South Africa	Awareness and education	Encouraging efficient wood and coal stove use
Tanzania	Land use and management	Reforestation and enhanced natural regeneration to maintain existing stocks
	Research and technology	Increasing efficiency of biomass stoves

India), and rehabilitation and regeneration of degraded ecosystems (e.g. India and Tanzania). In addition, they include a wide range of sustainable and responsible management practices and conservation measures. The needs in the other categories mostly relate to reducing the use of fuel wood by using more efficient stoves and alternative fuels, and creating greater awareness to the same purpose (e.g. in the Comoros and South Africa).

Table 8.11 *Biodiversity: adaptation needs*

Country	Measure	Need
Brazil	Institution building	Gene banks of existing flora and fauna; conservation units
China	Land use and management	Restoration of ecosystems; coral transplantation and restoration; recovery of soil; corridors
	Research and technology	Pest and disease control
	Institution building	
	Capacity building	Nature reserves
Comoros	Land use and management	Ecosystem protection
		Coastal zone management
	Research and technology	Monitoring coastal erosion and impacts on coral reefs
Egypt	Research and technology	Flood protection; improved water storage; impacts on coral reef ecosystems, coastal zones and water resources
	Institution building	Integrated coastal zone management
India	Land use and management	Linking protected areas, wildlife reserves and forests; natural regeneration in degraded forests; mixed-species forestry on degraded non-forest lands; fire prevention and management practices
	Institution building	More effective protected areas and wildlife conservation programmes; adaptation policies for forest ecosystems; *in situ* and *ex situ* conservation
Kenya	Land use and management	Improved forest management; ecosystem approach to land management; restoration and sustainable management of riparian forests; rehabilitation of most-damaged ecosystems
	Research and technology	Monitoring of ecosystem health
	Institution building	*Ex situ* conservation of local/indigenous plant varieties and animal breeds

Table 8.11 (*cont.*)

Country	Measure	Need
Malawi	Land use and management	Improved wetland and conservation measures; good agricultural and land use practices around water bodies and wetlands; improved forest management; conventional wildlife management techniques; improved fire management in game reserves; community-based ranching and breeding
	Institution building	Strengthening National Aquaculture Centre; fish gene bank
Nepal	Land use and management	Natural regeneration; prioritization of species vulnerable to climate change; habitat management for protected species; conservation of species and resources outside protected areas
	Research and technology	Identifying agro-ecological zones sensitive to climate change; increase knowledge of effects of climate change on species
South Africa	Land use and management	Extending protected areas to adjacent land with high topographical relief; vegetation and animal management policies, translocation and direct intervention
	Research and technology	Link climate change and species distribution; improve data on plant and animal diversity; species inventories and monitoring
	Institution building	Improve national conservation strategy; conservation-area network; focused *ex situ* conservation
Tanzania	Land use and management	Management plans for protected areas; community-based management programmes in areas surrounding national parks and game reserves; developing migration corridors and buffer zones

8.5.3 Adaptation needs

All of the case-study countries have formulated adaptation needs in relation to biodiversity, since they appear very well aware of their vulnerability and the consequences biodiversity loss will bring about (see Table 8.11). These needs fall within the categories of land use and management, institution building, and research and technology. Adaptation needs in the category of land use and management are highly similar to the mitigation needs in this sector, including, for example, the restoration of ecosystems (e.g. China and Kenya), improved management practices (e.g. the Comoros, Kenya and Malawi) and conservation measures (e.g. India, Malawi, Nepal, South Africa and Tanzania). In addition, most countries aim to develop stronger institutions and capacities for nature protection and conservation, such as gene banks and nature reserves (e.g. Brazil, China, India, Kenya, Malawi and South Africa). Furthermore, several countries (e.g. the Comoros, Egypt, Nepal and South Africa) have articulated needs for monitoring and research of impacts of climate change on species and vulnerable ecosystems.

8.6 Agriculture sector needs: a comparative assessment

8.6.1 Agriculture and climate change

The relationship between agriculture and climate change has two important aspects. On the one hand, it has been estimated that agriculture accounts for about 15% of global greenhouse gas emissions (World Bank, 2008: 200–1). Notably, agriculture provides a major contribution to the global emissions of nitrous oxide and methane through fertilizer application, manure management, biomass burning, rice production and enteric fermentation in livestock production. On the other hand, it is expected that climate change will have far-reaching consequences for agriculture and will disproportionately affect the poor, since there will be greater risks of crop failures and livestock deaths. Adapting agricultural systems to climate change is therefore considered urgent, because its impacts are already evident and the negative trends will continue even if emissions are stabilized at current levels.

8.6.2 Mitigation needs

All of the case-study countries, except the Comoros and Kenya, have articulated mitigation needs for agriculture. These needs mostly fall within the category of research and technology, including, for example, improved rice cultivation technology and rice varieties (e.g. China, Egypt, Malawi and Nepal), fertilizer-application technology (e.g. Egypt, India, Malawi and Tanzania), improved feed for livestock to reduce methane emissions (e.g. India, Malawi, Nepal and South Africa), and non-tillage technology

Table 8.12 *Agriculture: mitigation needs*

Country	Measure	Need
Brazil	Land use and management	Biofuel production; sustainable land management
China	Research and technology	Improved rice varieties; fertilizers; bio-digesters; livestock breeding; non-tillage technology; organic agricultural practices; straw-treatment technology
Comoros	–	–
Egypt	Research and technology	Improved management of rice production; bio-digestors; fertilizer management technology; altering fermentation patterns
India	Research and technology	Improved animal feed (to reduce methane emissions); improved fertilizers; agro-forestry technologies; biogas technology; monitoring of fertilizer use
Kenya	–	–
Malawi	Land use and management	Nutrient and water management in rice fields; fertilizer management; water conservation
	Research and technology	Improved management of, and feed for, livestock
Nepal	Land use and management	Greener manures; changing herd compositions and feed practices; agro-forestry
	Research and technology	Selection of appropriate rice cultivars; non-tillage technologies
South Africa	Land use and management	Manure management; reduced burning of agricultural residues
	Research and technology	Optimization of herd composition and feed intake of livestock
Tanzania	Research and technology	Fertilizer application technology; livestock husbandry

(e.g. China and Nepal). Other needs refer to land use and management, and are mostly about management of water, fertilizers and manure. See Table 8.12.

8.6.3 *Adaptation needs*

All of the case-study countries, especially China, India and Kenya, have formulated adaptation needs in relation to agriculture. They mostly fall within the categories of land use and management, including, for example, improved water management practices (e.g. India, Kenya, Nepal and South Africa), diversification of agricultural activities (e.g. the Comoros, India and Nepal) and sustainable animal husbandry (e.g. Kenya and Malawi). Other adaptation needs refer to research and technology, such as drought-tolerant crop varieties (e.g. Brazil, Egypt, India and Nepal),

Table 8.13 *Agriculture: adaptation needs*

Country	Measure	Need
Brazil	Research and technology	Drought-tolerant crop varieties; irrigation technologies; biotechnology
China	Research and technology	Biotechnology; seed variety technology; agricultural pest and disease control; irrigation technology; automated agro-technology; agro-processing technology
Comoros	Land use and management	Small-scale agriculture
	Research and technology	Agricultural pest and disease control; suitable crop varieties; meteorological forecasts
Egypt	Research and technology	Quick-maturing and drought-resistant crop varieties; water management technologies; land-management technologies
India	Land use and management	Drought-resistant and quick-maturing varieties of crops; zero tillage; diversification of agricultural activities; watershed management
Kenya	Land use and management	Improved agronomic practices; improved soil conservation practices; water management practices for agriculture; sustainable animal husbandry
Malawi	Land use and management	Improved land husbandry practices
	Research and technology	Improved crop varieties; early-warning weather systems; improved animal breeds and feeding habits
Nepal	Land use and management	Improved water conservation; cultivation of drought-resistant varieties; crop diversification
	Research and technology	Model and scenario development for determining adaptation measures
South Africa	Land use and management	Moisture management strategies; altered irrigation methods; improved tillage for conservation
Tanzania	Land use and management	Drought-resistant and quick-maturing varieties of crops in dry areas; compost-managing practices
	Research and technology	Improved irrigation technology; pest and disease management technologies; non-tillage technologies

irrigation technologies (e.g. Brazil, China and South Africa), agricultural pest and disease control (e.g. China, the Comoros and Tanzania), and early-warning weather forecasts (e.g. the Comoros and Malawi). Brazil and China have explicitly articulated the need for biotechnology. See Table 8.13.

8.7 Conclusions

This chapter has made a comparative appraisal of the various national reports prepared by 10 countries in the context of the climate change regime. The following overall conclusions apply only to the case-study countries, although this research may catalyse certain inferences about the climate mitigation and adaptation needs of other, similar, DCs.

Quality of and consistency between documents. In many countries the diverging authorship of individual and separate documents appears to lead to differing and inconsistent conclusions. Even within single documents, one often observes a dichotomy between individual sections. Telling examples are the NCs of Egypt and India. The bulk of the documents is so general and vague that it is often difficult to identify concrete measures. Furthermore, the definitions of key terms differ between national reports from different countries and the origin of data is not always clear.

Mitigation versus adaptation needs. The balance between mitigation and adaptation needs differs by sector. With regard to the energy and forestry sectors, most needs relate to mitigation. In relation to the biodiversity and agriculture sectors, most needs relate to adaptation. For the four sectors in total, the balance tips in favour of a higher number of mitigation needs.

Needs per sector. By far the most needs have been formulated in relation to forestry; otherwise the number of needs is balanced across the sectors of energy, biodiversity and agriculture. Taking the mitgation/adaptation distinction into account, most needs refer to mitigation in the forestry and energy sectors, and the least needs have been formulated about adaptation in those sectors.

Needs per country. By far the most needs have been formulated by China and, to a somewhat lesser extent, by India and Malawi. China has listed the most mitigation needs, whereas Malawi has articulated the most adaptation needs. This may be partly explained in terms of their efforts to compile additional national reports, such as NAPAs and/or TNAs. This is in contrast with Brazil, Egypt and South Africa, which, despite their status as major emitters as well as being vulnerable to the impacts of climate change, are less explicit about their needs. It seems that these countries have experienced difficulties in formulating their needs, although these difficulties may have arisen, partially or wholly, from considerations of sovereignty and security, especially in the field of energy.

Needs per category. The balance among the different types of needs (research and technology, land use and management, institution building, capacity building, social development, awareness and education) differs by sector (see Table 8.14). Overall, for the four sectors, the emphasis is on land use and management, and research and technology, and to a lesser extent on institution building. Far less attention is paid to

Table 8.14 *Dominant categories of needs per sector*

Sector	Mitigation	Adaptation
Energy	Research and technology	Research and technology
Forestry	Land use and management, and institution building	Land use and management, and institution building
Biodiversity	Land use and management	Land use and management, and institution building
Agriculture	Research and technology	Land use and management, and research and technology

the other categories, especially in relation to adaptation. The division among the main categories is largely similar for all of the case-study countries.

Mitigation needs

Energy. Most of the case-study countries have formulated mitigation needs, which mostly fall within the categories of research and technology, focusing on clean coal technology, energy efficiency and non-fossil fuels such as hydro, solar, thermal, wind power and biomass. The richer developing countries China and India have especially highlighted the need for energy technology, because they rely heavily on fossil fuels. Brazil has not developed any mitigation needs for energy because it aims to pursue its own renewables policies independently. Egypt and South Africa are still considering their options for mitigation.

Forestry. All of the case-study countries have formulated mitigation needs, with the emphasis on land use and management, and institution building measures. The countries in which net deforestation is taking place (Brazil, the Comoros, Kenya, Malawi, Nepal and Tanzania) have identified more mitigation needs than have those with stable or growing forest areas (China, Egypt, India and South Africa). Mitigation needs in the category of land use and management largely refer to reforestation and afforestation, and sustainable management practices. Those in the category of institution building mostly relate to joint and community forest management, reform of the property rights system, improved land use planning, and strengthening of monitoring and enforcement. Most of these countries did not focus on the role of market mechanisms as an instrument for forest management, although they increasingly do so in the political arena.

Biodiversity. All of the case-study countries, except Egypt, have articulated mitigation needs in the category of land use and management, typically entailing reforestation, restoration of degraded soils, reconstructing of slopes, and rehabilitation and regeneration of degraded ecosystems. In addition, they include a wide range of sustainable and responsible management practices and conservation measures.

Agriculture. Nearly all of the case-study countries have articulated mitigation needs, which mostly fall within the category of research and technology, including improved rice cultivation technology and rice varieties, non-tillage technology, fertilizer-application technology and improved feed for livestock to reduce methane emissions.

Adaptation needs

Energy. Most of the case-study countries have identified one or two adaptation needs, which mostly refer to climate risk and impact assessment and early-warning systems for floods and droughts.

Forestry. All of the case-study countries, except Kenya, have articulated adaptation needs, mostly falling within the categories of land use and management, and institution building. There is some overlap with the forestry mitigation needs since the proposed measures also relate to the need for improved forest management policies and practices, such as afforestation, agro-forestry, species selection and the establishment of forest seed banks for indigenous species.

Biodiversity. All of the case-study countries have formulated adaptation needs. These needs fall within the categories of land use and management, institution building, and research and technology, including, for example, the restoration of ecosystems, improved management practices and conservation measures. In addition, most of these countries aim to develop stronger institutions and capacities for nature protection and conservation.

Agriculture. All of the case-study countries have formulated adaptation needs. Several fall within the categories of land use and management, including, for example, improved water management practices, diversification of agricultural activities and sustainable animal husbandry. Other adaptation needs refer to research and technology, including drought-tolerant crop varieties, irrigation technologies, pest and disease control, and early-warning weather forecasts.

Mitigation, adaptation and development

Several case-study countries have pointed out that mitigation and development do not necessarily go hand in hand, especially in the energy sector, and that mitigation activities may harm their prospects for development. They thus implicitly believe in the 'U' shape of the environmental Kuznets curve (see Chapter 1). They are less uncomfortable about linking adaptation to development needs.

Acknowledgements

This chapter has drawn on a series of ERM masters' theses written at VU University Amsterdam by Pravesh Baboeram (energy), Milena Garita (biodiversity), Caro

Lorika (agriculture), Matthew Smith (forestry) and Hsin-Ping Wu (all issues). It is, furthermore, based on extensive student work and class discussions, involving Corinne Cornelisse, Grace Lamminar, Marilen Espinoza, Marit Heinen, Roy Porat, Ruben Zondervan, Belinda McFadgen, Remon Dolevo, Charles Owusu, Laura Meuleman, Ieva Oskolokaite, Emilie Hugenholtz, Hassan El Yaquine, Olwen Davies, Andrej Wout, Chad Rieben, Wouter Wester, Francesca Feller, Brenda Schuurkamp, Anna Harnmeijer, Jens Stellinga, Pieter Pauw, Yvette Osinga, Nguyen Thi Khanh Van, Joao Fontes, Sarianne Palmula, Laybelin Ogano Bichara, Viviana Gutierrez Tobon, Eline van Haastrecht, Coby Leemans, Efrath Silver, Michelle Beaudin and Jorge Triana.

References

Brazil, Government of (2004). *Brazil's Initial National Communication to the United Nations Framework Convention on Climate Change*. Brasilia: Ministry of Science and Technology.

China, Government of the People's Republic of (1998). *Technology Cooperation Framework – China*. Beijing: Government of the People's Republic of China.
   (2000). *China Wind Power – Study Report*. Beijing: Wind Power Expert Team.
   (2004). *The People's Republic of China Initial National Communication on Climate Change*. Beijing: State Development Planning Commission.

Comoros, Government of Union of the (2002). *Initial National Communication on Climate Change*. Moroni: Ministry of Development, Infrastructures, Post and Telecommunications and International Transports.
   (2006). *Technologies Needs Assessment in the Priority Areas*. Moroni: Ministry of Rural Development, Fishing, Handicraft and Environment.
   (2006). *National Action Programme of Adaptation to Climate Change (NAPA)*. Moroni: Ministry of Rural Development, Fisheries, Handicraft and Environment.

Easterling, W. E., Aggarwal, P. K., Batima, P. *et al.* (2007). Food, fibre and forest products, in *Climate Change 2007: Impacts, Adaptation and Vulnerability. Contribution of Working Group II to the Fourth Assessment Report of the Intergovernmental Panel on Climate Change*, ed. M. L. Parry, O. F. Canziani, J. P. Palutikof, P. J. van der Linden and C. E. Hanson. Cambridge: Cambridge University Press, pp. 273–313.

Egypt, Government of The Arab Republic of (1999). *Initial National Communication on Climate Change*. Cairo: Egyptian Environmental Affairs Agency.
   (2001). *Building Capacity for Egypt to Respond to UNFCCC Communications Obligations*. Cairo: Egyptian Environmental Affairs Agency.

FAO (2007). *State of the World's Forests 2007*. Rome: Food and Agriculture Organization.

FCCC Secretariat (2009). *Second Synthesis Report on Technology Needs Identified by Parties Not Included in Annex I to the Convention: Note by the Secretariat of the UNFCCC to the Subsidiary Body for Scientific and Technological Advice (SBSTA)*. UN Document FCCC/SBSTA/2009/INF.1, of 29 May 2009.

G-77 and China Summit Declaration (2005). *Doha Declaration*, Second South Summit 2005, G-77/SS/2005/1.

Gitay, H., Suárez, A., Watson, R. and Dokken, D. J., eds. (2002). *Climate Change and Biodiversity. IPCC Technical Paper V*. Geneva: Intergovernmental Panel on Climate Change.

India, Government of (2004). *India's Initial Communication to the United Nations Framework Convention on Climate Change*. New Delhi: Ministry of Environment & Forests.

IPCC (2007). Summary for policymakers, in *Climate Change 2007: Synthesis Report*, ed. Intergovernmental Panel on Climate Change. Cambridge: Cambridge University Press.

Kenya, Government of (2002). *Kenya's Initial National Communication*. Nairobi: Ministry of Environment and Natural Resources.

(2005). *Kenya's Climate Change Technology Needs and Needs Assessment Report under The United Nations Framework Convention on Climate Change*. Nairobi: Ministry of Environment and Natural Resources.

Malawi, Government of (2002). *Initial National Communication under the United Nations framework Convention on Climate Change*. Lilongwe: Ministry of Natural Resources and Environmental Affairs.

(2003). *Report on Malawi's Climate Technology Transfer and Needs Assessment*. Lilongwe: Ministry of Natural Resources and Environmental Affairs.

(2006). *National Adaptation Programme of Action (NAPA)*. Lilongwe: Ministry of Natural Resources and Environmental Affairs.

Nabuurs, G. J., Masera, O., Andrasko, K. *et al.* (2007). Forestry, in *Climate Change 2007: Mitigation. Contribution of Working Group III to the Fourth Assessment Report of the Intergovernmental Panel on Climate Change*, ed. B. Metz, O. R. Davidson, P. R. Bosch, R. Dave and L. A. Meyer. Cambridge: Cambridge University Press, pp. 541–80.

Nepal, Government of (2004). *Nepal Initial National Communication to the Conference of the Parties of the United Nations Framework Convention on Climate Change*. Kathmandu: Ministry of Population and Environment.

South Africa, Government of (2000). *South Africa's Initial National Communication under the United Nations Framework Convention on Climate Change*. Pretoria: Government of the Republic of South Africa, Department of Environmental Affairs and Tourism.

Tanzania, Government of United Republic of (2003). *Tanzania's Initial National Communication under the United Nations Framework Convention on Climate Change*. Dodoma: Vice President's Office.

(2007). *National Adaptation Programme of Action (NAPA)*. Dodoma: Vice President's Office.

(not dated). *Tanzania, Climate Technology Needs Assessment Report*. Dodoma: Vice President's Office.

World Bank (2008). *World Development Report 2008: Agriculture for Development*. Washington, D.C.: World Bank.

9

The supply of aid and the need–supply nexus

NICOLIEN VAN DER GRIJP AND JOYEETA GUPTA

9.1 Introduction

This chapter examines the supply of assistance to developing countries (DCs) on the basis of an examination of Country Strategy Papers (CSPs; see Section 6.3.3). These papers are compiled in cooperation between partner countries and the European Commission and draw, where available, on Poverty Reduction Strategy Papers (PRSPs) prepared for the World Bank and IMF. The assumption behind such papers is that building on 'national ownership' of ideas for development assistance is likely to enhance the effectiveness of aid (see Chapter 2).

More recently, as climate change has become an issue of higher priority and the links between development and climate change have become more obvious, international organizations have started to recognize the need to mainstream climate change into these documents (see Chapters 3 and 4; Table 3.2). UNEP–UNDP (2007) explicitly see the PRSPs as an instrument for linking poverty and environmental issues (see Section 3.3.2), and argue that such reports could be seen as effective entry points for international assistance.

Against this background, this chapter seeks to ask to what extent the supply of development aid addresses the needs for climate assistance. In comparing the supply with the needs, it reflects back on the demand for climate assistance as identified in Chapter 8. The current chapter reviews the CSPs agreed between the European Commission and the 10 case-study countries (see Section 8.2).

9.2 The supply of assistance

9.2.1 Introduction

This section reviews the supply of aid on environmental issues. It does so by undertaking four sectoral case studies, concerning energy, forestry, biodiversity and agriculture.

Mainstreaming Climate Change in Development Cooperation: Theory, Practice and Implications for the European Union, ed. Joyeeta Gupta and Nicolien van der Grijp. Published by Cambridge University Press. © Cambridge University Press 2010.

9.2.2 The supply of aid on environmental issues

Environmental issues scarcely fitted into the aid agenda in the pre-1990 period. Instead, in the run up to the UN Conference on Environment and Development in Rio de Janeiro in 1992, development cooperation projects and programmes were often critiqued for their negative environmental impacts (Werksman, 1993; Hicks *et al.*, 2008). With the end of the cold war and expectation of a peace dividend, attention turned towards environmental degradation as preparations for the Rio Conference took place. As mentioned earlier, of the 40 issues discussed at Rio, only 4 were short-listed for assistance via the Global Environment Facility, feeding speculation that resources would now be diverted from development goals to environmental goals. The DCs insisted that resources for the environment should be new and additional to ODA (see Section 4.4.4). Although a grand bargain was made at Rio to provide an Earth increment of 15% and to triple GEF funds, this expenditure 'failed to materialize' afterwards (Hicks *et al.*, 2008: 3). Even though the DCs then tried to integrate local and national issues into the mandate of the GEF, this was not accepted within the concept of incremental costs (Hicks *et al.*, 2008; Gupta, 1995). Ten years later, at the UN Conference on Financing the Environment, new promises were made. In 2005, the G8 again made promises at Gleneagles to raise funding for environmental issues (see Chapter 4) – promises that were perhaps meant to be broken!

When actually studying what part of aid goes to environmental issues, a new problem emerges. Data on environmental aid are difficult to find and classify, and definitions on what constitutes environmental assistance are diffuse. Hicks *et al.* (2008) show that environmental aid in the 1980s has been of the order of USD 1–7 billion, peaking in the mid 1990s (in constant USD 2000). 'Dirty aid' has been substantially more, receiving about ten times as much funding in the 1980s and falling to three times as much by 1999. Roberts *et al.* (2008) argue that a study of 115 000 aid projects randomly selected from the OECD portfolio for the period 2000–6 shows that funding for mitigation projects was about USD 1 billion per year. After five years of hovering around USD 1 billion annually, it increased in 2006 to USD 10 billion funding, inter alia, for building large hydroelectric power plants, which have been controversial. Roberts *et al.* (2008) classify 'adaptation' funding as 'minuscule' in comparison, and mostly focused on disaster planning and management.

9.2.3 The Country Strategy Papers: a new form of conditionality?

Country reporting schemes such as PRSPs were initially introduced as a means to engage DCs in determining their own strategies for reducing poverty and were seen as exemplifying national ownership. The PRSPs were expected to herald the end of conditionality and exogenous ideas of how poverty could be reduced. In practice, however, this has not (yet) happened. Analyses of the content of PRSPs indicate that the vast majority retain most of the main policy components of the former donor-led and donor-driven policy conditionality, encompassing the core components of the Washington Consensus (Riddell, 2007: 239). Even where the documents were largely a result of national activity, agents 'felt compelled to genuflect to the structural reform agenda' (Sobhan 2002: 63). Many PRSPs were even written by World Bank staff (Riddell, 2007). The PRSPs that were meant to end conditionality merely gave it a new lease of life! This raises the question of whether the CSPs agreed between the European Commission and its partner countries (CEC, 2001a–2008c) can be seen in a similar light.

9.2.4 The methodology

The CSPs are meant to draw on PRSPs, where available. These PRSPs describe the macro-economic structure of a country and structural and social policies that aim at promoting growth and reducing poverty. These documents should also highlight the financial needs of the country and how these needs are to be met. They are ostensibly prepared by governments in cooperation with the IMF and World Bank staff as well as local stakeholders. The preparation process is often partnered with the IMF/World Bank Joint Staff Assessments. A PRSP provides the basis for decisions on bank lending. There are five principles for these documents. They should be (1) country driven and owned, (2) result oriented and focused on the poor, (3) comprehensive in recognizing the multi-dimensional nature of poverty, (4) prepared in partnership with stakeholders and (5) embody a long-term perspective.

More than 60 low-income countries have participated in the drawing up of PRSPs and these assessments have helped to shape policy in these countries. The focus on poverty helps to complement the previous uni-directional focus on growth. However, an early IMF assessment (2002) concluded that poverty reduction policies need to be followed up by action and that the countries themselves are responsible for that. The assessment furthermore concluded that PRSPs helped countries gain ownership of an understanding of how poverty should be reduced in their countries, how to involve stakeholders in such a discussion and how to prioritize poverty, and helped to legitimize the process through engaging the donor community.

The main features of PRSPs and CSPs are highlighted in Table 9.1.

Not all of the 10 case-study countries are obliged to submit a PRSP (see Table 9.2).

Table 9.1 *Main features of the reporting frameworks*

	PRSP	CSP
Established in	1999	2001
Purpose	To reduce and eradicate poverty	To define areas of assistance by the EU, especially in the area of poverty reduction
Who compiles the report	Joint effort under the auspices of the IMF	Joint effort under the auspices of the EU
Who must do it	Low-income countries	EU partner countries
To whom must it be submitted	IMF and World Bank	EU
Proposed content	Must analyse how poverty can be reduced in the country	Must cover areas upon which development cooperation with the EU should focus

Table 9.2 *Case-study countries and documents submitted, including year of publication*

Country	PRSP	First-generation CSP	Second-generation CSP
Brazil	–	2002	2007
China	–	2002	2007
Comoros	2006	2002	2008
Egypt	–	2002	2007
India	–	2002	2007
Kenya	2005	2003	2008
Malawi	2002	2001	2008
Nepal	2003	2003	2007
South Africa	–	2003	2007
Tanzania	2006	2001	2008

9.3 Energy sector supply: a comparative assessment

9.3.1 Energy, climate change and development

Energy is an essential factor for achieving economic growth and fulfilling social needs, and is closely linked to human development. It is expected that in coming years the energy demand in developing countries will further rise as they achieve a certain level of development. However, if the growing demand for energy in developing countries is met by the use of fossil fuels, emissions of greenhouse gases will rise accordingly. This may stimulate development in the short term but could have negative impacts for development in the longer term, since developing countries are relatively vulnerable to the impacts of climate change. It is therefore necessary to aim at a decoupling of emissions from energy use.

In this respect, Gaye (2007) argues in a UNDP report that sufficient attention should be paid to poor people in developing countries and their energy needs, in order to improve their standard of living. Renewable energy technologies such as biomass, hydro, solar and wind power could possibly sustain the aim of development in DCs without increasing the level of greenhouse gas (GHG) emissions exponentially (Asif and Muneer, 2007). However, most low- or zero-GHG-emitting technologies are more expensive than fossil fuel based energy sources, which makes the question of who will pay for the renewable energy choices of developing countries important (van Ruijven *et al.*, 2008).

In the past, donors have often financed energy projects in developing countries. However, the share of ODA from OECD DAC countries and multilateral banks spent on the energy sector has been declining since the 1990s (Tharakan *et al.*, 2007). Between 1997 and 2005, a total amount of USD 6 billion was spent on energy projects through bilateral and multilateral support (Tirpak and Adams, 2008), which equals approximately 2% of all ODA. Japan has been by far the largest donor in energy-related aid, since about 66% of energy aid came from Japan; Germany had a share of 12%, and the French share was 3.4%. According to Tirpak and Adams (2008), there has been a certain shift in the type of projects financed by bilateral and multilateral funding during the period 1997–2005: the funding for renewable energy has increased, but it is still proportionally much less than that for fossil fuel based projects.

9.3.2 Mitigation and adaptation in CSPs

The supply of energy-focused EU assistance that is to be provided during the period 2007–13 is shown in Table 9.3. Clearly, the emphasis is on funding for mitigation, especially with respect to China and India. The projects funded are related to a wide range of sustainable energy options, including clean coal technology, energy efficiency, recovery of natural gas and renewable energy sources. In addition, money will be spent on institution and capacity building for CDM projects in China and India. Adaptation measures, however, have thus far been a neglected area.

9.3.3 Meeting the needs for climate assistance

On comparing the supply of assistance in the area of energy with the needs identified in Chapter 8, it is evident that there are some matches in relation to mitigation (see Table 9.4). This holds for the needs of the primarily fossil fuel dependent countries

Table 9.3 *Supply of assistance in the area of energy (2007–2013)*

	BR	CH	CO	EG	IN	KE	MA	NE	SA	TZ
Mitigation										
Sustainable energy in general	–	–	–	–	–	–	–	–	–	S
Clean energy technology in general	–	–	–	–	S	–	–	–	–	–
Clean coal technology	–	–	–	–	–	–	–	–	S	–
Recovery of natural gas	–	S	–	–	–	–	–	–	–	–
Carbon capture and storage	–	S	–	–	–	–	–	–	–	–
Energy efficiency	–	S	–	–	S	–	–	–	–	–
Renewable energy in general	S	S	–	S	S	–	–	–	–	–
Nuclear energy	–	–	–	–	–	–	–	–	–	–
Biofuels	–	–	–	–	–	–	–	–	–	–
Hydrogen	–	S	–	–	–	–	–	–	–	–
Wind energy	–	–	–	–	–	–	–	–	–	–
Solar energy	–	–	–	–	–	–	–	S	–	–
Hydropower	–	–	–	–	–	–	S	–	–	–
Geothermal energy	–	–	–	–	–	–	–	–	–	–
Reforestation for fuel wood	–	–	–	–	–	S	–	–	–	–
Improved cooking stoves	–	–	–	–	–	–	–	–	–	–
Institution and capacity building for the CDM	–	S	–	–	S	–	–	–	–	–
Adaptation										
Energy sector impact assessment on energy sector	–	–	–	–	–	–	–	–	–	–
Hydroelectricity impact assessment	–	–	–	–	–	–	–	–	–	–
Forestry impact assessment	–	–	–	–	–	–	–	–	–	–
Floods and droughts pre-warning system	–	–	–	–	–	–	–	–	–	–
Reduced dependence on single sources of energy	–	–	–	–	–	–	–	–	–	–
Afforestation	–	–	–	–	–	–	–	–	–	–

Notes: S, supply. The countries are abbreviated BR, Brazil; CH, China; CO, the Comoros; EG, Egypt; IN, India; KE, Kenya; MA, Malawi; NE, Nepal; SA, South Africa; and TZ, Tanzania.

China and India. Interestingly, development aid from the EU also focuses on mitigation measures in Brazil, Egypt and South Africa, although these countries have not been very explicit about their needs in their climate reports for the FCCC and are still in the phase of exploring options for mitigation. However, the EU has paid less attention to the mitigation needs of the countries that are dependent on non-fossil fuels (the Comoros, Kenya, Malawi, Nepal and Tanzania), despite the fact that several of them have articulated quite explicit needs, especially in relation to renewable energy.

Table 9.4 *Demand–supply nexus on energy*

	BR	CH	CO	EG	IN	KE	MA	NE	SA	TZ
Mitigation										
Sustainable energy in general	–	–	–	–	–	–	–	–	–	M
Clean energy technology in general	–	N	–	–	M	–	–	–	–	–
Clean coal technology	–	N	–	–	N	–	–	–	S	–
Recovery of natural gas	–	M	–	–	N	N	N	–	–	–
Carbon capture and storage	–	S	–	–	N	–	–	–	–	–
Energy efficiency	–	M	–	–	S	–	N	–	–	–
Renewable energy in general	S	S	–	S	S	–	–	–	–	N
Nuclear energy	–	–	–	N	–	–	–	–	–	–
Biofuels	–	–	–	–	–	–	N	–	–	N
Hydrogen	–	S	–	–	N	–	–	–	–	–
Wind energy	–	N	N	N	–	N	–	–	–	N
Solar energy	–	–	–	–	N	N	N	S	–	–
Hydropower	–	–	N	–	–	N	M	–	–	N
Geothermal energy	–	–	N	–	–	N	–	–	–	–
Reforestation for fuel wood	–	–	–	–	–	S	–	–	–	–
Improved cooking stoves	–	–	N	–	–	–	N	–	–	–
Institution and capacity building for the CDM	–	N	–	–	S	–	–	–	–	–
Adaptation										
Energy sector impact assessment	–	–	–	–	N	–	–	N	–	–
Hydroelectricity assessment	N	–	–	–	–	N	–	–	–	–
Forestry impact assessment	–	–	N	–	–	–	–	–	–	–
Floods and droughts pre-warning system	–	N	–	–	–	–	–	–	–	N
Reduced dependence on single sources of energy	–	–	–	–	–	–	N	–	–	–
Afforestation	–	–	–	–	–	–	N	–	–	–

Notes: N, need; S, supply; M, match between need and supply. The countries are abbreviated as in Table 9.3.

9.4 Forestry sector supply: a comparative assessment

9.4.1 Forestry, climate change and development

Forestry and development are intricately linked, since forests are rich ecosystems, housing much of the world's biodiversity and providing livelihoods and homes to many millions of people. People and forests have had an evolving relationship over time. Initially, forests were seen as resources to be exploited for short-term development goals. Subsequently, the importance of forests for supporting local livelihoods was seen as more important. The role of forests in environmental

Table 9.5 *Supply of assistance in the area of forestry (2007–13)*

	BR	CH	CO	EG	IN	KE	MA	NE	SA	TZ
Mitigation										
Global carbon markets	–	S	–	–	–	–	–	–	–	–
Property rights	–	–	–	–	–	–	–	S	–	–
Agro-forestry	–	–	–	–	–	–	–	–	–	–
Alternative income for forest dwellers and dependents	S	–	–	–	–	S	S	–	–	–
Increase of plantations	–	–	–	–	–	–	–	–	–	–
Increased use of wood for housing	–	–	–	–	–	–	–	–	–	–
Improved use of biomass and fuels	S	–	–	–	–	–	–	S	–	–
Adaptation										
Forest habitat corridors	–	–	–	–	–	–	–	–	–	–
Fire management	–	–	–	–	–	–	–	–	–	–
Pest and disease control	–	–	–	–	–	–	–	–	–	–
Improved forest management	S	–	–	–	–	–	S	–	–	–
Improved tree-species selection	–	–	–	–	–	–	–	–	–	–
Use of irrigation/fertilizers	–	–	–	–	–	–	–	–	–	–
Biotechnology	–	–	–	–	–	–	–	–	–	–
Other										
Forest research and monitoring	–	–	–	–	–	–	–	–	–	–
Improved governance	S	–	–	–	S	S	–	–	–	–
Increased awareness	–	–	–	–	–	–	–	–	–	–
Increased technical, institutional and administrative capacity	–	–	–	–	–	–	–	–	–	–

Notes: S, supply. The countries are abbreviated as in Table 9.3.

protection became the next issue to be prioritized. By the turn of the century, the importance of an integrated approach to forestry because of its multi-facetted impacts on development was being recognized.

Between 1974 and 2006, ODA spent on forestry varied widely. The OECD DAC statistics show that forestry ODA slowly increased through the 1980s to a peak in 1992, but since then has nearly halved. By far the most important donor with regard to forestry is Japan.

The history of assistance in the area of forestry can be classified into five phases (Persson, 2003). In the 1960s, the focus was on industrial forestry – harvesting forestry for development; in the 1970s the focus was on social forestry – to serve those dependent on forests; in the 1980s and 1990s the focus has been on environmental forestry and the creation and protection of protected areas and parks, followed by sustainable forestry focusing on integrated approaches and agro-forestry. The current discussions on Reducing Emissions from Deforestation and Degradation (REDD) may push the forestry discussions in the direction of managing forests to decrease greenhouse gas emission.

Table 9.6 *Demand–supply nexus on forestry*

	BR	CH	CO	EG	IN	KE	MA	NE	SA	TZ
Mitigation										
Global carbon markets	–	M	–	N	N	N	–	N	N	–
Property rights	N	–	–	–	N	–	N	M	N	N
Agro-forestry	N	–	N	–	–	N	N	N	–	N
Alternative income for forest dwellers and dependents	M	–	–	–	–	S	M	–	–	N
Increase of plantations	N	N	N	N	N	N	N	N	N	N
Increased use of wood for housing	–	–	N	–	–	N	–	–	–	N
Improved use of biomass as fuels	M	–	–	–	–	–	–	M	–	–
Adaptation										
Forest habitat corridors	–	–	–	–	–	–	–	N	–	N
Fire management	–	–	–	–	–	–	N	N	N	N
Pest and disease control	–	N	–	–	–	–	–	–	–	N
Improved forest management	M	N	N	N	N	N	M	N	N	N
Improved tree-species selection	–	N	N	N	N	–	N	N	N	N
Use of irrigation/fertilizers	–	–	–	–	–	–	–	–	–	–
Biotechnology	N	–	–	–	–	–	–	–	N	–
Other										
Forest research and monitoring	N	–	N	–	N	N	N	N	N	M
Improved governance	M	N	–	–	M	M	N	N	–	N
Increased awareness	–	N	–	–	N	N	N	–	N	–

Notes: N, need; S, supply; M, match between need and supply. The countries are abbreviated as in Table 9.3.

9.4.2 Mitigation and adaptation in CSPs

The supply of forestry-related EU assistance for the period 2007–13 is shown in Table 9.5. Evidently, the supply of assistance is focused on the countries with net deforestation, Brazil, Kenya, Malawi and Nepal, with the exception of Tanzania, however. The assistance provided is mostly in the area of improved forest management, alternative income for forest dwellers and dependents, improved use of biofuels, and improvement in general governance. Less attention has been paid to adaptive measures.

9.4.3 Meeting the needs for climate assistance

On comparing the supply of assistance in the area of forestry with the needs earlier identified, it can be seen that there are relatively many matches with respect to Brazil, Malawi and Nepal (see Table 9.6). In addition, it is evident there is hardly any supply without demand. The matches are mostly in the area of improved forest management and better governance. Although clear needs have been articulated in the area of improved selection of tree species, agro-forestry, plantations and research and monitoring efforts, the supply of assistance here is at a low level.

9.5 Biodiversity sector supply: a comparative assessment

9.5.1 Biodiversity, climate change and development

Climate change threatens to increase the pressure on the world's biodiversity, thereby putting at risk the livelihoods of thousands of communities in developing countries that depend on natural resources for income and food security (Gitay *et al.*, 2002). It is acknowledged that humans derive immense benefits from biodiversity and that ecosystem services provided by it are crucial to the economies of many countries (Costanza *et al.*, 1997; Díaz *et al.*, 2006), making life 'possible and worth living' (IPCC, 2002). Such ecosystem services include pest control, pollination, medicines and food, among others.

Thus far, there has been no separate category in development aid statistics for biodiversity. However, de facto, biodiversity issues are increasingly being included in agriculture, forestry and rural development categories (Lapham and Livermore, 2003), and such activities include conservation, sustainable use and/or fair and equitable sharing of the benefits (OECD, 2008). However, scarcely any development aid is allocated to biodiversity (Janni, 2000). Total expenditure on biodiversity has increased from negligible in the 1980s to a peak in 1991 and again in 2005, but has remained mostly below USD 200 million. Amongst EU members, Germany provides most assistance for biodiversity-related issues.

9.5.2 Mitigation and adaptation in CSPs

The supply of biodiversity-related EU assistance for the period 2007–13 is shown in Table 9.7. It is evident that the emphasis is especially on the protection and sustainable use of natural resources and on improved governance, and that less attention is being paid to mitigation measures.

9.5.3 Meeting the needs for climate assistance

On comparing the supply of assistance in the area of biodiversity with the needs, it is clear that there are quite a few matches, especially with respect to adaptation measures (see Table 9.8). Furthermore, Brazil and China are receiving relatively extensive assistance for the protection of biodiversity. However, the fact that almost

Table 9.7 *Supply of assistance in the area of biodiversity (2007–13)*

	BR	CH	CO	EG	IN	KE	MA	NE	SA	TZ
Mitigation										
Avoidance of deforestation	S	–	–	–	–	–	–	–	–	–
Fire management	–	–	–	–	–	–	–	–	–	–
Mixed species planting	–	–	–	–	–	–	–	–	–	–
Afforestation/reforestation	–	–	–	–	–	–	–	–	–	–
Decreased use of fuel wood	–	–	–	–	–	–	–	S	–	–
Improved land management	S	S	–	–	–	S	–	–	–	–
Adaptation										
New and improved management of protected areas	S	–	–	–	–	–	–	–	–	–
Conservation strategy	–	S	–	–	–	S	–	–	–	–
Improved connectivity	–	–	–	–	–	–	–	–	–	–
Protection/sustainable use of natural resources	S	S	S	S	S	S	S	S	S	S
Coastal zone management	–	–	–	–	–	–	–	–	–	–
Rehabilitation/restoration of ecosystems	–	–	–	–	–	S	–	–	–	–
Pest/invasive species control	–	–	–	–	–	–	–	–	–	–
Other										
Research and monitoring	–	–	–	–	–	–	–	–	–	–
Improved governance	S	S	S	S	S	S	S	S	S	S
Increased awareness	–	–	–	–	–	–	–	–	–	–
Increased technical, institutional and administrative capacity	–	S	S	–	–	–	S	–	S	–

Notes: S, supply. The countries are abbreviated as in Table 9.3.

all of the case-study countries expressed a need for research and monitoring activities has not been picked up in the supply process.

9.6 Agriculture sector supply: a comparative assessment

9.6.1 Agriculture, climate change and development

Agriculture and development have been linked throughout history because agriculture was one of the first main occupations of society. According to the World Development Report 2008 (World Bank, 2008), agriculture for both crop production and animal husbandry occupies approximately 40% of the total land area, and contributes 4% to the world's gross domestic product (GDP). The report distinguishes three distinct agricultural worlds: one agriculture-based, one transforming and one urbanized. In the agriculture-based countries, agriculture and its associated industries are essential to growth and to reducing mass poverty and food insecurity.

Table 9.8 *Demand–supply nexus on biodiversity*

	BR	CH	CO	EG	IN	KE	MA	NE	SA	TZ
Mitigation										
Avoidance of deforestation	M	N	S	S	S	S	S	S	S	S
Fire management	S	S	S	S	S	S	N	S	S	S
Mixed species planting	S	S	S	S	N	N	S	N	S	S
Afforestation/reforestation	S	N	S	S	N	S	S	N	S	S
Decreased use of fuel wood	N	S	N	S	S	N	N	M	N	N
Improved land management	M	M	N	S	N	–	N	N	S	S
Adaptation										
New and improved management of protected areas	M	N	S	S	N	S	S	N	N	N
Conservation strategy	N	M	S	S	N	M	N	N	N	N
Improved connectivity	S	N	S	S	N	S	S	N	S	S
Protection/sustainable use of natural resources	M	M	–	–	–	M	–	M	M	M
Coastal zone management	S	S	N	N	S	S	S	S	S	S
Rehabilitation/restoration of ecosystems	N	N	N	S	N	M	S	N	S	N
Pest/invasive species control	S	N	S	S	S	S	N	S	N	N
Other										
Research and monitoring	N	N	N	N	N	N	N	N	N	N
Improved governance	M	M	–	–	–	–	–	M	–	–
Increased awareness	S	S	S	N	N	S	S	S	S	S
Increased technical, institutional and administrative capacity	N	M	S	S	N	N	M	N	–	–

Notes: N, need; S, supply; M, match between need and supply. The countries are abbreviated as in Table 9.3.

The Comoros, Kenya, Malawi, Nepal and Tanzania belong to this category. For transforming countries, agriculture is no longer a major source of economic growth, but is still of crucial importance for rural populations. Brazil, China, Egypt, India and South Africa are members of this group. In urbanized (mostly developed) countries, agriculture can help to reduce remaining rural poverty. None of our case-study countries belongs to this group.

9.6.2 Mitigation and adaptation in CSPs

The supply of agriculture-related EU assistance for the period 2007–13 is shown in Table 9.9. Most efforts at assistance focus on trade promotion, water conservation measures, diversification of farmers' incomes, improved farming practices and combatting agricultural pests and diseases. Furthermore, it seems that practically no assistance is being provided with respect to mitigation activities.

Table 9.9. *Supply of assistance in the area of agriculture (2007–13)*

	BR	CH	CO	EG	IN	KE	MA	NE	SA	TZ
Mitigation										
Agro-forestry	–	–	–	–	–	–	–	–	–	–
Conservation tillage	–	–	–	–	–	–	–	–	–	–
Agronomy practices	–	–	–	–	–	–	S	–	–	–
Soil carbon markets	–	–	–	–	–	–	–	–	–	–
Manure and fertilizer technologies	–	–	–	–	–	–	–	–	–	–
Dietary additives	–	–	–	–	–	–	–	–	–	–
Improved animal breeding/ husbandry	–	–	–	–	–	–	–	–	–	–
Feeding practices	–	–	–	–	–	–	–	–	–	–
Bio-energy development	–	–	–	–	–	–	–	–	–	–
Rice field management	–	–	–	–	–	–	–	–	–	–
Adaptation										
Crop and livestock insurance	–	–	–	–	–	–	–	–	–	–
Trade promotion	S	S	S	S	S	S	S	S	S	S
Property rights	–	–	–	–	–	–	–	–	–	
Water conservation measures	–	–	S	–	S	S	S	–	–	S
Diversification of farmers' income	–	–	S	–	–	S	S	–	S	S
Crop/seed varieties	–	–	–	–	–	S	–	–	–	–
Improved farming practices	S	–	–	–	S	S	–	–	–	S
Food banks for crops and livestock	–	–	–	–	–	–	–	–	–	–
Combatting agricultural pests and diseases	S	S	S	S	–	–	–	S	–	S
Other										
Capacity building	S	S	S	S	–	–	S	S	S	–
Research and development	S	S	–	S	S	S	–	–	–	S
Climate change monitoring and information communication	–	–	S	–	–	S	–	–	–	S

Notes: S, supply. The countries are abbreviated as in Table 9.3.

9.6.3 Meeting the needs for climate assistance

On comparing the supply of assistance in the area of agriculture with the needs identified earlier, it can be seen that there are some matches in the areas of capacity building, research and development, and water conservation measures (see Table 9.10). There are no matches in relation to mitigation needs in the agricultural sector or the need for new varieties of seeds to ensure adaptation of agriculture to changing climatic circumstances.

Table 9.10 *Demand–supply nexus on agriculture*

	BR	CH	CO	EG	IN	KE	MA	NE	SA	TZ
Mitigation										
Agro-forestry	N	–	–	–	–	N	N	N	–	N
Conservation tillage	–	N	–	N	N	–	–	–	N	N
Agronomy practices	–	–	–	N	N	S	–	–	–	–
Soil carbon markets	–	–	–	–	–	–	–	–	–	–
Manure and fertilizer technologies	–	N	–	N	N	–	N	–	N	N
Dietary additives	–	–	–	–	N	–	N	N	N	N
Improved animal breeding/ husbandry	–	N	–	–	N	N	N	N	N	N
Feeding practices	–	N	–	–	N	–	N	N	N	N
Bio-energy development	N	–	–	–	–	–	–	–	–	–
Rice field management	–	N	–	N	N	–	N	N	–	N
Adaptation										
Crop and livestock insurance	–	–	–	–	–	–	–	–	–	–
Trade promotion	S	S	S	S	S	S	S	S	S	S
Property rights	N	–	–	–	N	–	–	–	–	–
Water conservation measures	N	N	M	N	M	M	M	N	N	M
Diversification of farmers' income	N	–	M	–	–	M	S	–	S	S
Crop/seed varieties	N	N	N	N	N	M	N	N	N	N
Improved farming practices	S	N	–	N	M	M	N	N	–	M
Food banks for crops and livestock	–	–	–	–	–	–	N	–	–	–
Combatting agricultural pests and diseases	S	M	M	S	–	–	–	S	S	N
Other										
Capacity building	M	M	M	M	M	M	M	S	M	S
Research and development	M	M	N	M	S	M	–	–	N	M
Climate change monitoring and information communication	N	N	M	N	N	M	N	–	N	S

Notes: N, need; S, supply; M, match between need and supply. The countries are abbreviated as in Table 9.3.

9.7 The CSPs revisited

The CSPs have clearly evolved as an instrument since their inception in 2001. Table 9.11 shows the differences between aid priorities in the first- and second-generation CSPs. The most significant difference is that the second generation tends to be more practical and takes environmental concerns more seriously into account (see Section 6.4). The early CSPs did not mention climate change at all. However, with the introduction of country environmental profiles, climate change issues have become more explicitly discussed in CSPs, by elaborating, for example, on the expected impacts of climate change on agricultural production and water resources.

Table 9.11 *Development assistance priorities in first- and second-generation CSPs*

Country	Priorities 2002–6	Priorities 2007–12
Brazil	(1) Sustainable and equitable economic growth through structural reform; (2) social development; (3) environmental conservation, especially in the Amazon region	(1) Enhancing bilateral relations to improve social inclusion and to achieve greater equality; (2) promote the environmental dimension of sustainable development
China	(1) Social and economic reform; (2) environment and sustainable development; (3) good governance and human-rights-related policies	(1) Assisting in dealing with global concerns and challenges over the environment, energy and climate change; (2) human resource development
Comoros	(1) Education; (2) decentralized cooperation	(1) Rehabilitation of transport infrastructure (roads); (2) support to education; (3) assistance to improve governance (finance, justice and decentralization) and institutional support
Egypt	(1) Implementation of the EU–Egypt Association Agreement; (2) economic transition; (3) balanced socio-economic development	(1) Democracy, human rights and justice; (2) competitiveness and productivity of the economy; (3) improve management of human and natural resources
India	(1) Education; (2) health; (3) environment	(1) Health and education; (2) implement the EU–India Partnership Action Plan
Kenya	(1) Agriculture and rural development; (2) transport and roads infrastructure; (3) Macro-economic budget support	(1) Regional economic integration by means of supporting (transport) infrastructure, agriculture and rural development; (2) capacity building
Malawi	(1) Agriculture and natural resources; (2) transport infrastructure	(1) Agriculture and food security; (2) transport infrastructure
Nepal	(1) Poverty reduction; (2) consolidation of democracy and conflict mitigation; (3) integration into the international economy	(1) Education; (2) stability and peace building; (3) trade facilitation
South Africa	(1) Equitable access and sustainable provision of social services; (2) equitable and sustainable economic growth; (3) deepening democracy; (4) regional integration and co-operation	(1) Pro-poor sustainable economic growth; (2) basic services for the poor; (3) good governance
Tanzania	(1) Macro-economic support	(1) Infrastructure, communications and transport; (2) trade and regional integration; (3) macro-economic support

9.8 Conclusions

This chapter provides two groups of conclusions, focusing on the supply of assistance and on the match between needs and demands.

In relation to the supply of assistance, this chapter concludes first that the supply of assistance is consistently focused on three areas: poverty alleviation, mostly through health and education services and infrastructure development; economic reform and aid for trade; and the promotion of good governance and capacity building. It is difficult to identify what can be classified as environmental aid and what is not. Second, CSPs of the second generation are increasingly focusing on environmental issues. The supply of aid related to energy focuses on clean coal technology, energy efficiency and renewable energy sources. The supply of aid on forestry focuses on improved governance and forest management, and provision of an alternative income for those dependent on the forests for their livelihoods. The supply of aid on biodiversity focuses on protection and sustainable use of natural resources, land use and management practices, and improved governance and capacity building. The supply of aid on agriculture focuses on trade promotion, water conservation, diversification of farmers' incomes, combatting agricultural pests and diseases, capacity building, and research and development.

With regard to the demand–supply nexus, it seems that there is a mismatch between the supply of development aid and the needs for climate assistance. First, CSPs are written at a higher level of abstraction than climate reports. They identify aid priorities but do not specify the specific projects that will be executed under these priorities. Second, the aid offered by the EU tends to focus on economic governance and/or reform. However, most recipient countries are looking for specific help in the areas of land use and management, and technology transfer. Of course, this raises the question of 'who' the legitimate source of 'demand' is – the state or stakeholders within the state? Third, in the area of energy, the EU has paid less attention to the mitigation needs of the countries that are (still) dependent on non-fossil fuels, despite the fact that several of them have articulated quite explicit needs, especially in relation to renewable energy. In the area of forestry, plantations, improved forest management, tree species selection, and research and monitoring are areas in which scarcely any assistance is being provided. In the area of biodiversity, needs for research and monitoring are scarcely being met. In the area of agriculture, mitigation needs have thus far totally been neglected.

In sum, these mismatches may reflect the fact that the DCs themselves do not ensure that their different reports communicate internally consistent information, or that the EU is strongly ideologically driven in its CSPs and focuses on good governance and participation in free-trade processes, which can be seen as a sort of disguised conditionality and is not in line with the original idea behind CSPs.

Acknowledgements

This chapter has drawn on a series of ERM masters' theses written at VU University Amsterdam by Pravesh Baboeram (energy), Milena Garita (biodiversity), Caro Lorika (agriculture), Matthew Smith (forestry) and Hsin-Ping Wu (all issues). It is, furthermore, based on extensive student work and class discussions, involving Corinne Cornelisse, Grace Lamminar, Marilen Espinoza, Marit Heinen, Roy Porat, Ruben Zondervan, Belinda McFadgen, Remon Dolevo, Charles Owusu, Laura Meuleman, Ieva Oskolokaite, Emilie Hugenholtz, Hassan El Yaquine, Olwen Davies, Andrej Wout, Chad Rieben, Wouter Wester, Francesca Feller, Brenda Schuurkamp, Anna Harnmeijer, Jens Stellinga, Pieter Pauw, Yvette Osinga, Nguyen Thi Khanh Van, Joao Fontes, Sarianne Palmula, Laybelin Ogano Bichara, Viviana Gutierrez Tobon, Eline van Haastrecht, Coby Leemans, Efrath Silver, Michelle Beaudin and Jorge Triana.

References

Asif, M. and Muneer, T. (2007). Energy supply, its demand and security issues for developed and emerging economies. *Renewable and Sustainable Energy Reviews*, **11**, 1388–413.

CEC (2001a). Malawi–European Community, Country Strategy Paper and Indicative Programme for the Period 2001–2007. Available online at http://ec.europa.eu/development.

(2001b). United Republic of Tanzania–European Community, Country Strategy Paper and National Indicative Programme for the Period 2001–2007. Available online at http://ec.europa.eu/development.

(2002a). Federative Republic of Brazil–European Community. Country Strategy Paper 2001–2006 and National Indicative Programme 2002–2006. REV30, 13 June 2002, CSP clean. Available online at http://ec.europa.eu/external_relations.

(2002b). Country Strategy Paper China. Commission Working Document. Available online at http://ec.europa.eu/external_relations.

(2002c). Egypt Country Strategy Paper 2002–2006 & National Indicative Programme 2002–2004. EURO-MED Partnership. Available online at http://ec.europa.eu/external_relations.

(2002d). EC Country Strategy Paper India, 10 September 2002. Available online at http://ec.europa.eu/external_relations.

(2002e). Union of the Comoro Islands–European Community (2002). Cooperation Strategy and Indicative Programme for the Period 2002–2007. DEV/071/2002-EN. Available online at http://ec.europa.eu/development.

(2003a). Kenya–European Community. Country Strategy Paper. Agreed at Nairobi, 15 October 2003. Available online at http://ec.europa.eu/development.

(2003b). Country Strategy Paper. Nepal and the European Community Co-operation Strategy 2002–2006, 13 November 2003. Available online at http://ec.europa.eu/external_relations.

(2003c). South Africa–European Community Country Strategy Paper and Multi-Annual Indicative Programme for the Period 2003–2005. Available online at http://ec.europa.eu/development.

(2007a). Brazil Country Strategy Paper 2007–2013. E/2007/889, 14 May 2007. Available online at http://ec.europa.eu/external_relations.

(2007b). China Strategy Paper 2007–2013. Available online at http://ec.europa.eu/external_relations.

(2007c). Egypt Country Strategy Paper 2007–2013, 4 January 2007. Available online at http://ec.europa.eu/external_relations.

(2007d). Cooperation between the European Union and South Africa. Joint Country Strategy Paper 2007–2013. Available online at http://ec.europa.eu/development.

(2007e). India Country Strategy Paper 2007–2013. Available online at http://ec.europa.eu/external_relations.

(2007f). Nepal Country Strategy Paper 2007–2013. Available online at http://ec.europa.eu/external_relations.

(2007g). Union des Comores–Communauté Européenne (2007). Document de Stratégie Pays et Programme Indicatif National pour la Période 2008–2013. Agreed in Lisbon, 9 December 2007. Available online at http://ec.europa.eu/development.

(2008a). Republic of Kenya–European Community. Country Strategy Paper and Indicative Programme for the Period 2008–2013. Agreed at Lisbon, 9 December 2007. Available online at http://ec.europa.eu/development.

(2008b). Republic of Malawi–European Community, Country Strategy Paper and Indicative Programme for the Period 2008–2013. Agreed at Lisbon, 9 December 2007. Available online at http://ec.europa.eu/development.

(2008c). United Republic of Tanzania–European Commission, Country Strategy Paper and Indicative Programme for the Period 2008–2013. Agreed at Lisbon, 9 December 2007. Available online at http://ec.europa.eu/development.

Costanza, R., d'Arge, R., de Groot, R. *et al.* (1997). The value of the world's ecosystem services and natural capital. *Nature*, **387**, 253–60.

Díaz, S., Fargione, J., Chapin, F. S., III and Tilman, D. (2006). Biodiversity loss threatens human well-being. *PLoS Biol*, **4**(8), e277. doi:10.1371/journal.pbio.0040277.

Gaye, A. (2007). Access to Energy and Human Development. *Human Development Report Office Occasional Paper to the Human Development Report 2007/2008: Fighting Climate Change. Human Solidarity in a Divided World*. New York, NY: United Nations Development Programme.

Gitay, H., Suárez, A., Watson, R. and Dokken, D. J., eds. (2002). *Climate Change and Biodiversity*. IPCC Technical Paper V. Geneva: Intergovernmental Panel on Climate Change.

Gupta, J. (1995). The Global Environment Facility in its North–South context. *Environmental Politics*, **4**(1), 19–43.

Hicks, R. L., Parks, B. C., Roberts, J. T. and Tierney, M. J. (2008). *Greening Aid? Understanding the Environmental Impact of Development Assistance*. Oxford: Oxford University Press.

IPCC (2002). *Climate Change and Biodiversity*. Geneva: IPCC.

Janni, O. (2000). *The EU and Biodiversity in Tropical Countries*. Milan: Fondazione Eni Enrico Mattei.

Lapham, N. and Livermore, R. (2003). *Striking a Balance: Ensuring Conservation's Place on the International Biodiversity Assistance Agenda*. Washington, D.C.: Conservation International.

OECD (2008). DAC Creditor Reporting Systems Aid Statistics. Available online at http://stats.oecd.org/WBOS/Index.aspx?DatasetCode=CRSNEW.

Persson, R. (2003). *Assistance to Forestry: Experiences and Potential for Improvement*. Bogor: Centre for International Forestry Research.

Riddell, R. (2007). *Does Foreign Aid Really Work?* Oxford: Oxford University Press.

Roberts, J. T., Starr, K., Jones, T. and Abdel-Fattah, D. (2008). *The Reality of Official Climate Aid. Oxford Energy and Environment Comment*. Oxford: Oxford Institute for Energy Studies.

Sobhan, R. (2002). Aid effectiveness and policy ownership. *Development and Change*, **33** (3), 539–48.

Tharakan, P. J., de Castro, J. and Kroeger, T. (2007). Energy sector assistance in developing countries: current trends and policy recommendations. *Energy Policy*, **35**, 734–8.

Tirpak, D. and Adams, H. (2008). Bilateral and multilateral financial assistance for the energy sector of developing countries. *Climate Policy*, **8**(2), 135–51.

UNEP–UNDP (2007). *Guidance Note on Mainstreaming Environment into National Development Planning*. Nairobi/New York, NY: United Nations Environment Programme and United Nations Development Programme.

van Ruijven, B., Urban, F., Benders, R. M. J. *et al.* (2008). Modeling energy and development: an evaluation of models and concepts. *World Development*, **36**(12), 2801–21.

Werksman, J. D. (1993). Greening Bretton Woods, in *Greening International Law*, ed. P. Sands. London: Earthscan, pp. 65–84.

World Bank (2008). *World Development Report 2008: Agriculture for Development*. Washington, D.C.: World Bank.

Part V

Conclusions

10

Prospects for mainstreaming climate change in development cooperation

JOYEETA GUPTA AND NICOLIEN VAN DER GRIJP

> Between the idea
> And the reality
> Between the motion
> And the act
> Falls the Shadow.
>
> *T. S. Eliot,* The Hollow Men

10.1 Mainstreaming climate change

The development and development aid issue has long been a matter for North–South struggles, unceasing demands and unkept promises. The climate change regime is the latest forum for North–South stress, where old grievances like colonialism and new ones like neo-imperialism merge in a complex pot of interdependence and yet distrust. Interdependence, because the climate change problem cannot be addressed by individual nations alone, distrust because the rules for sharing responsibility for causing the problem and solving it are vague. However, global dynamics and politics are changing rapidly. As the spheres of influence of countries change, as development processes alter rapidly, the global community may be entering an era of unforeseeable economic and political change with unpredictable consequences.

Against this background, the literature on and governance over six decades of official development cooperation and two decades of climate cooperation have been examined with a view to determining whether these two forms of cooperation can learn from each other and whether climate change cooperation should be incorporated into development cooperation. This concluding chapter synthesizes the arguments made in the previous chapters in a final integrated analysis. It may be relevant to state upfront that this chapter differentiates between mainstreaming climate change into development and mainstreaming climate change into development cooperation.

It is argued, first, that climate change is a quintessential North–South issue, although clearly North and South are contested terms (see Section 10.2). The rest of

Mainstreaming Climate Change in Development Cooperation: Theory, Practice and Implications for the European Union, ed. Joyeeta Gupta and Nicolien van der Grijp. Published by Cambridge University Press. © Cambridge University Press 2010.

the chapter has to be seen in the light of these introductory comments. It is then argued that mainstreaming is the last stage of incorporating climate change into development and/or development cooperation, that it includes other environmental issues, gender mainstreaming and disaster mainstreaming, and that it has clear advantages and disadvantages (see Section 10.3). Moving on to a discussion of mainstreaming climate change into development, the chapter argues that there are clear substantive links between the two issues and that diverse actors support these links. However, actually mainstreaming climate change into development policy is very challenging (see Section 10.4). The next section examines mainstreaming climate change in development cooperation, showing that diverse actors support this trend, but that there are reasons for and reasons against (see Section 10.5). Subsequently, the relevant EU policies and the policies of the EU Member States are evaluated. This section also looks at the need for and supply of assistance to developing countries (see Section 10.6). The North–South issues about who should pay for assistance, why and to whom are then discussed (see Section 10.7). Finally, the chapter integrates the information provided (see Section 10.8) and makes suggestions for the EU (see Section 10.9). The key argument of the chapter is that climate change should be mainstreamed in development at all levels of governance, but that, given the arguments in Section 10.5.4, climate change should not be mainstreamed in development cooperation. However, the development cooperation portfolio can be subjected to proofing for climate impacts and integrative efforts to minimize the emissions of greenhouse gases.

10.2 The North–South dimension of the development and climate issue

10.2.1 Introduction

This section argues that the climate change problem comes as yet another North–South grievance for the developing countries, where past patterns of behaviour are repeated (see Section 10.2.2). However, the discussion is nuanced by the changing natures of both the North and the South (see Section 10.2.3).

10.2.2 The North–South dimension

The North–South dimension is explained in terms of three issues – development and development cooperation, climate change and the link between these issues. Development and development cooperation have a clear North–South character for four reasons. First, since the end of World War II and colonialism, developing countries (DCs) have been asking for the New International Economic Order and the right to development. Although the former was adopted in 1974, and tried to address the structural imbalances in global relations from the perspective of the

DCs, it was never implemented. Even though the latter was adopted in 1986 and more than 172 countries are party to the follow-up Vienna Convention and Programme of Action, the content of this right is contested (see Section 4.4.2).

Second, since 1958, the issue of assistance from industrialized countries (ICs) to DCs has been on the political agenda, and has primarily been based on the 'capacity to pay' principle. This has been articulated as an aid percentage calculated as 0.7% of GNI of ICs. Over the years, there has been continuing political commitment to this target (see Table 4.6), but the actual implemention has been problematic (see Figures 4.1 and 4.2). The commitment has been reinforced by global decisions to support the achievement of the Millennium Development Goals (MDGs), but at the same time has been undermined by sweeping generalizations about the failure of aid (see Chapter 2). Furthermore, the resources generated fall short of what is needed for achieving the MDGs (see Table 4.8).

Third, while DCs are trying to link the right to development with the voluntary aid commitment of the ICs, the ICs are afraid that this may turn into a right to development assistance (Adam, 2006) or a right to everything (Kirchmeier, 2006); see Section 4.4.2. Finally, aid flows do not compensate for the grievances of the DCs with respect to the international order, including the trade and commerce regime.

Climate change, too, has a clear North–South character. First, most emissions to date have come from the ICs, and it is these emissions that will have impacts in the coming decades. Most will be felt hardest in the countries of the developing world because of their geographical location and lower resilience, and because the impacts will exacerbate their existing problems (see Section 1.2.3). Second, the ICs have not kept their side of the original North–South deal (see Section 5.3), namely to reduce their own emissions drastically in order to make room for the emissions of the South to increase. If the climate problem is to be kept within 'safe limits', the room to emit greenhouse gases may be exhausted in two to three decades (UNDP, 2007). This has increased the pressure on the DCs to reduce the rate of growth of their emissions. However, these so-called safe limits are seen as unsafe by the small island states and the least-developed countries, who call for limiting the global mean temperature rise to 1.5 °C above pre-industrial levels (Oxfam, 2009).

Third, the resources for climate change assistance fall short of what is needed to address the climate change problem. While studies show that the assistance required for mitigation may be of the order of USD 200 billion (Opschoor, 2009) and that for adaptation about USD 70–100 billion annually (Opschoor, 2009; EACC, 2009), and that these estimates may be conservative (Behrens, 2009), the cumulative assistance since 1992 has been no more than about USD 500 million (see Table 5.4 and Section 5.4). Furthermore, while the use of market mechanisms for providing assistance to DCs has generated resources, it might not deliver on the sustainable development component, on the additionality of emission reductions and on

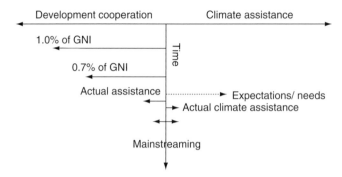

Figure 10.1 Promised versus committed resources.

technology transfer. By offsetting emissions in the ICs, the instrument reduces the pressure on them to reduce their own emissions (see Section 5.5). Fourth, the past emphasis of the ICs has been on defining climate change primarily as a sectoral, technocratic issue, rather than as a development issue, and emphasizing mitigation rather than adaptation (see Section 1.3), which has also been seen as problematic and continues to have a 'legacy' effect.

In combination, the climate change and development issues become a highly sensitive North–South issue. First, the *right* to development of 1986 is recognized as a *need* in the preamble and has become the 'right to and should promote sustainable development' principle in the Climate Convention. The right to traditional forms of development is thus not recognized as such. There has been an attempt within the UN Human Rights Council to reinvigorate this discussion through an examination of climate change and human rights (see Box. 5.1).

Second, the ICs have not delivered on their promise to provide 0.7% of GNI for about four decades. The promised new and additional resources for climate change appear to be equally elusive, being more in USD millions cumulatively than the USD billions needed annually. Figure 10.1 shows that, although resources were promised, the actual resources made available have been much less than what was promised.

Third, although market mechanisms have been created to promote GHG-friendly investments in the South, they basically offset IC emissions, and have no impact on normal technology transfer to the South or on rules of international trade.

10.2.3 The changing nature of North and South

In the development world, the North is referred to as the OECD countries and the South as the G-77 countries. Aid has been provided primarily by 23 of the 30 OECD countries (see Table 4.5).[1] In the climate change world, the ICs are seen as Annex I countries, and the rich among them have obligations towards the DCs as Annex II

---

[1] South Korea became a DAC member in 2010.

countries. The remaining approximately 150 non-Annex I countries are seen as potential recipients of aid.

However, these divisions no longer represent reality, since three major changes have taken place. First, the end of Communism led to new configurations in global politics and, while some of the richer former Communist countries became part of the 'North', the 'industrialized', 'developed' world, the poorer countries became, by default, DCs. During the cold war, the countries of the Communist bloc also provided assistance to DCs to increase their sphere of influence. This has decreased in the post-cold-war era. Second, some of the DCs had become over time very rich on a per capita basis and industrialized, but for political reasons remained part of the G-77. This category includes such countries as Singapore, Antigua and Barbuda, Barbados, the Bahamas, Qatar, Kuwait and the United Arab Emirates. Some relatively poor countries became part of the industrialized world, for example, through membership in the EU (e.g. Malta and Cyprus). These new entrants have sought assistance from the ICs and in some ways are competitors for aid rather than donors. Third, in recent years, the two giant economies – China and India – have started to develop rapidly, although this has scarcely made a dent on their per capita income (see Box 10.1). While systems of 'graduation' are being proposed and contested in the climate negotiations, this is clearly an important issue for the future. Nevertheless, there is a clear core of poor countries and there is a clear core of rich countries, and the analysis in this book focuses on these two cores.

10.3 Mainstreaming climate change

10.3.1 Introduction

Before delving into whether mainstreaming climate change is a good idea, this section discusses the theory of mainstreaming, the stages of incorporation of climate change concerns and the pros and cons of mainstreaming before identifying some key challenges.

10.3.2 The theory of mainstreaming

Defining mainstreaming

Climate policymakers and researchers have used the term 'mainstreaming' loosely to mean incorporation of climate change considerations into development processes. However, it is argued in this book that 'mainstreaming' has a very specific meaning. On the basis of the mainstreaming literature, Chapter 3 states that

Mainstreaming of climate change into development and/or development cooperation is the process by which development policies, programmes and projects are (re)designed,

---

Box 10.1 The changing role of China and India: from latecomer to frontrunner?

Introduction

Any effort to address climate change will need the constructive engagement of the large polluters – the USA, China, Russia, India and Japan – the top five emitters of greenhouse gases. The USA, however, has not committed to a target; the Russians are still capitalizing on their post-Perestroika emission reductions; and Japan is struggling to meet its 0.6% target under the Kyoto Protocol. The two DCs here are China and India.

Caveats

Before expecting China and India to take action, one should note that the unilateral behaviour of such countries as the USA and Russia is worrying. This suggests that some countries can avoid making a legal commitment, thereby creating a dangerous precedent. As a whole the ICs had promised to take significantly more action than they have done so far, prior to asking DCs to take action. Moreover, on a per capita basis, China's emissions rank 99th and India's 140th on a global basis (Baumert *et al.*, 2005). Their per capita emissions are not only lower than those of the ICs, but also significantly lower than those of other DCs.

China and India: internationally on the offensive

In an effort to create room for growth of their own emissions without sacrificing on the efforts to stabilize global greenhouse gas concentrations, the G-77 suggested at the 2009 climate negotiations in Bonn that the ICs reduce their emissions by more than 40% by 2020 and small island states by 45% in the coming two decades; an idea that was rejected immediately by the ICs, with the USA not offering much more than a stabilization target. At the same time, China and India were avoiding any suggestion that they should take on quantitative responsibilities.

China and India: domestically more active, but not proactive enough

Within China and India several climate-relevant policies have been initiated, but not necessarily implemented. Think tanks are active in both countries. China is more focused than India, but India has more pressing competitive issues on the agenda. However, both will be seriously affected by the impacts of climate change. Prior to the Copenhagen conference of 2009, China offered to reduce its carbon intensity by 40%–45% and India offered a 25% reduction by 2020. These are, however, not legally binding commitments.

China and India: proactive policy may be a matter of self-preservation

It is perfectly justified for both countries to point fingers at the recalcitrant ICs. However, in the final analysis their own domestic populations will be hurt most. Hiding behind the per capita argument may be morally astute vis-à-vis the ICs, but is not really helpful. Most of the emissions in both countries come from the rich; most of those vulnerable to climate impacts will be the poor, creating a North–South problem within these countries. Hence, some suggest sharing responsibility among the global high-emitting individuals (Chakravarty *et al.*, 2009). A governmental preoccupation with avoiding responsibility will just mirror what the North is doing to the South. On the other hand, perhaps technological and institutional solutions are easier to develop in China and India because of their massive student populations and (potential) civil societies. Perhaps mobilizing local people to search for sustainable options may pay off in the end when these two countries are not only able to ensure that the global community does not cross the 'dangerous' threshold in climate change, but also can market modern ideas to the world. The latecomers to development may become the frontrunners in the twenty-first century, if they put their minds to it. China and India seem now to be actively considering this option!

(re)organized, and evaluated from the perspective of climate change mitigation and adaptation. It means assessing how they impact on the vulnerability of people (especially the poorest) and the sustainability of development pathways – and taking responsibility to re-address them if necessary. Mainstreaming implies involving all social actors – governments, civil society, industry and local communities – in the process. Mainstreaming calls for changes in policy as far upstream as possible.

The stages of incorporation

The incorporation of climate change into the development and development cooperation process can occur along a spectrum of stages. In the early stages, some ad hoc experimental activities are carried out. This is followed by a search for win–win projects and programmes that are useful both for climate change and for development. Then there is a stage at which all development projects are screened for their possible vulnerability to climate change impacts. In the next stage, the development portfolio is examined from the perspective of integrating climate change mitigation into the existing projects and programmes. In these first four stages, trade-offs among economic, social and environmental values are continually made. In the final stage, the development paradigm and portfolio are redesigned, keeping the climate change perspective central (see Figure 10.2). Mainstreaming implies that the issue being mainstreamed becomes the overriding objective.

The scope of mainstreaming

Furthermore, climate mainstreaming is integral to environmental mainstreaming, and climate mainstreaming is closely related to dealing with gender issues and disaster risk management. All these fields are also key elements of both

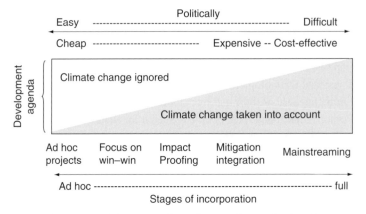

Figure 10.2 The stages of incorporating climate change in development and/or development cooperation.

development *and* development cooperation (see Chapter 3). Hence, mainstreaming climate change should be complementary to these three mainstreaming goals.

Design issues in mainstreaming

The mainstreaming of climate change into development and/or development cooperation implies using a CEGD lens – a Climate, Environment, Gender and Disaster lens – to establish the scope and design of development processes. Such a process could perhaps build on lessons from disaster mainstreaming (Benson and Twigg, 2007: 14) by using a seven-step approach: raising awareness; creating an enabling environment; development of tools, training and technical support; changes in operational practice; measuring progress; and learning and experience sharing at all stages. If the argument that climate change and development are closely interlinked is accepted, such a lens would be useful for redesigning development both in ICs and in DCs and for examining all global relations (see Figure 10.3).

In relation to the issue of whether climate change should be mainstreamed in development cooperation, a third element becomes critical: namely the lessons learned from development cooperation. Chapter 2 postulated that a 'clumsy BASICS' approach may have to be taken to ensure that development cooperation is successful. 'Clumsy' implies that development cooperation cannot be an elegant, efficient, pre-planned, drawing-board blueprint that is then implemented, but rather that it is a process in which, through continuous discussion with stakeholders and contextual learning, different views and understandings of relevant power politics are balanced to reach a solution that may work in a specific area. BASICS refers to Broader assessments; Alignment of approaches to countries and actors; avoiding

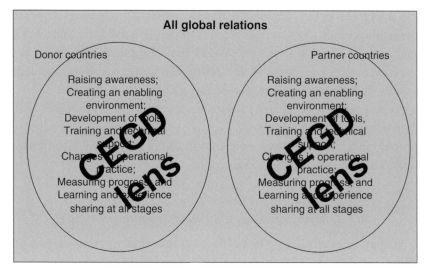

Figure 10.3 Mainstreaming climate change in development.

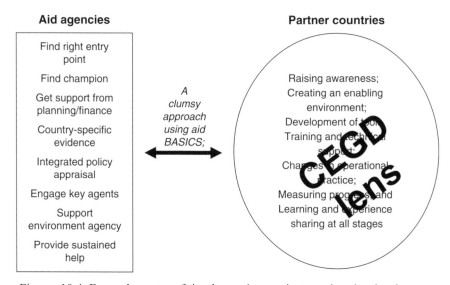

Figure 10.4 Four elements of implementing mainstreaming in development cooperation.

Simplicity, Imbalances and rigid Conditionality; and engaging Stakeholders (see Sections 2.4 and 10.5.2).

In line with demand-driven aid, UNEP's principles for mainstreaming environment (see Table 4.2) could be applied to donors. These principles are that development aid agencies should find the right entry point, find a 'champion', ensure the commitment of the planning or finance team, provide country-specific evidence, perform integrated policy appraisals, engage key sector agencies, consider the environment agency's capacity and acknowledge the need for sustained support. One may also be tempted to use the World Bank's transmission channels to the mainstreaming approach (Mani *et al.*, 2008); see Table 4.3. However, the latter approach is strongly anchored in neo-liberal ideology and does not question the ability of this approach to achieve sustainable development in completely different country contexts (cf. Gorz, 1994).

Mainstreaming in development cooperation would then imply a combination of the four ideas above, as shown in Figure 10.4.

10.3.3 Mainstreaming: challenges from the literature

Mainstreaming is an advantageous concept because it takes the issue being main-streamed seriously; does not attempt ad hoc, limited, once-only, sticking-plaster-type solutions; and reorganizes society and its institutions in an effort to address the issue. In the long term, mainstreaming should also become cost-effective and sustainable as a new equilibrium is reached.

However, mainstreaming is problematic because it is seen as politically challen-ging; questioning the vested interests of industry and the energy lobby; requiring the redesign of current production, consumption and lifestyle patterns; and unnecessary because societies should be able to improve themselves incrementally. Once vested interests understand what mainstreaming actually is, they lobby against changing the status quo, and mainstreaming efforts retreat into integration efforts. In the process trade-offs are made, often in favour of economic interests, and the issue may become invisible. The process appears to be hugely expensive because of the institutional change it appears to involve. The result is a 'mainstreaming lite' version – which trivializes the concept of mainstreaming.

This is why, in some of the mainstreaming discussions, a two-pronged strategy is recommended: (1) mainstream the subject under discussion (e.g. climate change in development and/or development cooperation); and (2) have a focused policy that promotes the subject (e.g. a focused climate change strategy), to ensure that the first effort does not fall short of full mainstreaming and does not become invisible before being addressed.

10.4 Climate change as a development issue

10.4.1 Introduction

Having defined and explored mainstreaming as a concept, this section examines whether mainstreaming climate change into development makes sense.

10.4.2 The substantive links

Section 1.4 argued that there are clear substantive links between climate change and development. This is because development for the poor and the rich may increase their greenhouse gas emissions; the poor, however, may have fewer choices. A range of policy interventions can enable the adoption of appropriate and affordable steps that may reduce these emissions. In relation to adaptation, 'maldevelopment' may affect the ability of people to adapt, while development opportunities may enhance the resilience of people to cope with climate change. Furthermore, climate change may have impacts on the ability of people to develop and on their infrastructure and ecosystems.

Although these substantive links are undisputed in the literature (IPCC-3, 2007), the key challenge is with respect to sequencing. What comes first, development or sustainable development? While the environmental Kuznets curve initially suggested that development precedes sustainable development as depicted in the inverted-U curve (see Figure 1.1), scholars subsequently argued that the inverted-U curve was more like an N-shaped curve (see Figure 1.2). The idea of an inverted-U curve initially suggested that DCs could grow first and only then would they have the resources to invest in sustainable development. It further suggested that the ICs would automatically invest in sustainable development. This would, of course, imply that DC emissions would grow. In order to avoid this, the suggested solution was affordable technology transfer from rich to poor.

However, the new insights show that (a) the theory is no longer valid, (b) IC emissions are not decreasing consistently with what is needed, (c) climate-friendly technology transfer is proceeding very slowly under the climate change regime, and (d) business-as-usual technology transfers have increased – but are probably transferring older and more affordable technologies to the South. This implies that incremental changes in ICs in combination with technology transfer to DCs are unlikely to address the climate change problem rapidly enough to ensure that it stays within 'safe' boundaries. This justifies the call for mainstreaming climate change in development.

10.4.3 Mainstreaming in development: challenges

Mainstreaming climate change in development is not an easy task. The USA sees climate change policies as seriously affecting its economic situation and that is why climate change mitigation has thus far received much less attention than it merits. Most ICs are incrementally developing their policies and hoping that such strategies will cause the least possible economic, political and social disruption to their societies. Most DCs too are being cautious, insofar as they are trapped in an out-of-date growth model and trying to keep up with the Joneses. Furthermore, at international level, vested interests seek to maintain the status quo and invest in marginal improvements rather than structural change. This is the biggest challenge associated with the climate change problem.

Climate change is yet another environmental problem that challenges the existing system of development. Dealing with climate change is not just a matter of tweaking the development system at the margins through environmental impact assessments and climate proofing, but calls for re-examining what development should actually look like, given the problem of climate change as a subset of many other environmental and social problems; see Chapter 1 and IPCC-3 (2007). Sustainable development may be the answer to this problem, but defining sustainable development remains intellectually difficult and politically challenging.

10.5 Climate change and development cooperation

10.5.1 Introduction

If mainstreaming climate change in development is a challenging, but possibly necessary, solution in the direction of sustainable development, does this automatically mean that mainstreaming climate change in development cooperation is a good idea? This section discusses the lessons learned from development and climate cooperation (see Section 10.5.2), examines the substantive links (see Section 10.5.3), and elaborates on the challenges for mainstreaming in development cooperation (see Section 10.5.4).

10.5.2 International cooperation

Development cooperation

Development cooperation, in its 60 years of existence, has had an interesting history. Motivated by diverse considerations (altruism, enlightened self-interest, political and strategic interests, security concerns, economic and, more recently, environmental reasons; see Table 2.3), influenced by the changing theoretical insights and practical experiences of each decade (see Table 2.2), development cooperation has been curiously controversial. While some argue that the overall effectiveness of development cooperation is disputable, this claim is based on a few macro-assessments of individual indicators rather than on a meta-analysis of the enterprise as a whole. Others argue that in some fields – such as health and agriculture – development cooperation has been very successful; at project level and in relation to project goals, development cooperation has been successful; and there are hopes that sector-wide and budget-related support will be more successful in the future, because of the underlying premise that these forms of assistance will support governments in the implementation of their own policies.

Governance on development cooperation has been scattered among UN agencies, although formally united in recent years by the UN Development Group. The 1945 UN Charter sets the norm; the UN General Assembly provides the forum where the general agreements are made; other UN bodies provide fora for issues with respect to the right to development and quantitative targets on development cooperation; the development banks focus on investment for development; the United Nations Development Programme focuses on development projects; and other UN agencies also deal with myriad development issues. However, there is no clarity about who the global donors are and why they should be donors, and who the partners should be and why they should be partners, except within the context of the OECD DAC (see Section 10.7). There has been a multitude of experiences, which have been integrated and adopted in the 2005 Paris Declaration on Aid Effectiveness (see Section 4.3.4), one of the first fora where donors and partners have come together to develop a common framework.

Lessons from aid history can be summed up in the acronym 'clumsy BASICS'. The term 'clumsy' (Verweij and Thompson, 2006; see Chapter 2) refers to the fact that pre-planned ideas based on theory and sketched on a drawing board are unlikely to have the expected results on the ground. The process is clumsy insofar as it encompasses a complex set of ideas and approaches, which have to be continuously balanced and fine-tuned, and sometimes overhauled, in order to develop successful programmes and projects. 'BASICS' refers to five ideas that need to be considered, but are not prescriptions to be followed blindly. 'B' is for 'broader assessments' of aid processes and approaches that take a transdisciplinary approach to analysing the results rather than converting the results into single criteria that have been determined in advance; this gives confidence in the aid process and justifies its continuation by ensuring that fine-tuning based on assessments is carried out. Such assessments would also give a much larger role to the partner countries and the stakeholders in partner countries in assessing these projects. 'A' is for 'aligning' aid type with partner country, organization and person, whereby long-term budgetary, programme and sectoral assistance is provided to competent persons in governments within countries with good governance, short-term project-related assistance is provided to competent persons and organizations within countries with poorer governance, and a community-based, basic-needs approach is applied to the poorest countries. 'S' is for 'simplicity avoidance' and realizing that change in specific places occurs within very specific context-relevant power relationships and that development cooperation has to find a route through these relationships in order to identify workable solutions. 'I' is for 'imbalances' or distortions to the local economy in terms of income, diversion of expertise and resources from one field to another, policy substitution effects and other distortions that should be avoided. 'C' is for rigid 'conditionalities', including conditions on partner country behaviour, tied aid and technical assistance. However, while these can have counter-productive results, they also create a domestic constituency in donor countries; and removing them may remove the domestic constituency. 'S' is for engaging and mobilizing 'stakeholders'.

Ultimately, the *literature tells us more about what doesn't work than about what does work*. What becomes increasingly clear is that planning and instrumental approaches may fail to work. Instrumental accountability approaches may also fail to yield the projected results. A more open, less instrumental, less planned (*à la* Easterly, 2007), less structured, clumsier (*à la* Verweij and Thompson, 2006) and more seeking approach may work.

Climate change cooperation

Climate change cooperation is younger in pedigree than development cooperation. The key lessons in climate change cooperation thus far can be summed up in the acronym LAME. 'L' is for the 'leadership' paradigm that was the driving force behind the regime – which is now limping as the USA, Russia, Norway and Australia lag far behind other ICs, and as the ICs lag behind the commitment needed to address the problem rapidly and transfer resources to DCs. 'A' is for aid levels that are currently in the millions, while the help needed is of the order of billions. 'M' is for market mechanisms that are problematic because of their character as an instrument that offsets the emissions of the North rather than as an instrument that rapidly catalyses change in the South. 'E' is for emission mitigation and adaptation technology transfer that is proceeding much slower than initially envisaged, while foreign direct investment continues to market older technologies (see Chapter 5).

10.5.3 *The substantive links and other driving forces*

Mainstreaming climate change in development cooperation appears to make sense in a number of ways.

- Logically, since (i) climate change impacts will affect all sectors of society and all projects and programmes being developed; and (ii) all projects, programmes and initiatives in society may exacerbate the climate change problem.
- Financially, since (i) there are not enough resources for meeting the MDGs and, worse still, the MDGs may be negatively affected by climate change; and (ii) there are not enough resources for meeting the climate change goals.
- Practically, since (i) the existing development cooperation agencies already have an institutional framework for implementing policies in DCs and could easily engage in climate cooperation; and (ii) there is probably a domestic constituency in the ICs that

would support measures on climate change, even if they would not support measures on development.

- From a DC perspective, since (i) these countries prioritize development; and (ii) these very goals may be affected by climate change itself through its physical impacts or through global climate change policy and its implications for mitigation policy.
- From a stakeholder perspective, insofar as it (i) brings the development and environmental communities (government, non-governmental, scientific) together; and (ii) unites their experiences and theories.
- From existing reporting and accountability trends, where (i) climate cooperation is calling for National Communications and National Adaptation Programmes of Action; and (ii) development aid is calling for the preparation of Poverty Reduction Strategy Papers and Country Strategy Papers, which also examine environment–poverty linkages.

Mainstreaming climate change in development cooperation has support from many different actors for different reasons, e.g. it helps existing aid agencies secure greater funding, and it helps development banks cope with the environmental critique (see Table 3.2).

10.5.4 Mainstreaming in development cooperation: challenges

However, there are six arguments against mainstreaming.

- Different paths to development. A key argument against mainstreaming climate change in development cooperation from the perspective of transferring Western ideologies, technologies, and lifestyles to the South is that this may exacerbate the sustainable development challenge. While such an approach was consistent with the inverted-U shape of the environmental Kuznets curve, it is no longer consistent with the N-shaped curve of scientific evidence (Caviglia-Harris *et al.*, 2009). While Metz *et al.* (2000) focused on avoiding the past patterns of the ICs, and the UNDP (2007) warns that nine Earths would be needed to meet the needs of people measured in terms of Western standards, possibly completely different pathways are needed both for ICs and for DCs (see Figure 10.5). As the IPCC puts it, and as has been cited in Chapter 1, probably DCs must navigate 'through an uncharted and evolving landscape' (IPCC-3, 2007: 693).
- Diversion – development cooperation in the context of promises. Another argument against mainstreaming is that aid promises have been at 0.7% of GNI and these have not been met (see Section 4.3.4). Climate cooperation was to be funded as part of the leadership paradigm through 'new and additional' funding (see Section 5.5), but the current flows fall considerably short of what is needed (see Section 5.4). Worse, mainstreaming may appear to fudge the two flows through an overly optimistic presentation of the way in which the two problems of climate change and development cooperation are linked. Some clearly see that mitigation will need extra assistance and hence limit mainstreaming to adaptation activities, arguing that '[i]ncorporating climate considerations into traditional development aid is the most effective way to assist with adaptation,

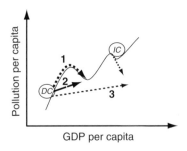

Figure 10.5 DCs and ICs need to move through an unchartered landscape.

and the Task Force recommends that adaptation assistance be delivered primarily through such an approach' (Independent Task Force US, 2008: 7). However, this is a very pernicious argument. It is a repetition of the older argument that adaptation is a local problem and, hence, that the international community should not pay for adaptation, rather than that the impacts have been caused by past industrialization and those responsible should pay for it. Under the new international pressure to pay for adaptation measures in DCs, the ICs are now seeking ways to relabel existing money for adaptation goals.

- Diversion – development cooperation in the context of resources needed. From a resource perspective, the argument can be put differently. About USD 100 billion annually is at present being spent on development cooperation, although in 2008 it reached USD 120 billion. Chapter 4 argued, from the literature, that an additional USD 50–135 billion is needed to meet the MDGs (Clemens *et al.*, 2007). In 1992, an additional USD 125 billion was needed from ICs to support policy work under Agenda 21 (Agenda 21 and Rio Declaration, 1992). The climate change literature suggests that up to USD 270 billion is needed annually to address mitigation and adaptation issues (see Table 10.1). There is some degree of overlap of the issues covered by the MDGs, climate goals and Agenda 21. Nevertheless, the order of numbers is so different that there is a very real fear that mainstreaming climate change into development cooperation may imply a hijacking of the limited existing development resources for climate purposes, especially since the latter issue is seen as more urgent for the ICs.
- Diversion – the Triple Triangle of beneficiaries. Although there are considerable overlaps and links between climate change and development, the DC beneficiaries of such cooperation differ considerably. The ostensible beneficiaries of development cooperation at present are the poorest at the bottom of the triangle; the DC beneficiaries of climate mitigation action are the industrial classes and large agriculture; while the DC beneficiaries of action on adaptation are across the board – from poor to rich (see Figure 10.6).
- Development cooperation as a small part of international transfers. Global cooperation relations between countries lie in the areas of trade, investment, environment and aid. Aid and environmental flows are a fraction of global financial flows (see e.g. Table 1.2) and to that extent have a limited impact on DCs in general, but a larger impact on the smaller, poorer DCs that scarcely receive investment flows and/or flows through the

Table 10.1 *The resources needed for cooperation*

	Existing development cooperation	Additional aid above existing ODA needed for MDGs	Resources needed for Agenda 21 goals	Resources needed for climate change assistance	Total
USD billion	~100–120	50–135	125	Up to 270	Up to an additional 430
Comment	~0.35% of GNI of donors	Based on literature review in Clemens *et al.* (2007); see Table 4.8	Chapter 33 of Agenda 21 (1992)	Based on literature review (Opschoor, 2009; Behrens, 2009; EACC, 2009); see Table 5.4	There are clearly overlaps between the categories

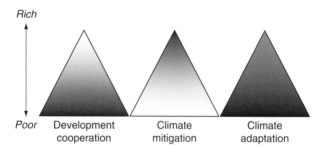

Figure 10.6 The diverging beneficiaries of development cooperation and climate change.

Clean Development Mechanism. Trade and investment are much more significant areas of cooperation – and in both worlds there is considerable discussion about the degree to which the views of the DCs and environmental and social impacts are taken into account.

- Mainstreaming may amount to a new disguised conditionality. There is a risk that the incorporation of a mainstreaming agenda into development cooperation may evolve into a kind of conditionality imposed by the ICs. If so, then, given the relatively poor experience with rigid conditionalities in the past (see Section 2.3.4), the omens for success are bad. However, clearly, if the money is earmarked for climate change, there should be some general principles that guide spending. But, extrapolating from earlier arguments in the negotiations, DCs would probably argue that, if the money is compensatory in nature for the impacts caused to them, then it is up to them how the money is spent. Where the resources are generated as part of a cooperative process, all parties concerned should democratically decide how the resources are spent in a joint management and implementation system.

10.6 Climate change, development cooperation and the European Union

10.6.1 Introduction

Having submitted that there are substantial arguments both in favour of and against mainstreaming climate change in development cooperation, this section examines how the EU and its Member States are dealing with the issue.

10.6.2 EU level: integrating climate change

The EU is taking measures to incorporate climate change into its development cooperation strategy. A number of conclusions about its strategy can be derived. First, it controls up to 20% of the total aid flows from its Member States, and, including the resources of the Member States, is responsible for about 60% of global OECD aid. Thereby it is the single largest donor in the global arena and has structural power. It can thus potentially have a considerable influence on development cooperation policy globally.

Second, its development cooperation is motivated by (a) the altruistic goal of achieving the MDGs; (b) increasingly important security goals in the post-9/11 era; and (c) ideological concerns of promoting a democratic framework, perhaps more with the intention of liberalizing economic institutions (aid for trade) than tampering with politics itself. For example, the European Consensus on Development focuses on poverty reduction and supporting DCs in defining their own development and on promoting democracy. The EU is probably less influenced by other strategic, post-colonial, or economic interests, as it averages out these concerns of its 27 Member States. This can be extrapolated from the study result that bilateral donors tend to promote their own interests more (McGillivray, 2004). Its experience is less long-standing than that of other donors, but it has a shorter learning curve since it has benefited from the experiences of others. It can thus, potentially, serve as an example of a 'good' donor. The experiences of the coming years will be decisive for a more definitive assessment of whether EU development cooperation policy can influence climate policy in DCs.

Third, since 1990, the EU has been steadily trying to incorporate environmental and climate change policy into its sectoral policies, including development cooperation policy. At rhetorical level, at least, the EU appears just as committed to incorporating climate change and other environmental concerns into its sectoral policy as into its development cooperation strategy, thus showing a degree of consistency in these policy arenas. Although it is pushing its Member States to adopt stringent climate targets that will impact on their development, it is easier for it to modify its own development cooperation policy, which is within its own mandate.

Fourth, unlike the OECD DAC (see Section 4.3.5), the EU aims to incorporate both climate mitigation and adaptation. In practice, some of this incorporation is taking place via the EU Country Strategy Papers (CSPs). These papers and other related reports may enable DCs to assess constructively how best to develop sustainably. Although the first-generation CSPs failed to include climate change concerns, the newer papers increasingly discuss climate-related issues (see Section 9.7). Other existing development cooperation projects are already subject to a number of checklists; however, climate change issues are not yet being systematically checked or integrated into environmental impact assessments.

Fifth, the EU's policy does not go beyond a policy commitment to climate-proof and/or integrate climate change into its development cooperation policy. While it has used the term 'mainstreaming', the content of the EU's action falls far short of that in terms of the definition used in this book. The EU is also committed to the integration of environmental, gender and disaster issues, but linking these effectively still has to be done.

Sixth, in 2008, the Heads of State meeting at the Monterrey follow-up conference in Doha (Doha Declaration, 2008: para. 42; see Chapter 4) noted that the EU collectively had agreed to provide 0.56% of GNP for ODA by 2010 and 0.7% by 2015, and to channel at least 50% of collective aid increases to Africa. The EU does not enter into discussions regarding 'new and additional' assistance in its internal deliberations, and it is not easy to understand what its collective view on this is.

Thus, the EU as the single largest donor, averaging out national interests and by virtue of not having any strong ideological roots, may be in a position to influence global aid policy and could consider using this position to improve global aid. Internally, the EU has still a long way to go to climate-proof and integrate its own development and development cooperation strategies. It is in the process of designing criteria to take climate change adaptation and mitigation, disaster management and gender issues explicitly into account in development programmes and projects. However, if the EU believes in national ownership of development strategies and policies, it has a limited role in actually climate-proofing or integrating climate change into aid policies. If it focuses on the latter, this may amount to a new conditionality. The EU has to walk a fine line between these two extremes.

Table 10.2 *EU Member States and the OECD DAC*

Member States belonging to the OECD DAC	Member States not belonging to the OECD DAC	Other OECD DAC members	Member States not adhering to the Paris Declaration
Austria, Belgium, Denmark, Finland, France, Germany, Greece, Ireland, Italy, Luxembourg, the Netherlands, Portugal, Spain, Sweden, UK	Bulgaria, Cyprus, Czech Republic, Estonia, Hungary, Latvia, Lithuania, Malta, Poland, Romania, Slovakia, Slovenia	Australia, Canada, Japan, New Zealand, Norway, South Korea, Switzerland, USA	Bulgaria, Latvia, Lithuania, Malta, Poland

10.6.3 Member State level: divergent

The EU consists of 27 Member States, which provide up to 80% of EU aid. The research reveals first that the EU membership has diverging responsibilities in the area of development cooperation. Of the 27 EU Member States, 15 are OECD DAC donor states and 21 have signed the Paris Declaration on Aid Effectiveness (see Table 10.2). Although development cooperation is an area of shared competence between the EU and its Member States (the fact that aid strategies do not uniformly apply to all Member States shows that there is not a harmonized strategy within the EU regarding the development cooperation policy of its Member States), the aid strategies of the Member States vary widely.

Second, most Member States have not yet met their 0.7% development cooperation norm. Only Denmark, Luxembourg, the Netherlands and Sweden (see Figure 4.2) have done so. While the Netherlands and Sweden have accepted the idea of 'new and additional' resources, and provide more than the commitment of 0.7% of GNI as assistance, Denmark, the UK, France, Italy and Germany have committed to increasing their ODA *via* increased expenditure on climate change activities.

Third, at Member State level there is a wide difference in motivations, goals and commitments concerning climate change. An extreme example is that some Member States are allowed to increase their emission levels of greenhouse gases under the Kyoto Protocol (e.g. Portugal by 27%, Spain by 15%, Ireland by 13% and Sweden by 4%) (Kyoto Protocol, 1997) and under the 2008 effort-sharing decision in the EU (e.g. Bulgaria by 20% and Romania by 19%). This shows that understanding on how to incorporate climate change into their own development process is still low and subject less to scientific endeavour than to political bargaining processes. Nevertheless, the case-study countries are trying to incorporate climate change into national development policies.

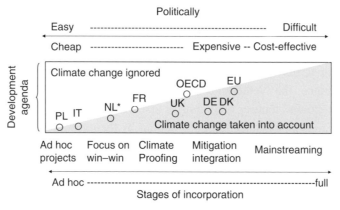

Figure 10.7 Policies of different countries on incorporation. N.B. The asterisk next to NL denotes that the Netherlands has opted to allow partner countries to develop their own strategies.

Fourth, in contrast to the comprehensive strategic approach to climate change taken at EU level, at Member State level each country has a different approach to incorporating the climate change agenda into its development cooperation strategy (see Figure 10.7). For instance, the use of terminology in the documents differs (e.g. Denmark and Germany have different meanings for climate proofing) and the situation is still in a state of flux. Some Member States prepare ad hoc climate-relevant projects, others are climate screening their development cooperation port-folio. Of the five Member States studied, Denmark and Germany take climate change seriously in their aid strategies. Denmark has clear objectives, specific actions, indicators and entry points. Germany has operationalized its ideas and subjects its portfolio to an assessment. The other countries are debating how best to incorporate climate change into their strategy. Likewise the emphasis given by individual countries to specific values in aid differs.

Fifth, many EU Member States are developing and using climate change incor-poration tools. Sometimes the tools are used in an ad hoc manner – such as the application of the 'quick scans' in the Netherlands. The German 'climate check' tool is systematically applied and seems to be a model that others can also use. If the screening process is too complex, it becomes difficult to implement.

Sixth, the motives of Member States vary considerably (see Chapters 2 and 7). While some (the UK and France) give aid to past colonies, others (Sweden) give aid based on specific criteria (Alesina and Weder, 2002). These two approaches can, of course, go together. Some countries label debt cancellation as aid; others, such as Denmark, France, Sweden, the Netherlands and the Czech Republic, include the costs of hosting refugees as aid; Germany, France and Portugal include the costs of hosting foreign students as aid (Joint European NGO Report, 2006). Some are seen as altruistic (e.g. Ireland, Denmark, Austria and Finland), others as selfish (e.g. the UK, Italy and France) (Berthélemy, 2006; cf. Nunnenkamp and Thiele, 2006).

Seventh, in terms of favourite partner countries, each donor has its reasons and while France, the Netherlands and the UK focus on the poor countries and past colonies, Poland focuses on countries that are in transition. Preferences change all the time. Small island states are conspicuous by their absence from consideration as partner countries.

10.6.4 Case studies

For the purpose of studying how climate change is being implemented in cooperation documents, the reports of and with respect to 10 DCs (namely Brazil, China, the Comoros, Egypt, India, Kenya, Malawi, Nepal, South Africa and Tanzania) were studied. These reports analysed the needs of these countries for climate assistance in the areas of energy, forestry, agriculture and biodiversity as revealed in their National Communications, National Adaptation Programmes of Action and Technology Needs Assessments. Subsequently, a study of the EU Country Strategy Papers (CSPs) was undertaken to examine the kind of assistance being offered to these DCs.

Certain caveats need to be taken into account. First, most of the reports have been written by many authors and there is often no consistency within (e.g. Egypt and India) and between the different reports. There is a greater degree of coherence in the needs of the smaller countries. Second, these reports were not always written with the explicit purpose of identifying needs for assistance. Third, the terminology is not always uniform.

Subject to these caveats, the needs of DCs are more explicitly articulated in the area of mitigation, rather than adaptation, except perhaps in relation to the biodiversity and agricultural sectors. Not surprisingly, China and India are most explicit about their needs for assistance, while Brazil, Egypt and South Africa are less explicit. Most of these countries request assistance especially with respect to land use and management, and research and technology, while assistance for governance issues is less emphasized. In the area of energy, the needs focus on research and technology for clean coal, fuel switching and energy efficiency, with China and India calling for modern technologies. On forestry, countries in which net deforestation is occurring (Brazil, the Comoros, Kenya, Malawi, Nepal and Tanzania) have clearer needs (e.g. land use and management largely refer to reforestation and afforestation, and sustainable management practices) than the others. With respect to biodiversity, with the exception of Egypt, all focus on needs with respect to land use and management, including reforestation, restoration of degraded soils, reconstructing of slopes, rehabilitation and regeneration of degraded ecosystems and sustainable and responsible management practices and conservation measures. On agriculture, needs are more in the area of research and technologies relating to issues such as rice cultivation technology and rice varieties, non-tillage technology, fertilizer-application technology and improved feed for livestock to reduce methane emissions.

On adaptation, energy needs tend to focus on climate risk and impact assessment and early-warning systems for floods and droughts. With respect to forestry, apart from Kenya, all have listed a variety of needs with respect to land use and management, and institution building. Some of these needs also serve mitigation goals such as improvement of forest management policies and practices, agro-forestry, species selection and the establishment of forest seed banks for indigenous species. On biodiversity, adaptation needs are similar. Agricultural needs also focus on land use and management, water management, diversification of agricultural activities, sustainable animal husbandry, drought-tolerant crop varieties, irrigation technologies, pest and disease control, and early-warning weather forecasts. While mitigation and adaptation activities may be very different in nature in the energy sector, there may be some synergies between mitigation and adaptation policies in the agriculture, forestry and biodiversity sectors.

An examination of the supply of assistance shows that, during the last 50 years, agricultural, forestry and energy assistance has been on the development agenda, whereas environment and biodiversity issues have only recently received attention. The CSPs reveal a reliance on ideas of good governance, capacity building and aid for trade, and support for traditional areas such as education, health and transport infrastructure. This approach has dominated the assistance. The second generation of CSPs has taken environmental issues on as well. In relation to our case-study sectors, the supply of aid on energy has focused on a wide range of sustainable energy options. Aid on forestry and biodiversity has focused on governance and forest management issues, including alternative livelihoods for those dependent on forestry. Aid on agriculture has focused on trade promotion, water conservation, diversification of farmers' incomes, combatting agricultural pests and diseases, capacity building and research and development.

There has not always been a match between needs and supply. This reveals that, although there is considerable discussion on country ownership of ideas that aid should support, there is a gap between the rhetoric of climate mainstreaming and the fact. The EU tends to focus on economic governance and/or reform, market instruments and trade, even though many of the countries have explicit needs for climate-related assistance. Although in some fields there is assistance needed in research and technology, e.g. drought-resistant seeds, it is in these practical areas that there is less support. For example, on forests, research and technology needs include improved seeds, while assistance is mostly in the form of governance help. The same is the case for biodiversity. Some of these mismatches may reflect the facts that (a) the documents serve different goals, (b) the PRSPs and CSPs have yet to incorporate climate change into the national planning process, and (c) the DCs themselves do not have a consistent vision of their needs.

10.7 North–South issues revisited

10.7.1 Introduction

There is a tendency to focus on a handful of developed countries and ask them to pay, and to critique their behaviour simply because they are part of a system of donor registration. This deflects attention from the real issue – why should countries pay, who should pay and to whom?

10.7.2 Why should developed countries pay?

Part of the underlying logic of development cooperation (see Chapters 2 and 4) and climate change cooperation (see Chapter 5) has been the understanding that the richer world has the scientific knowledge about how to develop at its command *and* that it is significantly richer and, therefore, better able to assist developing countries. Both development cooperation and climate change cooperation are also closely linked to a perception of responsibility for past actions, altruistic commitment to addressing global problems and alleviating poverty as a destabilizing force.

Much of development cooperation finds its roots also in cold-war politics and ideological interests, serious commitment to addressing global problems, security concerns and post-colonial interests. Historically, development cooperation grew from the voluntary willingness of the ICs and in response to the needs of the DCs. This led to a political commitment to adopt a quantitative target and discussions on the right to development in which participants sought to change the voluntary character of aid to a more 'compulsory' nature based on the principle of the ability to pay, akin to systems within social democracies.

Climate change cooperation, on the other hand, is based on the need for the ICs to show leadership in addressing the problem, since the bulk of past greenhouse gases were emitted by them. It is based both on their responsibility for past and present emissions and on their ability to pay. This, at any rate, is how IC responsibility has been phrased in the Climate Convention.

There are many arguments as to why the ICs should be seen as responsible for past and present emissions and thus as having a duty to reduce their own emissions and being liable to compensate others for the harm caused to them, as well as why a right to development justifies increases in emissions in DCs (Gupta, 1997; Baer *et al.*, 2008; Oxfam, 2007); see Chapter 5. The shift from liability to leadership and now to a limping leadership is, in itself, a critical argument as to why the DCs are politically quite sceptical about the ICs' willingness to provide assistance to DCs and to reduce their own emissions.

10.7.3 Who should pay?

Research shows that the issue of who pays ODA is an area of fuzzy governance. There are no criteria for countries that pay or should pay. The disbursement of ODA began in the Western world as a voluntary activity and there were a range of motivations behind this. It is coordinated by the OECD DAC. However, not all OECD members are members of the DAC. Similarly, not all OECD members are Annex I/B parties (e.g. South Korea and Mexico) under the Climate Convention and Kyoto Protocol. This reflects probably on the nature of the OECD and its role as a think tank. Similarly, although the EU has a development cooperation policy, not all EU Member States are either members of the OECD DAC or subscribers to the Paris Declaration on Aid Effectiveness (see Table 10.2).

A related issue is that there are donors outside the OECD DAC. China and some countries in the Middle East have been active donors with a completely different style (see Box 1.1). The governance process on development cooperation appears to have no link with these bodies. The question is when does one, or should one, become eligible to become a donor? This is a critical question since the needs of the poorer countries keep increasing, while the group of donor countries remains relatively limited and the pressures on them keep increasing. This is also problematic for the ICs insofar as some are exempted while others are not.

Although a critical issue over time has been the issue of donor coordination, and the OECD has tried to respond to this need, the question of whether the OECD is the appropriate authority to coordinate donors may be posed. While the UNDP coordinates the UN Development Group, it does not have the actual steering power to coordinate global development cooperation. Furthermore, in the development cooperation world, the partner countries have more of a subsidiary than a partnership role – although this changes where 'ownership' of development projects and programmes shifts. Do partners have a say in OECD DAC and EU policy making? Although these countries have signed up to the Paris Declaration, does that give them an effective voice in development cooperation?

In the climate change regime, some countries listed in Annex I are not obliged to pay. Table 10.3 shows that some EU and non-EU ICs have no obligations to pay despite past industrialization and pollution; and that some DCs have graduated into the EU – Malta and Cyprus!

Table 10.3 *ICs with and without financial obligations*

ICs with obligations		ICs without obligations	
EU	Non-EU	EU	Non-EU
Austria, Belgium, Denmark, European Economic Community, Finland, France, Germany, Greece, Ireland, Italy, Luxembourg, Netherlands, Portugal, Spain, Sweden, UK	Australia, Canada, Iceland, Japan, Norway, New Zealand, Switzerland, USA	Bulgaria, Czech Republic, Estonia, Hungary, Latvia, Lithuania, Slovakia, Poland, Romania New EU: *Malta, Cyprus*	Belarus, Russian Federation, Turkey, Ukraine New ICs: *Croatia, Liechtenstein, Monaco, Slovenia*

Table 10.4 *G-77 and non-Annex I countries*

G-77 countries (130)	Non G-77 countries by group (23)				
	New OECD (2)	CEITs (11)		AOSIS (6)	Misc. (4)
133 – 3 members (Palestine is not an independent state; Yugoslavia is not allowed to participate; Romania is in Annex 1)	Mexico, Korea (Republic of)	Albania, Armenia, Azerbaijan, Georgia, Kazakhstan, Kyrgyz Republic, Macedonia (Former Yugoslav Republic of), Moldova, Tajikistan, Uzbekistan, Yugoslavia (Federal Republic of)		Cook Islands, Kiribati, Nauru, Niue, Palau, Tuvalu	Andorra, Holy See, Israel, San Marino

Source: Gupta (2008).

10.7.4 Who should receive assistance?

The Group of 77 (G-77), which was born in 1964, today counts about 130 members. However, from the climate change perspective, about 150 countries are technically seen as non-Annex I/B countries and hence eligible for assistance (see Table 10.4).

The OECD list of recipient countries is provided in Table 10.5. The least-developed 50 countries are uncontroversially among the group that qualifies for both climate aid and development cooperation. However, there remain some issues

Table 10.5 *The OECD list of recipient countries*

LDCs	Other low-income countries	LMI countries and territories	UMI countries and territories
Afghanistan, Angola, Bangladesh, Benin, Bhutan, Burkina Faso, Burundi, Cambodia, Central African Republic, Chad, Comoros, Congo (DR), Djibouti, Equatorial Guinea, Eritrea, Ethiopia, Gambia, Guinea, Guinea-Bissau, Haiti, Kiribati, Laos, Lesotho, Liberia, Madagascar, Malawi, Maldives, Mali, Mauritania, Mozambique, Myanmar, Nepal, Niger, Rwanda, Samoa, São Tomé and Príncipe, Senegal, Sierra Leone, Solomon Islands, Sudan, Tanzania, Timor-Leste, Togo, Tuvalu, Uganda, Vanuatu, Yemen, Zambia	Côte d'Ivoire, Ghana, Kenya, Korea, DR, Krygyz Republic, Nigeria, Pakistan, Papua New Guinea, Tajikistan, Uzbekistan, Vietnam, Zimbabwe	Albania, Algeria, Armenia, Azerbaijan, Bolivia, Bosnia and Herzogovina, Cameroon, Cape Verde, China, Colombia, Congo, Republic, Dominican Republic, Ecuador, Egypt, El Salvador, Georgia, Guatemala, Guyana, Honduras, India, Indonesia, Iran, Iraq, Jordan, Macedonia, Marshall Islands, Micronesia, Moldova, Mongolia, Morocco, Namibia, Nicaragua, Niue, Palestinian Administered Areas, Paraguay, Peru, Philippines, Sri Lanka, Swaziland, Syria, Thailand, *Tokelau*, Tonga, Tunisia, Turkmenistan, Ukraine, *Wallis and Futuna*	*Anguilla*, Antigua and Barbuda, Argentina, Barbados, Belarus, Belize, Botswana, Brazil, Chile, Cook Islands, Costa Rica, Croatia, Cuba, Dominica, Fiji, Gabon, Grenada, Jamaica, Kazakhstan, Lebanon, Libya, Malaysia, Mauritius, *Mayotte*, Mexico, Montenegro, *Montserrat*, Nauru, Oman, Palau, Panama, Serbia, Seychelles, South Africa, *St Helena*, St Kitts/Nevis, St Lucia, St Vincent and Grenadines, Suriname, Trinidad and Tobago, Turkey, Uruguay, Venezuela

Source: http://www.oecd.org/dataoecd/62/48/41655745.pdf.
Notes: names in italics refer to extra-territorial regions of ICs, not to independent states. LMI, lower middle income; UMI, upper middle income.

on the margin. The OECD includes its own member states as potential recipients of aid, or partner countries (e.g. Mexico); offshore IC territories (e.g. Montserrat and Mayotte); and groups that are not countries and technically speaking potential recipients of climate change assistance – although there is no formal listing of the DCs in the climate change regime (e.g. Montserrat, etc.). It excludes countries that are technically eligible for assistance under the climate change regime (e.g. Israel,

Holy See), and includes countries that are technically excluded for assistance under the climate change regime (e.g. Turkey).

Furthermore, the climate change regime itself is complicated. It includes a number of poorer countries among the ICs, and leaves a number of quite rich countries among the DCs, e.g. Qatar, Brunei and Singapore (Gupta, 2008). These are issues that create complications about whether merging the two discussions, with their different histories in terms of donors, responsibilities for providing assistance and partner countries, can be effectively undertaken.

10.8 Integrated analysis

10.8.1 Introduction

This section shows the divergence in views underlying the seeming convergence of rhetoric (see Section 10.8.2), and the paradoxes within the key issues (see Section 10.8.3).

10.8.2 Divergence underlying convergence

Seeming convergence on development, development aid and mainstreaming climate change in both

This book reveals a convergence in views of different countries and actors on the need to recognize the right to development of DCs; the 0.7% commitment of ICs; the need to implement the Paris Declaration on Aid Effectiveness, not only in terms of providing a regular and reliable amount of aid but also in terms of supporting the lessons learned in development cooperation; the need for leadership on climate change; the need to provide new and additional assistance to DCs for climate-relevant measures; the need to mobilize all actors, including private actors, through the use of market mechanisms; the need for appropriate technology transfer from ICs to DCs; and the need to link development with climate change. The bottom line is that climate change and development are so intimately linked to each other that addressing one without the other is not likely to work. As a logical corollary, and given the reasons in Section 10.5.3, the global community should try also to 'mainstream' climate change into development cooperation.

Underlying divergence

Underneath the seeming convergence of views between ICs and DCs there is an underlying divergence in views.

Although the right to development of DCs has been accepted, extensions to link it to the rights of states as opposed to the rights of peoples are disputed;

making quantitative commitments to development cooperation from the ICs legally binding is contested since this could potentially convert their voluntary generosity into a monitorable legal commitment; quantitative commitments in the climate change regime are contested because of their implications for emission reductions in the ICs and assistance for adaptation and mitigation costs in DCs.

Although both ICs and DCs have in their own and joint fora repeatedly advocated that the 0.7% norm for development assistance should be respected, there is a gap between norm recognition and norm implementation.

Athough both ICs and DCs have adopted the Paris Declaration on Aid Effectiveness, implementation of the concepts in this Declaration is complicated and challenging, and not all donors are able to do so. The implication of this Declaration for other groups of donors is also not clear.

Although ICs and DCs have repeatedly accepted and used the rhetoric of the need for ICs to show leadership on climate change, the leadership of ICs is flawed insofar as the large IC polluters are avoiding major commitments; some of the smaller ICs have targets to increase their emissions; and, although the EU is doing its best to push the negotiations forwards, convincing Member States is not all that easy. The EU, however, has an unconditional target of reducing its emissions by 20% in 2020.

Although there was a stated consensus to provide 'new and additional assistance' to DCs for climate-relevant measures, there are major differences in the way these terms are interpreted. At the Conference of the Parties in Copenhagen in December 2009, a Copenhagen Accord was 'noted', not 'adopted', which stated that developed countries commit to providing new and additional resources 'approaching USD 30 billion for the period 2010–2012' and aim to 'mobilize jointly USD 100 billion a year by 2020' (Copenhagen Accord, 2010). This legally non-binding statement offers some hope, but the resources, although significantly greater than those allocated at present, fall short of what is needed.

Although market mechanisms were supposed to help catalyse major change in DCs, the sustainable development returns on the mechanisms remain elusive; the actual 'additionality' of the emission reductions is contested; the technology transfer component in these mechanisms is limited; the ability to catalyse change in the large countries is limited, while these mechanisms tend to bypass the smaller countries; and the emission reductions are used to offset emission increases in the North, leading to limited net reduction.

Although there was consensus on the need for appropriate technology transfer from ICs to DCs and many administrative steps have been taken to promote this, technology transfer under the climate regime has been limited, and normal flows of technology via foreign direct investment continue unabated.

Although there are clear substantive reasons why climate change should be mainstreamed in development in poor and rich countries, in the international arena vested interests resist the structural changes this will call for at local through to global levels.

Although many actors are converging on the need to mainstream climate change in development cooperation and there are six good reasons for mainstreaming, there are six good reasons why mainstreaming should not be undertaken (see Section 10.5.4).

This book moves towards the conclusion that mainstreaming climate change in development cooperation is not a good idea. This conclusion would not be valid if (a) large resources commensurate with both the development and the climate change problem were available, (b) if all countries were actually mainstreaming climate change into their own development processes, and (c) if all flows of investment into DCs were actively mainstreaming climate change.

Between the idea and the reality falls the shadow of distrust

Inevitably, T. S. Eliot comes to mind, for between the consensus and the reality falls the shadow. There are many gaps between the rhetoric and the action of the ICs, between the rhetoric of mainstreaming climate change in development cooperation and mainstreaming climate change in domestic development; and there is a partnership gap between the DCs and ICs on development cooperation. The gap between the rhetoric and the implementation in both the development field and the climate change field brings up the issue of the trust between ICs and DCs: 'Some of the factors accounting for the present lack of progress in achieving a climate agreement are rooted in a deep deficit of trust between developed and developing countries' (Opschoor, 2009: 38); and 'Those who do not trust one another to keep to commitments can rarely negotiate successfully, especially on something as complex as a post-Kyoto climate framework' (Commission on Climate Change and Development, 2009: 9).

10.8.3 Governance and lessons learned: the paradoxes

This assessment has brought to light a number of issues. Different conclusions possibly apply at different levels of analysis.

- Development cooperation – governance and lessons learned. The assessment on development cooperation and its governance reveals that governance is scattered throughout the UN system and links are often not effectively made; there is no clarity about who should be donors and who should be recipients, or partner countries, except in very general terms, why and in relation to what criteria; the most organized body on development cooperation is the OECD DAC which unites donor countries; there have been more lessons learned about how not to structure development cooperation than about how aid can be structured; and the lessons learned are often counterintuitive and difficult to implement. Clumsy BASICS may provide some guidelines for aid professionals. The *ownership paradox* of aid is that, if aid is successful only when there is 'ownership' and 'commitment' to implementation in partner countries, this reduces the need for substantive policy in the donor world, for tied aid, technical assistance and conditionality, but also implies that the constituency that supports providing aid in ICs may no longer be supportive!

- Climate cooperation – governance and lessons learned. The assessment on climate cooperation and governance reveals that, even in the much more centralized negotiating arena of the climate change negotiations, it is easier to establish norms than to implement them and short-term self-interest is more dominant in influencing results than long-term global interests. Here there is no confusion regarding who should pay, but curiously some ICs are exempted. The leadership paradigm does not show statesmanship; the new and additional financial resources are more a symbolic show of goodwill and a sop to avoid future liability than reflective of what is needed to address the problem; the market mechanism is flawed, although it can be improved; and technology-transfer flows under the Convention are rendered ineffective by the flow of normal technology transfer under foreign direct investment. The *climate paradox* is that valuable time is lost in the negotiations arguing about who is right and who is wrong, who should take measures and how measures should be taken. The impacts of climate change are already visible and ultimately the burden will fall heavily on the DCs, and they have to be ready to face this burden because power structures will not change overnight, as 50 years of history show. It is for them a matter of survival!

- Mainstreaming climate change in development. There are many substantive reasons for mainstreaming climate change in development, but action is not forthcoming. The *mainstreaming paradox* is that it is a very attractive idea with the potential to solve problems in theory but cannot deal with the problems of vested interests, power politics and short-term self-interests reflected in short-term democracies, and path-dependence. As a result, mainstreaming becomes captive to the process and gets reduced to a series of check-lists, leading to 'mainstreaming lite'.

- Mainstreaming climate change in development cooperation. Despite the convergence in view about the need to mainstream climate change in development cooperation, given the low flows of resources in development cooperation, given the resources needed for addressing the MDGs (which aim only to halve the number of people without access to basic resources!), given that all other flows continue in sublime ignorance of the climate change issue, given that most ICs have not mainstreamed climate change into their own development process, climate change should not be mainstreamed into development cooperation. The *paradox* here is that mainstreaming may divert resources unless climate change goals are not ends in themselves but subservient to development cooperation goals, in which case it is not mainstreaming, but climate proofing or climate integration!

- Policies within the EU. The EU is an idealistic body with expansionist tendencies encouraging new members to accept its *acquis communautaire* and aims to tackle North–South issues single-handedly in a hostile environment of other ICs where post-Communist countries, including Russia, non-colonial countries, including the USA and Australia, and new Member States all see little obligation towards the South. The balancing act in trying to win the trust of the South, while avoiding having the bulk of the responsibility fall on a handful of EU countries, is a challenging task. Recognizing the theoretical importance of mainstreaming, the EU has set in motion a compromise decision to try to mainstream climate change into development cooperation, but does not go much further than integration. The *EU paradox* is that it has leadership ambitions and structural power, but limited leadership quality. Not many are yet following, not even many of its Member States.

10.9 Recommendations for the European Union

The section below includes (a) substantive and qualitative recommendations on mainstreaming climate change, and (b) recommendations for the EU with respect to various actors.

10.9.1 Substantive recommendations on mainstreaming climate change

Mainstreaming climate change in development is a good idea

This book concludes that mainstreaming climate change in development (not development cooperation) is a good and challenging idea because of the strong substantive links between climate change and development. Such mainstreaming requires existing development ideologies, policies, instruments and practices to be scrutinized from the perspective of climate change and reflexively redesigned to meet climate change goals.

Such mainstreaming should not be done at the cost of other mainstreaming exercises and should take gender issues, other environmental issues and disaster-related issues into account (see Section 10.2.1). Such mainstreaming approaches should also not be implemented in a way that excludes or marginalizes the poor or makes achievement of the MDGs more difficult (see Section 10.3.1). In other words, the climate change agenda should not rewrite the existing mainstreaming agendas, but should be part of a much larger reflexive approach to existing development patterns. Because of the very serious nature of climate change, mainstreaming climate change should be used as a lever to revisit the development paradigm in terms of all its major shortcomings.

Such mainstreaming is essentially a political and social process; it is contextual in nature; technology and economic instruments are subservient to the political and social reform that it calls for.

Such mainstreaming needs to be undertaken in ICs, DCs and investment, trade and banking regimes if it is to be effective. However, since mainstreaming has a habit of making the issue being mainstreamed invisible ('mainstreaming lite'), it is also vital to keep the climate change regime strong and intact (see Section 10.2.1).

Mainstreaming climate change in development cooperation
is not a good idea

Mainstreaming climate change in development cooperation, under the present political circumstances, given the history of both development and climate change cooperation, is not a good idea.

First, it is not a good idea because mainstreaming climate change in development cooperation is merely a symbolic commitment to mainstreaming in all development processes. Without reforming trade, investment and banking regimes further, mainstreaming in development cooperation will not achieve much. If climate change is mainstreamed in the other fields, ODA could remain focused on poverty reduction.

Second, current development patterns in the ICs leave much to be desired, and aid processes focused on transferring ideologies, instruments and technologies from the North to the South are likely to create greater problems.

Third, mainstreaming climate change in development cooperation without increasing the resources for both areas would be politically extremely insensitive to DCs and would further exacerbate the existing distrust between ICs and DCs.

Fourth, mainstreaming climate change in development cooperation without taking into account the resources needed would imply a diversion of resources from development goals to climate change goals, which would be politically insensitive and substantively ineffective.

Fifth, mainstreaming climate change in development cooperation would imply a different set of beneficiaries, other than those orginially intended for development cooperation. This would imply a major shift in priority for donor countries.

Finally, mainstreaming climate change into development cooperation might become the new rigid conditionality!

If mainstreaming in development cooperation is inevitable, then ...

Since mainstreaming climate change into development makes sense and a large group of stakeholders is trying to push for mainstreaming climate change in development cooperation, the process leading towards mainstreaming climate change in development cooperation may be inevitable. However, if one is to make it a success, mainstreaming must adopt an appropriate CEGD (climate, environment, gender, disaster) lens to assess development and suggest changes; be accompanied by a dedicated policy for CEGD policy; ensure *clumsy tripartite decision-making* involving stakeholders, private parties and governments both of countries disbursing ODA and of partners to design context-relevant, locally owned policies; increase ODA to 0.7% and raise new and additional resources above this amount; generate resources commensurate with what is needed; assess the beneficiaries and prioritize the poorest; build on the lessons learned from ODA

Table 10.6 *Conditions of success for mainstreaming climate change into development cooperation*

Issue	Conditions of success
Will mainstreaming become 'mainstreaming lite'?	1. Adopt a CEGD (climate, environment, gender, disaster) lens to assess development and suggest changes. 2. Include also a dedicated policy for CEGD policy.
Will mainstreaming transfer Western modalities to the South?	3. Ensure *clumsy tripartite decision-making* involving stakeholders, private parties and governments both of ODA countries and of partners to design context-relevant, locally owned policies. Avoid focus on formulae, efficiency, rationality and conditionality and accept clumsy solutions.
Will mainstreaming bypass the ODA commitment and the 'new and additional' argument?	4. Additionality: increase ODA to 0.7% and raise new and additional resources above this amount.
Will mainstreaming be a symbolic gesture in comparison with the efforts needed?	5. Commensurate resources: resources generated should be commensurate with what is needed or problems will not be addressed.
Will mainstreaming divert resources to the rich in poor countries?	6. Beneficiary assessment: to the extent possible, mainstreaming should prioritize the poorest.
Will mainstreaming learn from the lessons of aid?	7. Promote *BASICS*: • Broader goals and evaluations (not just effects on GDP or emission reduction; • Alignment between tools and countries/ actors; • Simplicity avoidance (e.g. look at contextual power relations); • Imbalance avoidance (avoid creating distortions in culture, policy and economy); • Conditionality avoidance (avoid being dogmatic about conditions); • Stakeholder engagement and mobilization.
Will mainstreaming be a symbolic gesture contradicted by other international flows?	8. Coherence and consistency: trade, investment and other development regimes must also mainstream CEGD.

Source: Based on Gupta *et al*. (2010).

Table 10.7 *Practical recommendations to improve development cooperation strategies*

Level	Recommendations
Political	Stimulate stronger and ongoing high-level endorsement in donor and partner countries of mainstreaming goals.
	Strengthen financial commitment by dedicating a specific long-term budget.
	Ensure a parallel climate change budget to prevent 'retreat into invisibility'.
Policy	Aim for common terminology, creating clarity about what is meant by terms like 'mainstreaming', 'integration' and 'climate-proofing'. This could build on the definitions provided in this book.
	Define clear objectives, targets and timetables.
	Formulate clear criteria of what counts as climate-related aid.
	Develop markers or indicators for measuring progress in climate mainstreaming.
Planning	Create a proper institutional setting for matching DC needs and aid supply.
	Present climate-related issues in a more structured way and explicitly in Regional and Country Strategy Papers.
	Make Regional and Country Strategy Papers 'climate-proof', and also other partnership documents, such as Economic Partnership Agreements.
	Assist DCs in identifying needs through bottom-up processes.
Implementation	Communicate the issue of climate change clearly and simply to development agencies, partner countries and embassies, avoiding the perception that it is another burden.
	Systematically apply impact assessment methodologies and other integration tools.
	Take a systematic approach towards climate screening, including a process for follow-up.
	Add a section specifically about climate in the *Environmental Integration Handbook for EC Development Cooperation*.
	Use development aid especially for pilot projects focused on innovative approaches.
	Develop other practical tools and guidelines.
At all levels	Monitor and evaluate aid projects and approaches in a structured and systematic way in order to learn what works and what does not.

and take 'clumsy BASICS' into account; and ensure consistency and coherency in other policy fields such as trade, investment and other development regimes (see Table 10.6).

In addition, the EU could consider the following strategies to improve the quality of its own decision-making process (see Table 10.7).

Table 10.8. *General recommendations to the European Union*

Suggestion	Arguments
Global level	
1. Promote the establishment of a Global Development Assistance Committee	The existing OECD DAC represents only OECD members; at UN level there is no such body; only competing bodies – the G8, the UN Development Group, the Banks, etc.
2. Promote discussion on who should be donors and who partners in the climate change regime	Donors are limited to only the richer industrialized countries, but could include past polluters from the countries with economies in transition. They could also include the richer developing countries, and possibly the rich in developing countries. Some climate victims may fall out of the boat since they are not traditional aid recipients or partner countries.
3. Revisit the offsetting role of the CDM	The CDM reduces pressure on the industrialized countries to reduce their own emissions. It slows down the process of emission reduction in the North!
Multilateral level	
1. Promote assessment of the role of the OECD DAC in influencing policy	Such assessments are scarce and could provide ideas about how to improve the OECD DAC itself as well as how to design a global DAC.
2. Insist that all new OECD members become DAC members and Annex II countries in the Climate Convention	This will treat like countries alike.
3. Influence the multilateral banks to accelerate the greening process	The literature shows that aid still funds 'dirty' projects.
European Union level	
1. Insist that all new EU members take on development cooperation responsibilities	This will treat like countries alike.
2. Influence Member State policy through a role-model approach	This will enhance consistency and coherence between Union and Member State policy.

10.9.2 Recommendation for the EU with respect to different actors

Whether or not the EU decides to mainstream climate change in development cooperation, it could play a more dominant role in global governance on development cooperation. This role is summed up in Table 10.8.

First, the EU and its first 15 Member States provide 59.5% of total global net OECD DAC ODA, amounting to about 0.39% of their combined GNI (see Chapter 6). If they increase this to 0.7% by 2015, they will most probably be the single largest donor in the global community, even taking into account other donors such as China and the Middle East. This gives the EU structural power in this field (Grubb and Gupta, 2000). The question is whether the EU should use its structural power to modify global politics on development, development cooperation and climate change. This chapter argues that the EU could consider doing that, since this helps to share responsibilities with other donors, it is vital that there is some degree of coordination with other donors to ensure that activities are not contradictory in impact, and it is uniquely placed to do so since it is generally more altruistic than its Member States.

At the global level, the EU could consider promoting the establishment of a global Development Assistance Committee of donors and recipients going beyond OECD donors (possibly within the Economic and Social Council). The existing OECD DAC could serve as a model, although the role of partners needs to be seriously examined. As a half-way step, the EU could consider actually engaging seriously with non-OECD donors and trying to understand their perspectives. It could also try to feed into the two other forums – namely the United Nations Development Group and the G8. The task of such a forum could be to identify who should be a donor and who a recipient, the role of partner countries in aid policy and the purpose of aid. Within the Climate Convention, the EU could suggest that some more thought be given to who should be considered as donors and who as recipients/partners. For example, on the basis of the 'polluter pays' principle, many of the economies in transition would become eligible to be funders. Under the 'ability to pay' principle, many of the newly rich DCs would become eligible as funders. The EU could also exert pressure to ensure that the CDM is improved and that the offsetting role of the CDM is revisited.

At the multilateral level, the EU could consider an evaluation of the activities of the OECD DAC and how it has assessed and shaped members' policies. A comprehensive assessment could provide the OECD DAC with some insights into how to improve its own role, but could also generate ideas about what a future global DAC should look like. The EU could further insist that all new OECD DAC members should take on responsibilities appropriate to a developed country in the climate change field, as well as responsibilities appropriate to a developed country in the development field. Another step that the EU could consider is putting pressure on the development banks to reform their ongoing investments in fossil fuels.

At the EU level, it could seriously be considered whether EU membership should bring with it responsibilities with respect to the rest of the world in terms of both ODA and climate change. To the extent that new Member States claim that they have present economic hardships and a low capacity to pay, perhaps they could receive a lower target. Since many have contributed to greenhouse gas emissions during the pre-recession period, they should definitely pay under the common but differentiated responsibilities principle with respect to climate change. Although the EU does not necessarily have any control over a Member State's bilateral ODA strategy, it may wish to influence that. If the EU wishes to make such a mainstreaming approach more substantive, it will have to take action at political, policy, planning and implementation levels.

10.10 A postscript

10.10.1 For the developing world

Fifty years of history in the aid world demonstrates that powerful countries and powerful people will not change the status quo to restructure the global economy and politics to make room for equitable solutions; neither will they translate moral obligations into a legal system of rights and obligations. Fifty years of history in the aid world shows that symbolic assistance may be provided, but much more for geopolitical, strategic and economic interests than for altruistic and moral reasons. Theoretical and empirical evidence may call for 'ownership' approaches, but, if this leads to a loss of constituency in the developed world, aid will decrease, not increase.

Twenty years of climate change history shows that, even where there is a clear case of a polluter and a victim, at least in the short term (there are current victims because of past pollution), the resources raised have been a fraction of those which are raised for development cooperation. Moralizing statements have become a recent fad, but there is neither statesmanship nor a citizen following.

In the June 2009 negotiations in Bonn, DCs made a last-ditch effort to insist that the ICs take responsibility by reducing their emissions by 40%–45% by 2020, thereby making room for the growth in emissions of developing countries. This was rejected. Although the press release heralding the Copenhagen Conference on Climate in December 2009 stated that a historic conference kicks off 'with strong commitment to clinch an ambitious climate change deal and an unprecedented sense of urgency to act', the final deal was vague, weak and only 'noted' by the participants, primarily because especially poor and vulnerable developing countries were extremely disappointed by the outcome. The DCs should reflect on what the last 50 years have taught them: that their moral and legal appeals for greater development assistance fall on deaf ears in the bulk of the ICs. Only a handful remain still sensitive to these calls. The power of numbers has hardly made a dent except in the adoption of a consensus text that accommodates their views but does not implement them. This has led to the growing gap between norm acceptance and norm implementation. In the meantime, anthropogenic climate change and its impacts have the potential completely to disrupt DC societies. On average, one in 19 of their citizens, compared with one in 1500 in the North, will face a climate disaster. Developing like the North is no solution, since this would require nine times our Earth in resources (UNDP, 2007). This is a moment for rethinking strategy: how can DCs completely bypass the ICs in a sustainable development process by unleashing their own intellectual humanpower and focusing all their attention on addressing their short-term problems, problems that will become worse each year? In 1937, Romein wrote a paper on the 'Law of the handicap of a jump start' (Romein, 1937) arguing that firstcomers can become backseaters, and vice versa: there is potential for latecomers to become frontrunners.

10.10.2 For the reader

Open-minded academic research tends to be highly critical of existing development patterns and inequalities. But where does such critique lead? Consciousness of the current shortcomings in development and development cooperation is an essential first step for understanding the future of mainstreaming climate change in development cooperation. Is it likely to be a short-term fad, evaporating in the gales of the financial crisis, or will the financial crisis give an extra window of opportunity to rewrite the rules of global sustainable development governance? Such rules should ensure that not just development cooperation, but also the development process itself, as well as international trade and investment regimes, all work in tandem to address the common global problems of this century. Humans have only one Earth, and the financial crisis shows how interlocked humans and their institutions are. The solutions are therefore also interlocked, and may allow a transition from a global economy to a global society.

Acknowledgements

This chapter has benefited considerably from the discussions within the ADAM P3B project about mainstreaming climate change in development cooperation with Joanne Bayer, Anthony Patt, Michael Thompson, Lennart Olsson, Åsa Persson and Richard Klein and the comments from Onno Kuik, Frans Oosterhuis, Eric Massey and Harro van Asselt. Hans Opschoor critically reviewed a previous version of this chapter. Earlier drafts have been presented at workshops dedicated to this chapter at the UNESCO-IHE Institute for Water Education, Institute for Social Studies and Bothends and incorporate the comments of those present: Pieter van der Zaag, Jeltje Kemerink, Marloes Mul, Ton Bresser, Yunus Mohamed, Frank Jaspers, Amaury Tilmant, Willy Douma, Harrie Oppenoorth, Eco Matser, Harrie Clemens, Daniëlle Hirsch, Wiert Wiertsema and Annelieke Douma. This chapter has been reviewed by Mike Hulme.

References

Adam, E. (2006). Preface to Kirchmeier, F. (2006). *The Right to Development: Where Do We Stand?* Dialogue on Globalization Occasional Papers No. 23. Geneva: Friedrich-Ebert-Stiftung.

Agenda 21 and Rio Declaration (1992). *Report on the UN Conference on Environment and Development, Rio de Janeiro, 3–14 June 1992.* UN Document. A/CONF.151/26/Rev. 1 (Vols. 1–III).

Alesina, A. and Weder, B. (2002). Do corrupt governments receive less foreign aid? *American Economic Review*, **92**, 1126–37.

Baer, P., Athanasiou, T., Kartha, S. and Kemp-Benedict, E. (2008). *The Greenhouse Development Rights Framework: The Right to Development in a Constrained World.* Berlin: Heinrich Boll Foundation.

Baumert, K., Herzog, T. and Pershing, J. (2005). *Navigating the Numbers: Greenhouse Gas Data and International Climate Policy.* Washington, D.C.: World Resources Institute.

Behrens, A. (2009). Financial impacts of climate change mitigation. *Climate Change Law Review*, **3**(2), 179–87.

Benson, C. and Twigg, J. (2007). *Tools for Mainstreaming Disaster Risk Reduction: Guidance Notes for Development Organizations.* Geneva: ProVention Consortium.

Berthélemy, J.-C. (2006). Bilateral donors' interest vs. recipients' development motives in aid allocation: do all donors behave the same? *Review of Development Economics*, **10**(2), 179–94.

Caviglia-Harris, J. L., Chambers, D. and Kahn, J. R. (2009). Taking the "U" out of Kuznets: a comprehensive analysis of the EKC and environmental degradation. *Ecological Economics*, **68**(4), 1149–59.

Chakravarty, S., Chikkatur, A., de Coninck, H. *et al.* (2009). Sharing global CO_2 emission reduction among one billion high emitters. *Proceedings of the National Academy of Sciences*, **106**, online pre-publication in the *PNAS Early Edition* on 6 July 2009.

Clemens, M. A., Kemp, C. J. and Moss, T. J. (2007). The trouble with the MDGs: confronting expectations of aid and development success. *World Development*, **35**(5), 734–51.

Commission on Climate Change and Development (2009). *Closing the Gaps: Report of the Commission on Climate Change and Development.* Stockholm: Swedish Ministry of Foreign Affairs.

Copenhagen Accord (2010). *Decision Noted by the Parties to the Fifteenth Conference of the Parties to the Climate Change Convention*, FCCC/CP/2009/L.7, 18 December 2009.

Doha Declaration (2008). *Doha Declaration on Financing for Development: Outcome Document of the Follow-up International Conference on Financing for Development to Review the Implementation of the Monterrey Consensus.* Doha, Qatar, 29 November–2 December 2008. Available online at http://www.un.org/esa/ffd//doha/documents/Doha_Declaration_FFD.pdf.

EACC (2009). *The Costs to Developing Countries of Adapting to Climate Change: New Methods and Estimates, The Global Report of the Economics of Adaptation to Climate Change Study.* Washington, D.C.: World Bank. Available online at http://siteresources.worldbank.org/INTCC/Resources/EACCReport0928Final.pdf.

Easterly, W. (2007). Was development assistance a mistake? *American Economic Review*, **97**(2), 328–32.

Gorz, A. (1994). *Capitalism, Socialism, Ecology.* London: Verso Books.

Grubb, M. and Gupta, J. (2000). Introduction: climate change, leadership roles and the European Union, in *Climate Change and European Leadership: A Sustainable Role for Europe*, ed. J. Gupta and M. Grubb. Dordrecht: Kluwer Academic Publishers, pp. 3–14.

Gupta, J. (1997). *The Climate Change Convention and Developing Countries: From Conflict to Consensus?* Dordrecht/Boston, MA: Kluwer Academic Publishers.

  (2008) *Engaging Developing Countries in Climate Change Negotiations*. Study for the European Parliament's Temporary Committee on Climate Change (CLIM), IP/A/CLIM/IC/2007–111. Institute for European Environmental Policy (IEEP) and Ecologic, Briefing number 631–715. Brussels/London/Berlin: Institute for European Environmental Policy and Ecologic.

Gupta, J., Persson, A., Olsson, L. *et al.* (2010). Mainstreaming climate change in development cooperation: conditions for success, in *Adaptation and Mitigation of Climate Change*, ed. M. Hulme and H. Neufeldt. Cambridge: Cambridge University Press.

Independent Task Force on Climate Change (2008). *Confronting Climate Change: A Strategy for US Foreign Policy*. Washington, D.C.: Council on Foreign Relations.

IPCC-3 (2007). *Climate Change 2007: Mitigation Contribution of Working Group III to the Fourth Assessment Report of the Intergovernmental Panel on Climate Change*. Cambridge: Cambridge University Press.

Joint European NGO report (2006). *EU Aid: Genuine Leadership or Misleading Figures? An Independent Analysis of European Governments' Aid Levels*. Brussels: Concord.

Kirchmeier, F. (2006). *The Right to Development: Where Do We Stand?* Dialogue on Globalization Occasional Papers No. 23. Geneva: Friedrich-Ebert-Stiftung.

KP (1997). *Kyoto Protocol to the United Nations Framework Convention on Climate Change*. Signed 10 December 1997, in Kyoto; entered into force 16 February 2005. Reprinted in (1998) *International Legal Materials*, **37**(1), 22.

Mani, M., Markandaya, A. and Ipe, V. (2008). *Policy and Institutional Reforms to Support Climate Change Adaptation and Mitigation in Development Programs: A Practical Guide*. Washington, D.C.: World Bank.

McGillivray, M. (2004). Descriptive and prescriptive analyses of aid allocation: approaches, issues, and consequences. *International Review of Economics and Finance*, **13**, 275–92.

Metz, B., Davidson, O. R., Martens, J.-W., van Rooijen, S. N. M. and McGrory, L. v. W., eds. (2000). *Methodological and Technological Issues in Technology Transfer*. Cambridge: Cambridge University Press.

Nunnenkamp, P. and Thiele, R. (2006). Targeting aid to the needy and deserving: nothing but promises? *World Economy*, **29**(9), 1177–201.

Opschoor, H. (2009). *Sustainable Development and a Dwindling Carbon Space*. Public Lecture Series 2009, No. 1. The Hague: Institute of Social Studies.

Oxfam (2007). *What's Needed in Poor Countries, and Who Should Pay?* Oxford: Oxfam International.

  (2009). *UN Climate Negotiations in Bonn 1–12 June 2009: Background Briefing*. Oxford: Oxfam International. Available online at http://www.oxfam.de/download/Background_brief_Bonn_talks.pdf.

Romein, J. (1937). De dialectiek van de vooruitgang, in *Het onvoltooid verleden*, J. Romein. Amsterdam: Querido. Available online at http://www.dbnl.org/tekst/_for003193501_01/_for003193501_01_0124.htm.

UNDP (2007). *Fighting Climate Change: Human Solidarity in a Divided World. UNDP Human Development Report 2007–2008*. New York: Palgrave Macmillan.

Verweij, M. and Thompson, M., eds. (2006). *Clumsy Solutions for a Complex World: Governance, Politics and Plural Perceptions*. Basingstoke: Palgrave.

Index